Y0-BTA-511

CONVEXITY THEORY AND ITS APPLICATIONS IN FUNCTIONAL ANALYSIS

L.M.S. MONOGRAPHS

Editors: D. EDWARDS and H. HALBERSTAM

1. Surgery on Compact Manifolds *by* C. T. C. Wall, F.R.S.
2. Free Rings and Their Relations *by* P. M. Cohn
3. Abelian Categories with Applications to Rings and Modules *by* N. Popescu
4. Sieve Methods *by* H. Halberstam and H.-E. Richert
5. Maximal Orders *by* I. Reiner
6. On Numbers and Games *by* J. H. Conway
7. An Introduction to Semigroup Theory *by* J. M. Howie
8. Matroid Theory *by* D. J. A. Welsh
9. Subharmonic Functions, Volume 1 *by* W. K. Hayman and P. B. Kennedy
10. Topos Theory *by* P. T. Johnstone
11. Extremal Graph Theory *by* B. Bollobás
12. Spectral Theory of Linear Operators *by* H. R. Dowson
13. Rational Quadratic Forms *by* J. W. S. Cassels, F.R.S.
14. C^* Algebras and their Automorphism Groups *by* G. K. Pedersen
15. One-Parameter Semigroups *by* E. B. Davies
16. Convexity Theory and its Applications in Functional Analysis *by* L. Asimow and A. J. Ellis

Published for the London Mathematical Society
by Academic Press Inc. (London) Ltd.

CONVEXITY THEORY AND ITS APPLICATIONS IN FUNCTIONAL ANALYSIS

L. ASIMOW
University of Wyoming, Laramie, Wyoming, USA

A. J. ELLIS
University College of Swansea, Swansea, Wales, UK

1980

ACADEMIC PRESS

A Subsidiary of Harcourt Brace Jovanovich, Publishers

London New York Toronto Sydney San Francisco

ACADEMIC PRESS INC. (LONDON) LTD.
24/28 Oval Road
London NW1

United States Edition published by
ACADEMIC PRESS INC.
111 Fifth Avenue
New York, New York 10003

British Library Cataloguing in Publication Data

Asimow, L
Convexity theory and its applications in
functional analysis.
—(London Mathematical Society. Monographs;
16 ISSN 0076–0560).
1. Convex sets 2. Functional analysis
I Title II. Ellis, A J III. Series
517′.7 QA640 80–40648

ISBN 0–12–065340–0

PRINTED IN GREAT BRITAIN BY
J. W. ARROWSMITH LTD, WINTERSTOKE ROAD, BRISTOL

To Marilyn Asimow and Jennifer Ellis

Preface

With the appearance in 1966 of *Lectures on Choquet Theory*, Phelps [**172**], the representation theory of compact convex sets became accessible to a large, and as it develops, enthusiastic audience. The geometric appeal of the Choquet simplex is irresistible; it has led quite naturally to an exploration of related geometric structure in more general state spaces. This enterprise, while somewhat interesting in its own right, would be of little note but for the remarkable insights it has provided in appreciating various analytical aspects of the underlying function spaces. This feature was already apparent in the comprehensive treatment in 1971 of Choquet Theory in the book by Alfsen [**5**]. Since that time the geometric study of compact convex sets has rewarded its devotees not only with generalizations of individual theorems in functional analysis, but with the prospect of a unified geometric theory that yields a clearer understanding of a reasonable variety of classical results.

Our object here is to promote this geometric perspective. We hasten to make the standard disclaimer concerning the lack of totality in our selection of topics. Granting the necessarily idiosyncratic nature of our subject matter we have attempted to illustrate the means by which a fairly elementary geometric theory, based on partially ordered Banach spaces and duality, can be applied in a systematic fashion to concrete function spaces (real and complex) and unital Banach algebras.

In Chapter 1 we have gathered together the functional analytic preliminaries that constitute our basic tools. The reader with only a basic knowledge of abstract real analysis will find there statements of the main results as well as short but self-contained treatments of the Krein–Šmulyan Theorem, the basic Choquet Theory and the Bishop–Phelps Theorem.

Chapter 2 gives the basic duality results, lattice theory and concrete representation theorems for order unit spaces and Banach lattices of type *M* and *L*. Since much of the book treats subspaces of continuous functions on compact Hausdorff spaces we felt the classical representation theorems of abstract ordered Banach spaces were especially pertinent.

We continue the treatment of real affine function spaces in Chapter 3 by examining in detail the case where the state space is a Choquet simplex. In Chapter 4 we show how the study of real $A(K)$ spaces can be employed even for complex-valued function spaces by means of a complex state space. Chapter 5 gives a survey of the application of the theory to the study of (non-commutative) Banach algebras.

Our thanks are due to Caroline Johnson and Paula Melcher for their expert typing of the manuscript. We are very grateful to Dr K. F. Ng for reading the manuscript and making many helpful suggestions. We have also benefited from useful discussions of parts of the book with Dr T. B. Andersen. The second author wishes to thank the University College of Swansea for sabbatical leave during the Autumn Term of 1976, enabling him to visit the University of Wyoming.

July 1980

L. Asimow
A. J. Ellis

Contents

CHAPTER 1

Preliminaries

We bring together in this chapter most of the general material in functional analysis that will be relied on in subsequent chapters. We have simply stated some of the standard results and have provided complete proofs of others. The distinction is based mainly on which theorems are pretty much standard in a graduate level analysis course (say, for example Royden [**185**] or Rudin [**187**]) and which are left for more advanced courses.

Thus we omit proofs of the Separation Theorem and the bipolar calculations (which are used repeatedly throughout the remainder of the text) and the Krein–Milman Theorem. We do prove the Krein–Šmulyan Theorem (also an invaluable tool in our later study) in Section 2. In Section 3 we present a technical lemma which incorporates most of the iteration procedures in Banach spaces (including the Open Mapping Theorem) that arise later on.

In Sections 4–6 we introduce state spaces and prove the Choquet–Bishop–de Leeuw Theorem on representing measures. We conclude with a proof of the Bishop–Phelps Theorem.

1. SEPARATION AND POLAR CALCULUS

Let (E, F) be a pair of (real or complex) linear spaces in duality, with each total over the other; i.e. if $(a, x) = 0$ for all $x \in F$ then $a = 0$ and if $(a, x) = 0$ for all $a \in E$ then $x = 0$. Let (E, F) be endowed with the weakest topologies for which the linear functionals $(\cdot, x)$ and $(a, \cdot)$ are continuous on E and F for each a and each x. Then E and F are locally convex linear topological spaces with each being the (dual) space of continuous linear functionals over the other.

1.1. THEOREM (SEPARATION THEOREM). *If A and B are disjoint convex subsets of E with A compact, and B closed then there exists an $x \in F$ such that* $\max\{\text{re}\,(a, x)\colon a \in A\} < \inf\{\text{re}\,(b, x)\colon b \in B\}$.

Let X be a subset of E or F. We denote the *convex hull* of X by co X and the *closed convex hull* by $\overline{\text{co}}\, X$. We write bal X for $\{\alpha x: x \in X \text{ and } |\alpha| = 1\}$ and cob X for the *convex balanced hull* of X.

We write ρ_X, or $\rho(X)$, for the *support function* of X, which is defined on the dual space by

$$\rho_X(a) = \sup\{\text{re}\,(a, x): x \in X\}$$

if $X \subset F$ (and dually for $X \subset E$). Note that ρ_X may be an extended real-valued function.

If $0 \in A$ and A is convex we define the *Minkowski functional* of A by

$$p_A(\text{or} \quad p(A))(a) = \inf\{r \geq 0: a \in rA\}.$$

If (E', F') is another dual pair and $T: E \to E'$ is a continuous map then the adjoint $T^*: F' \to F$ is given by

$$(a, T^*y) = (Ta, y).$$

(The continuity of T guarantees $T^*y \in F$.)

For $A \subset E$ we define the *polar*

$$A^0 = \{x \in F: \text{re}\,(a, x) \leq 1 \quad \text{for all } a \in A\}.$$

If $A \subset F$ then A^0 is the subset of E defined similarly.

1.2. Theorem (Polar Calculus).

(i) *A^0 is a closed convex set containing* 0 *and*

$$A^{00} = \overline{\text{co}}\,(A \cup \{0\}).$$

(ii) *$A \subset B$ implies $B^0 \subset A^0$.*

(iii) *$(\lambda A)^0 = (1/\lambda)A^0$ for $0 < \lambda < \infty$.*

(iv) *$(A \cup B)^0 = A^0 \cap B^0$.*

(v) *If $T: E \to E'$ is continuous with adjoint $T^*: F' \to F$ then for $A \subset E$,*

$$(T(A))^0 = (T^*)^{-1}(A^0).$$

(vi) *If $0 \in A$ then $p(A^0) = \rho(A)$.*

Proof. Part (1) follows from the Separation Theorem and (ii)–(v) are straightforward. For (vi) we have first that $\rho(A)(x) = \infty$ if and only if $x \notin (\lambda A)^0 = (1/\lambda)A^0$ for all $0 < \lambda < \infty$, if and only if $p(A^0)(x) = \infty$. If $0 \leq \rho(A)(x) < \lambda$ then $x/\lambda \in A^0$ so that $p(A^0)(x) \leq \lambda$ and hence $p(A^0) \leq \rho(A)$. If $p(A^0)(x) < \lambda$ then $x/\lambda \in A^0$ so that $\rho(A)(x) \leq \lambda$. Thus $\rho(A) = p(A^0)$.

We note that (i) and (iv) together imply

(vii) If A, B are closed convex sets containing 0 then $(A \cap B)^0 = \overline{\text{co}}\,(A^0 \cup B^0)$. If, in addition, A is a *cone* ($\lambda A \subset A$ for all $\lambda > 0$) then so is A^0

and

$$(A \cap B)^0 = \overline{\mathrm{co}}\,(A^0 \cup B^0) = (A^0 + B^0)^-.$$

We say x is an *extreme point* of X if whenever $x = \lambda y + (1-\lambda)z$ ($y, z \in X$ and $0 < \lambda < 1$) then $y = x = z$. We denote the set of extreme points of X by ∂X or ext X.

Similarly, the convex subset F or X is called a *face* if whenever $x \in F$ is a convex combination of y and z in X then y and z belong to F.

1.3. THEOREM (KREIN–MILMAN). *If X is a compact convex set in E (or F) then $X = \overline{\mathrm{co}}\,\partial X$. If $X = \overline{\mathrm{co}}\,Y$ and Y is compact then $\partial X \subset Y$.*

In almost all cases we will be taking E to be a normed linear space with dual space $F = E^*$. We write

$$A_r = \{a \in A : \|a\| \leq r\}.$$

Thus the (closed) unit ball is denoted E_1 and the unit ball of E^* is $E_1^* = E_1^0$.

The dual topologies of our previous discussion are called the weak and weak* (w^*) topologies respectively. Since convex sets in E have the same closure in the weak and norm topologies we generally will not distinguish these cases.

1.4. THEOREM. (ALAOGLU). *A subset X of E^* is w^*-compact if and only if X is w^*-closed and $\|\cdot\|$-bounded.*

1.5. THEOREM. *Let A and B be closed convex sets containing 0 with A compact.*

(i) *The set $A + B$ is closed.*

(ii) *The set* $\mathrm{co}\,((A + P_B) \cup B) = \overline{\mathrm{co}}\,(A \cup B)$, *where*

$$P_B = \{x \in B : p(B)(x) = 0\}.$$

Proof. We show (ii): Let (x_α) be a net in $\mathrm{co}\,(A \cup B)$ with $x_\alpha \to x$. Let

$$x_\alpha = \lambda_\alpha y_\alpha + (1-\lambda_\alpha) z_\alpha; \qquad 0 \leq \lambda_\alpha \leq 1, \quad y_\alpha \in A, \quad z_\alpha \in B.$$

Since A is compact, by passing to a subnet, we can assume

$$y_\alpha \to y \in A \quad \text{and} \quad \lambda_\alpha \to \lambda.$$

If $\lambda < 1$ then eventually $\lambda_\alpha < 1$ so that

$$z_\alpha = (x_\alpha - \lambda_\alpha y_\alpha)/(1-\lambda_\alpha) \to (x - \lambda y)/(1-\lambda) \in B.$$

Thus $x \in \mathrm{co}\,(A \cup B)$. If $\lambda = 1$ and $0 < r \leq 1$ is given then eventually $1 - \lambda_\alpha < r$ so that

$$(1-\lambda_\alpha) z_\alpha = x_\alpha - \lambda_\alpha y_\alpha \in rB.$$

Hence $x-y\in rB$ for all $r>0$ so that $p(B)(x-y)=0$. In this case

$$x=y+(x-y)\in A+P_B.$$

2. KREIN–ŠMULYAN THEOREM AND COROLLARIES

We briefly discuss various topologies on the dual, E^*, of a normed linear space E for which the elements of E remain continuous as linear functionals on E^*.

Definition. A non-empty collection $\mathscr{A}$ of sets in E is called a *topologizing family* for E^* if

(i) $A\in\mathscr{A}$ implies A is norm-bounded,
(ii) $A, B\in\mathscr{A}$ implies there exists $C\in\mathscr{A}$ such that $A\cup B\subset C$,
(iii) $A\in\mathscr{A}$ implies $\alpha A\in\mathscr{A}$ for scalars (real or complex) α,
(iv) A finite implies $A\in\mathscr{A}$.

2.1. THEOREM. *If $\mathscr{A}$ is a topologizing family in E then*

$$\mathscr{A}^0\doteq\{A^0\colon A\in\mathscr{A}\}$$

is a neighbourhood base at 0 *for a locally convex topology on E^* for which the elements of E are continuous linear functionals on E^*.*

Proof. We just note here that the standard von Neumann axioms are satisfied. For example, to check that given A^0 there is a B^0 such that $\lambda B^0\subset A^0$ for all $|\lambda|\leq 1$ take, using (ii) and (iii),

$$B\supset 2[\pm A\cup\pm iA].$$

Then $B^0\subset\frac{1}{2}[A^0\cap(-A^0)\cap(iA^0)\cap(-iA^0)]$. Thus $x\in B^0$ implies $\pm x,\ \pm ix\in(\frac{1}{2})A^0$ so that for $|\lambda|\leq 1$, $\lambda x=(\mathrm{re}\,\lambda)x+i(\mathrm{im}\,\lambda)x\in(\frac{1}{2})A^0+(\frac{1}{2})A^0\subset A^0$.

Three important examples of topologizing families are

(1) $\mathscr{A}=\{A\subset E\colon A \text{ finite}\}$ which yields the w^*-topology on E^*,
(2) $\mathscr{A}=\{A\subset E\colon A \text{ norm-compact}\}$,
(3) $\mathscr{A}=\{A\subset E\colon A \text{ is norm-bounded}\}$ which yields the strong, or norm, topology on E^*.

We call the topology of Example (2) the *bounded-weak** (*bw**) topology on E^*. The importance of the bw^*-topology derives from its close connection with the w^*-topology.

2.2. THEOREM. *The set $V\subset E^*$ is bw^*-open if and only if $V\cap E_r^*$ is relatively w^*-open in E_r^* for each $r\geq 0$.*

We give a proof based on the following two lemmas.

2.3. LEMMA. *Let $V \subset E^*$ and $x \in E^*$ be given. Let $\|x\| = s$. Then the following are equivalent:*

(i) *$V \cap E_r^*$ is a relative w^*-neighbourhood of x in E_r^* for all $r \geq s$,*

(ii) *$(V - x) \cap E_r^*$ is a relative w^*-neighbourhood of 0 in E_r^* for all $r \geqslant 0$.*

Proof. Note that

$$(x + W) \cap E_r^* \subset x + W \cap E_{r+s}^* \quad \text{and} \quad x + W \cap E_r^* \subset (x + W) \cap E_{r+s}^*.$$

If (i) holds and $r \geq 0$, choose W, a w^*-neighbourhood of 0, such that $(x + W) \cap E_{r+s}^* \subset V$. Then $x + W \cap E_r^* \subset V$ so that $W \cap E_r^* \subset V - x$. The argument for (ii) implies (i) is quite similar.

2.4. LEMMA.

(i) *If V is a bw^*-neighbourhood of 0 then $V \cap E_r^*$ is a relative w^*-neighbourhood of 0 in E_r^* for all $r \geq 0$.*

(ii) *If $V \cap E_r^*$ is a relative w^*-open neighbourhood of 0 in E_r^* for all $r \geq 0$ then V is a bw^*-neighbourhood of 0.*

Proof. For (i) choose A norm-compact in E with $A^0 \subset V$ and let $r > 0$ be given. Choose A_0 finite such that

$$A \subset A_0 + E_{(1/2)r}.$$

Then $A \subset \text{co}\,[2A_0 \cup E_{1/r}]$ so that $E_r^* \cap (2A_0)^0 \subset A^0 \subset V$. For (ii) we shall construct a sequence $(A_n)_{n=0}^{\infty}$ such that

(a) each A_n is finite,

(b) $A_n \subset E_{1/n}$ for $n \geq 1$,

(c) $(A_0 \cup \cdots \cup A_n)^0 \cap E_{n+1}^* \subset V$.

Take A_0 finite and $A_0^0 \cap E_1^* \subset V$. Assume $A_0, \ldots, A_{n-1}$ $(n \geq 1)$ are chosen satisfying (a), (b), (c). Let

$$K = E_{n+1}^* \cap (A_0 \cup \cdots \cup A_{n-1})^0 \cap V^c.$$

Note that K is w^*-compact in E_{n+1}^*, and $K \cap E_n^* = (A_0 \cup \cdots \cup A_{n-1})^0 \cap E_n^* \cap V^c \subset V \cap V^c = \varnothing$.

Using a standard compactness argument, we choose a w^*-closed convex neighbourhood U of 0 such that $W = E_n^* + U$ is a w^*-closed neighbourhood of E_n^*, disjoint from K. Thus W^0 is a finite-dimensional set in the interior of $E_{1/n}$ and we can choose a finite set A_n such that $W^0 \subset \text{co}\, A_n \subset E_{1/n}$. Thus (a), (b) hold and

$$\varnothing = W \cap K \supset A_n^0 \cap E_{n+1}^* \cap (A_0 \cup \cdots \cup A_{n-1})^0 \cap V^c.$$

Thus

$$A_n^0 \cap (A_0 \cup \cdots \cup A_{n-1})^0 \cap E_{n+1}^* = (A_0 \cup \cdots \cup A_n)^0 \cap E_{n+1}^* \subset V,$$

so (c) is satisfied. Now take

$$A = \text{norm-cl}\left(\bigcup_{n=0}^{\infty} A_n\right).$$

Then A is compact and

$$A^0 \cap E^*_{n+1} \subset (A_0 \cup \cdots \cup A_n)^0 \cap E^*_{n+1} \subset V.$$

Hence

$$A^0 = \bigcup_{n=0}^{\infty} (A^0 \cap E^*_{n+1}) \subset V.$$

Proof of Theorem 2.2. Let V be bw^*-open and $x \in V$ with $\|x\| = s$. Then $0 \in V - x$ is bw^*-open so that by Lemma 2.4(i), $(V-x) \cap E^*_r$ is a relative w^*-neighbourhood of 0 in E^*_r for all $r \geq 0$ and hence, by Lemma 2.3, $V \cap E^*_r$ is a relative w^*-neighbourhood of x in E^*_r for all $r \geq s$.

For the converse choose $x \in V$. We show $V - x$ is a bw^*-neighbourhood of 0. If $(V-x) \cap E^*_t$ is relatively w^*-open in E^*_t for all $t \geq 0$ then Lemma 2.4(ii) applies. But $y \in (V-x) \cap E^*_t$ implies $x + y \in V$ so that $V \cap E^*_r$ is a w^*-neighbourhood of $x+y$ for all $r \geq \|x+y\|$ and hence (Lemma 2.3, (i) implies (ii)) $[V-(x+y)] \cap E^*_r$ is a w^*-neighbourhood of 0 in E^*_r for all $r \geq 0$.

Finally, this says (Lemma 2.3 (ii) implies (i)) $(V-x) \cap E^*_r$ is a w^*-neighbourhood of y in E^*_r for all $r \geq \|y\|$. But $y \in E^*_t$ so that $t \geq \|y\|$.

We can now state some of the very useful characterizations of w^*-continuous functionals and w^*-closed convex sets. The first corollary is a direct consequence of Theorem 2.2.

2.5. Corollary. *The function f is bw^*-continuous on E^* if and only if $f|E^*_r$ is relatively w^*-continuous for all $r \geq 0$.*

Let $\phi: E \to E^{**}$ be the canonical embedding. Then ϕ is an isometry and, since E^{**} is a Banach space, the completion of E can be identified as norm-cl $\phi(E)$ in E^{**}.

Corollary (Banach–Grothendiek Theorem). *If $a \in E^{**}$ then a is bw^*-continuous on E^* if and only if $a \in$ norm-cl $\phi(E)$.*

Proof. If a is bw^*-continuous then, by Corollary 2.5, for each $r > 0$ there is an $A \subset E$ with A finite and $A^0 \cap E^*_r \subset \{x : \text{re}\,(a, x) \leq 1\}$. Then taking polars in E^{**} gives (since $A^{00} = \text{co}\,\phi(A) \subset \phi(E)$)

$$x \in [A^0 \cap E^*_r]^0 = \text{co}\,(\phi A \cup E^{**}_{1/r}) \subset \phi E + E^{**}_{1/r}.$$

Hence $x \in$ norm-cl $\phi(E)$.

For the converse let $(a_n)_{n=1}^{\infty} \subset E$ with $\|\phi(a_n) - a\| \to 0$. For $r > 0$ let $X = E_r^*$ and let $\theta: E \to C(X)$, taking $b \to (\phi b)|_X$. Then θ is an isomorphism onto a subspace of $C(X)$ and $\|\theta b\| = r\|b\|$. Hence $\sup_X |\phi(a_n) - a| = r\|\phi(a_n) - a\| \to 0$ so that $a \in C(X)$. Thus Corollary 2.5 yields the bw^*-continuity of a.

2.6. COROLLARY. *If E is a Banach space then the following are equivalent:*

(i) *a is a w^*-continuous linear functional on E^*,*
(ii) *$a \in \phi E$,*
(iii) *a is bw^*-continuous.*

2.7. COROLLARY (KREIN–ŠMULYAN THEOREM). *Let E be a Banach space and let K be a convex subset of E^*. Then K is w^*-closed if and only if $K \cap E_r^*$ is w^*-closed for all $r \geq 0$.*

Proof. Let $V = K^c$ so that the condition implies $V \cap E_r^*$ is relatively w^*-open for all $r \geq 0$ and hence V is bw^*-open. Thus K is bw^*-closed and hence is an intersection of bw^*-closed half-spaces. But the previous corollary guarantees these half-spaces are in fact w^*-closed and therefore, K is w^*-closed.

3. GAUGE LEMMA AND COMPLETENESS

It is convenient to formulate here a general iteration scheme in Banach spaces, which will allow us to conclude a variety of duality results later on. This is expressed in terms of *gauges*, i.e. Minkowski functionals of convex sets containing 0 and possessing certain convergence properties.

Definition. If p is a positively homogeneous, sub-additive functional on a normed linear space E such that $\sum_{n=1}^{\infty} p(x_n) < \infty$ implies $x = \sum_{n=1}^{\infty} x_n$ exists and $p(x) \leq \sum_{n=1}^{\infty} p(x_n)$ then p is called a *gauge.*

It is the Minkowski functional of the set

$$B = \{x \in E : p(x) \leq 1\}.$$

If p has the weaker property that

$$p(x) \leq \sum_{n=1}^{\infty} p(x_n) \quad \text{whenever } x = \sum_{n=1}^{\infty} x_n \text{ exists}$$

then p is called a *pre-gauge.*

We say the convex set B (containing 0) is a *gauge set*, or a *pre-gauge set* if p_B has the corresponding property. It may happen that B is a proper subset of $\{x : p_B(x) \leq 1\}$.

3.1. PROPOSITION. *Let A and B be closed convex sets containing 0 in the normed linear space E and let $T: E \to F$ be a bounded linear map to the normed space F.*

(i) *A and B are pre-gauge sets.*
(ii) *If B is complete and bounded then B is a gauge set.*
(iii) *If B is a gauge set then $A+B$ and $\text{co}\,(A \cup B)$ are pre-gauge sets.*
(iv) *If B is a gauge set then $T(B)$ is a pre-gauge set.*

Proof. Note first that the Minkowski functional p_A is lower-semi-continuous if and only if A is closed, since

$$\{x \in E: p_A(x) \le \alpha\} = \alpha A \quad \text{for } 0 < \alpha < \infty$$

and

$$\{x \in E: p_A(x) = 0\} = \bigcap_{\alpha > 0} \alpha A.$$

Thus, if $x = \lim s_n$, where $s_n = \sum_{i=1}^n x_i$, then $p_A(x) \le \liminf p_A(s_n) \le \liminf \sum_{i=1}^n p_A(x_i) = \sum_{n=1}^\infty p_A(x_n)$. This shows (i). If B is complete and bounded then, for some $M > 0$, $\|\cdot\| \le M p_B$; thus $\sum p_B(x_n) < \infty$ implies $\sum \|x_n\| < \infty$ and hence $x = \sum x_n$ exists. For (iii), let $C = \text{co}\,(A \cup B)$ and let $x = \sum_{n=1}^\infty x_n$ with $\sum_{n=1}^\infty p_C(x_n) < \infty$. Let α_n be such that $0 \le p_C(x_n) < \alpha_n < p_C(x_n) + \varepsilon/2^n$. Then $x_n \in \alpha_n\, \text{co}\,(A \cup B)$ so that $x_n = \alpha_n \lambda_n a_n + \alpha_n (1 - \lambda_n) b_n \doteq y_n + z_n$ where $p_A(y_n) + p_B(z_n) = \alpha_n \lambda_n p_A(a_n) + \alpha_n (1 - \lambda_n) p_B(b_n) \le \alpha_n$. In particular we have $\sum p_B(z_n) < \infty$ so that $z = \sum z_n$ with $p_B(z) \le \sum p_B(z_n)$. It follows that $y = \sum y_n = \sum (x_n - z_n) = x - z$ exists and hence $p_A(y) \le \sum p_A(y_n) < \infty$. Thus, if $p_A(y) < \lambda$ and $p_B(z) < \mu$, then

$$x = (\lambda + \mu)[(\lambda/(\lambda + \mu))(y/\lambda) + (\mu/(\lambda + \mu))(z/\mu)] \in (\lambda + \mu)\, \text{co}\,(A \cup B).$$

Hence $p_C(x) \le \lambda + \mu$. Since λ, μ are arbitrary,

$$p_C(x) \le p_A(y) + p_B(z) \le \sum (p_A(y_n) + p_B(z_n)) \le \sum \alpha_n \le \sum p_C(x_n) + \varepsilon.$$

The argument for $A+B$ is similar. Finally, if $C = T(B)$, $y = \sum y_n$ and $\sum p_C(y_n) < \infty$ then choose α_n such that

$$p_C(y_n) < \alpha_n < p_C(y_n) + \varepsilon/2^n.$$

Then $y_n \in \alpha_n C$ so that $y_n = T(x_n)$ with $p_B(x_n) \le \alpha_n$. Hence $x = \sum x_n$ exists and $p_B(x) \le \sum p_B(x_n) \le \sum p_C(y_n) + \varepsilon$. But $y = \sum y_n = \sum T(x_n) = T(x)$ and $p_C(y) = p_C(Tx) \le p_B(x)$ so (iv) follows.

It will frequently happen that the gauge, or pre-gauge sets that arise in practice are *not* closed, as in Proposition 3.1(iii) and (iv). Nevertheless, with some additional hypotheses it turns out the Minkowski functionals of the sets and their closures are the same. To formulate this we require the notion of the distance from a point x to a set A via a gauge set B, defined by

$$d(A, B)(x) = \inf\{r \geq 0 : x \in A + rB\}.$$

We will abbreviate this by d_B, or just d, if the sets A and B are understood. We also require the closure of d, given by

$$\bar{d}(A, B) = \inf\{r \geq 0 : x \in (A + rB)^{-}\}.$$

The Gauge Lemma gives sufficient conditions for $d(A, B) = \bar{d}(A, B)$ on a domain set D.

3.2. LEMMA (GAUGE LEMMA). *Let A, B, D be convex sets containing* 0 *in the normed linear space E and satisfying the following hypotheses:*

(i) *$D + \langle B\rangle \subset D$, where $\langle B\rangle$ is the linear subspace spanned by B,*
(ii) *$\bar{d}(A, B) \leq \alpha d(A, E_1)$ on D for some $\alpha > 0$,*
(iii) *any one of*
  (a) *A is closed and B is a gauge set,*
  (b) *A is compact and B is a pre-gauge set, or*
  (c) *A is complete and B is a bounded pre-gauge set.*

Then $d(A, B) = \bar{d}(A, B)$ on D.

(Note that $d(A, E_1)$ in (ii) is the usual distance from a point to A.)

Proof. Denote $d(A, E_1)$ by δ. We have $\bar{d} \leq d$ by definition. Let $x_0 \in D$ with $\bar{d}(x_0) < r_0 < r < \infty$. Choose a sequence $(r_n)_{n=1}^{\infty}$ of positive numbers with

$$r_1 > r_0 \quad \text{and} \quad \sum_{n=1}^{\infty} r_n < r.$$

Since $\bar{d}(x_0) < r_0$ we have

$$x_0 = a_1 + b_1 + c_1; \qquad a_1 \in A, \qquad p_B(b_1) < r_1 \quad \text{and} \quad \|c_1\| < r_2/\alpha.$$

Then, by (i), $x_0 - b_1 \in D$ and $x_0 - b_1 = a_1 + c_1$ implies $\delta(x_0 - b_1) < r_2/\alpha$. Hence

$$x_0 - b_1 = a_2 + b_2 + c_2; \qquad a_2 \in A, \qquad p_B(b_2) < r_2 \quad \text{and} \quad \|c_2\| < r_3/\alpha.$$

Continuing by induction we obtain sequences (a_n), (b_n) and (x_n) such that

$$a_n \in A, \quad p_B(b_n) < r_n, \quad \|c_n\| < r_{n+1}/\alpha \quad \text{and} \quad x_0 = a_n + (b_1 + \cdots + b_n) + c_n.$$

Thus, if (iii) (a) holds then there exists

$$b = \sum b_n \quad \text{with } p_B(b) \leq \sum p_B(b_n) < r.$$

Hence a_n converges to $a \in A$ and so $x_0 \in A + rB$. If (b) holds then a subsequence (a_m) converges in A and so $\sum b_n = b$ exists with $p_B(b) < r$. In case (c) we have

$$\|a_{n+1} - a_n\| = \|b_{n+1} + c_{n+1} - c_n\| \leq Mr_{n+1} + (1/\alpha)(r_{n+1} + r_n),$$

where M is a bound for B. Thus (a_n) is Cauchy so that (a_n) converges to $a \in A$ and consequently $b = \sum b_n$ exists with $p_B(b) < r$.

In making applications of the Gauge Lemma we shall adopt the convention of writing $A \leftrightarrow \cdot$, $B \leftrightarrow \cdot$, etc. where we fill in the blanks with the names of the sets currently playing the roles of A, B, D. The meaning of d, $\bar{d}$ and δ will follow accordingly. We say B is *D-regular* if $p(B) = p(\bar{B})$ on D.

3.3. COROLLARY. *Let D be a closed convex cone and let B be a pre-gauge set with $\pm B \subset D$. If $D_1 \subset \alpha\bar{B}$ for some $\alpha > 0$ then B is D-regular and $D_1 \subset \alpha' B$ for all $\alpha' > \alpha$.*

Proof. Apply the Gauge Lemma with $A \leftrightarrow \{0\}$. Then $d, \bar{d}, \delta$ are $p(B), p(\bar{B})$ and $\|\cdot\|$. Thus $D_1 \subset \alpha\bar{B}$ says

$$\bar{d} \leq \alpha\delta$$

so that the conclusion follows.

3.4. COROLLARY. *If D is a complete subspace and B a pre-gauge in D such that $D = \bigcup_{n=1}^{\infty} n\bar{B}$ then $D_1 \subset \alpha' B$ for all α' greater than some α. Also, B is D-regular.*

Proof. Since D is a subspace, $\pm B \subset D$. Since $0 \in \operatorname{cor}(\bar{B})$ in D, the Baire Category Theorem shows $D_1 \subset \alpha\bar{B}$ for some α. Thus Corollary 3.3 applies.

Let E, F be Banach spaces with T a bounded linear map from E onto a dense subspace of F. Let $T^*: F^* \to E^*$ be the adjoint and denote $X \doteq T^*(F_1^*)$. Then X is w^*-compact, balanced and convex. Let N be the linear span

$$\langle X \rangle = \bigcup_{n=1}^{\infty} nX.$$

We now give some of the standard completeness results: for example, the equivalence of (i) and (ii) is the Open Mapping Theorem and the equivalence of (i) and (v) is the Closed Range Theorem.

3.5. THEOREM. *The following are equivalent:*

(i) *$T: E \to F$ is onto;*
(ii) *$F_1 \subset T(E_s)$ for all s greater than some r;*
(iii) *$(T(E))_1 \subset T(E_s)$ for all $s > r$;*
(iv) *T^* has w^*-closed range N;*
(v) *T^* has norm-closed range N;*
(vi) *$N_1 \subset rX$, some $r > 0$.*

Proof. Let $B = T(E_1)$. Then Corollary 3.4 applies (with $D \leftrightarrow F$) and hence (i) implies (ii). That (ii) implies (iii) is trivial. (iii) implies, for $\alpha < \beta < 1$, $r < s$

$$F_\alpha = [\mathrm{cl}\ T(E)]_\alpha \subset \mathrm{cl}\,[T(E)_\beta] \subset \mathrm{cl}\,[T(E_{\beta s})].$$

Thus $F_1 \subset r\bar{B}$ so Corollary 3.3 yields (ii) and hence, (i).

Now (iv) implies (v) and (v) yields (vi) by applying the equivalence of (i) and (ii) in the dual spaces. If (vi) holds then $N_1 = (rX) \cap E_1^*$ is w^*-compact. Therefore, by the Krein–Šmulyan Theorem, N is w^*-closed so that (iv), (v) and (vi) are equivalent. Finally, (vi) says precisely that

$$(T^*)^{-1}(E_1^*) \subset rF_1^*$$

which, by the polar calculus, is equivalent to

$$D_1 \subset r\bar{B}.$$

However, as we have seen, this is equivalent to (ii), using Corollary 3.3.

4. SUBSPACES OF $C(X)$ AND AFFINE FUNCTIONS

In this section we introduce affine function spaces and discuss their representations as real subspaces of $C(X)$. To begin, we take X as a compact Hausdorff space and let $C(X)$ denote the (real or complex-valued) continuous functions on X. Then $C(X)$ is a Banach space with the usual sup-norm: $\|f\| = \sup\{|f(x)|: x \in X\}$.

We first discuss *state spaces* in the more general setting of subspaces of complex-valued functions. However, the results are all valid in the real space as well and we shall restrict ourselves to subspaces of $C_{\mathbb{R}}(X)$ in the last part of this section.

Thus, let M be a (complex) subspace of $C(X)$ such that M separates the points of X and contains the constant functions. Let M^* be the dual Banach space. If $L \in M^*$ we say $L \geq 0$ providing $L(f) \geq 0$ whenever f is a non-negative real-valued function in M.

4.1. PROPOSITION. *If $L \in M^*$ satisfies $L(1) = 1 = \|L\|$ then, for each $f \in M$, $L(f) \in \overline{\mathrm{co}}\, f(X)$.*

Proof. We observe that $\overline{\mathrm{co}}\, f(X)$ in $\mathbb{C}$ is the intersection of the closed discs $D(\alpha, r)$ (centre α, radius r) which contain $f(X)$. For each such disc we have

$$\|f - \alpha 1\| = \sup\{|f(x) - \alpha|: x \in X\} \leq r$$

so that

$$|L(f) - \alpha| = |L(f) - \alpha L(1)| = |L(f - \alpha 1) \leq \|L\|\,\|f - \alpha 1\| \leq r.$$

Hence $L(f) \in D(\alpha, r)$.

We say M is *self-adjoint* if $f \in M$ implies the complex-conjugate $\bar{f} \in M$. In this case re f and im f belong to M and $L \in M^*$ with $L \geq 0$ implies re $L(f) = L(\text{re } f)$.

4.2. PROPOSITION. *If $L \in M^*$ then $L(1) = \|L\|$ implies $L \geq 0$. If M is self-adjoint then $L \geq 0$ implies $L(1) = \|L\|$.*

Proof. If $L(1) = \|L\| > 0$ and $f \geq 0$ then, by Proposition 4.1,

$$L(f)/\|L\| \in \overline{\text{co}}\, f(X) \subset R^+.$$

Hence $L(f) \geq 0$.

Let $L \geq 0$ and $\|f\| \leq 1$, with M self-adjoint. Choose $\alpha \in \mathbb{C}$ such that $|\alpha| = 1$ and $|L(f)| = \alpha L(f)$. Then

$$|L(f)| = L(\alpha f) = \text{re } L(\alpha f) = L(\text{re } \alpha f) \leq L(1).$$

Hence $\|L\| \leq L(1) \leq \|L\|$.

Definition. The *state space* of M, denoted S_M, equals

$$\{L \in M^* \colon L(1) = 1 = \|L\|\}.$$

Since S_M is the intersection of a w^*-closed hyperplane with the unit ball M_1^* we see that S_M is w^*-compact and convex. We define $\phi : X \to M^*$ (the evaluation map) by $\phi(x) = L_x$ with $L_x(f) = f(x)$. Then it is easy to see that ϕ is continuous and $\phi(X)$ is a compact subset of S_M.

4.3. THEOREM. *The state space $S_M = \overline{\text{co}}\, \phi(X)$.*

Proof. Let $S_0 = \overline{\text{co}}\, \phi\, (X)$. If $f \in M$ then $f(X) \subset \{L(f);\ L \in S_0\} \doteq f(S_0)$ and hence, for $L_0 \in S_M$,

$$L_0(f) \in \overline{\text{co}}\, f(X) \subset f(S_0).$$

But then re $L_0(f) \leq \max\,\{\text{re } L(f) \colon L \in S_0\}$ so that, by the Separation Theorem, $L_0 \in S_0$.

The Krein–Milman Theorem yields the following.

4.4. COROLLARY. ext $S_M \subset \phi(X)$.

The subset $\phi^{-1}(\text{ext } S_M)$, denoted $\partial_M X$, is called the *Choquet boundary* of M. In general it need not be closed or even a Borel set. As we shall see, however, it does have the property of being a *boundary*, meaning that each $f \in M$ attains its norm on $\partial_M X$.

4.5. THEOREM. *If $f \in M$ then there is an $x_0 \in \partial_M X$ such that*

$$|f(x_0)| = \|f\|.$$

Proof. Choose $x \in X$ such that $\|f\| = \alpha f(x)$, with $|\alpha| = 1$. Let

$$T = \{L \in S_M : L(\alpha f) = \|f\|\}.$$

Then $L_x \in T$ and hence T is a non-empty compact convex set. Moreover, since T is the subset of S_M where the continuous linear functional αf attains its maximum, it is easy to see that

$$\text{ext } T \subset \text{ext } S_M.$$

Thus, the Krein–Milman Theorem guarantees an $x_0 \in \partial_M X$ with $\phi(x_0) = L_{x_0} \in \text{ext } T$.

4.6. COROLLARY. *The unit ball M_1^* equals $\overline{\text{cob}}\, \phi(\partial_M X)$.*

Proof. Given $f \in M$ there is an $x_0 \in \partial_M X$ with

$$|f(x_0)| = \|f\|.$$

Thus for $L \in M_1^*$, $\text{re } L(f) \leq \|f\| = \alpha f(x_0)$ $(|\alpha| = 1) = \text{re } \alpha L_{x_0}(f)$. Thus $[\overline{\text{cob}}\, \phi(\partial_M X)]^0 \subset (M_1^*)^0$ and hence

$$M_1^* \subset \overline{\text{cob}}\, \phi(\partial_M X) \subset M_1^*.$$

Suppose K is a compact convex subset of a locally convex space F. It is convenient to assume the duality relationship of Section 1 between $F^* = E$ and F. The set K remains compact in the weak topology on F induced by E. We define the space $A(K)$ as the (Banach) subspace of $C_{\mathbb{R}}(K)$ consisting of the continuous real-valued *affine* functions on K. By affine, we mean $f(\lambda x + (1-\lambda)y) = \lambda f(x) + (1-\lambda) f(y)$, where $x, y \in K$ and $0 \leq \lambda \leq 1$.

We again have the evaluation map

$$\phi : K \to S_{A(K)} \doteq S \subset A(K)^*.$$

We let

$$\theta : E \to A(K); \qquad (\theta a)(x) = (a, x).$$

We will assume first that the function $1 = \theta a$ for some $a \in E$. This just amounts to having K contained in a closed hyperplane of F disjoint from 0. Let N denote the linear subspace of F spanned by K. Thus

$$N = \bigcup_{n=1}^{\infty} n \text{ co}\,(K \cup -K).$$

4.7. THEOREM.

(i) *The evaluation map* ϕ *is an affine homeomorphism of* K *onto* S;

(ii) *The map* ϕ *extends (uniquely) to a linear isomorphism (also denoted* ϕ*) of* N *onto* $A(K)^*$, *identifying* $\mathrm{co}\,(K \cup -K)$ *with* $A(K)_1^* = \mathrm{co}\,(S \cup -S)$;

(iii) $\theta^* = \phi^{-1}$;

(iv) θE *is dense in* $A(K)$.

Proof. It is easy to see that ϕ is affine, one-to-one and continuous. Thus Theorem 4.3 shows that $S = \overline{\mathrm{co}}\,\phi(K) = \phi(K)$. Since the spaces are real, (ii) follows from Corollary 4.6, where

$$\overline{\mathrm{cob}}\,\phi(\partial_{A(K)}K) = \mathrm{co}\,(S \cup -S).$$

For (iii), $(a, \theta^* \circ \phi z) = (\theta a, \phi z) = (a, z)$. Hence $\theta^* \circ \phi = \mathrm{id}: N \to N$. Finally,

$$(\theta E)^0 = (\theta^*)^{-1}(0) = \phi(0) = \{0\}.$$

This proves $(\theta E)^{00} = (\theta E)^- = A(K)$.

If K is not a closed hyperplane of F to begin with, we can let $K_0 = K \times \{1\} \subset F \times \mathbb{R}$. Clearly K_0 is an affine homeomorphic copy of K and $F \times \{1\}$ is a closed hyperplane in the duality $(E \times \mathbb{R}, F \times \mathbb{R})$ defined by

$$((a, r), (x, s)) = (a, x) + rs.$$

4.8. COROLLARY. *If* K *is a compact convex subset of* F *(locally convex) and* $E = F^*$ *then* $\theta E + \mathbb{R} \cdot 1$ *is dense in* $A(K)$.

Proof. Let $\theta_0: E \times \mathbb{R} \to A(K \times \{1\})$. Then Theorem 4.7(iv) shows functions of the form

$$f(x) = (a, x) + r = ((a, r), (x, 1)) \quad \text{are dense in } A(K).$$

In essence, Theorem 4.7 shows that K can be realized as the state space of a subspace ($A(K)$) of continuous functions. This is quite convenient and we shall frequently assume, without further explicit mention, that K is so situated. Conversely, if K is a state space to begin with, then the next result shows nothing changes.

4.9. THEOREM. *If* M *is a closed subspace of* $C_{\mathbb{R}}(X)$, *containing* 1 *and separating points, then* M *is isometrically isomorphic to* $A(S_M)$.

Proof. Since $S = S_M$ is in a closed hyperplane in the duality (M, M^*) we have θM dense in $A(S)$. But in this case Corollary 4.6 shows θ is an isometry, and hence onto, since M is closed.

5. REPRESENTING MEASURES

Let X be a compact Hausdorff space and let $\mathscr{C}_0$ denote the collection of all compact G_δ's (countable intersections of open sets) in X. Then $\mathscr{C}_0$ is a set *lattice* (closed under finite intersections and unions) and it follows easily from this that the collection $\mathscr{A}$, consisting of sets that are finite disjoint unions of differences, $A\backslash B$ $(A, B \in \mathscr{C}_0)$, is an *algebra* of sets. If $\mathscr{B}_0$ is the smallest σ-algebra containing $\mathscr{A}$ (and hence the σ-algebra spanned by $\mathscr{C}_0$) then elements of $\mathscr{B}_0$ are referred to as *Baire* sets. It can also be described as the smallest σ-algebra for which the continuous functions on X are measurable. A finite (non-negative) measure μ on $\mathscr{B}_0$ is called a *Baire measure*. By building up from $\mathscr{C}_0$ it is easy to see that for any $B \in \mathscr{B}_0$, $\mu(B) = \sup\{\mu(A) : A \in \mathscr{C}_0 \text{ and } A \subset B\} = \inf\{\mu(V) : X\backslash V \in \mathscr{C}_0 \text{ and } B \subset V\}$. Hence we say a Baire measure is *regular*.

The Baire sets are a sub-class of the *Borel* sets, $\mathscr{B}$, comprising the smallest σ-algebra containing *all* open sets. A (finite non-negative) Borel measure μ is said to be *regular* if, for any $E \in \mathscr{B}$,

$$\mu(E) = \sup\{\mu(K) : K \text{ compact and } K \subset E\}$$

and

$$\mu(E) = \inf\{\mu(U) : U \text{ open and } E \subset U\}.$$

In contrast to the Baire measures, Borel measures are not automatically regular.

We say a regular Borel measure of total mass one is a probability measure on X and denote the set of such measures by $M_1^+(X)$. If $\mu \in M_1^+(X)$ then μ corresponds to a state $L_\mu \in S$, the state space of $C(X)$, given by

$$L_\mu(f) = \int_X f \, \mathrm{d}\mu.$$

That the converse is also true is the content of our next theorem. For a proof of this fundamental result, see for example, Rudin [**184**].

5.1. Theorem (Riesz Representation Theorem). *If $L \in S$, the state space of $C(X)$, then there exists a unique probability measure $\mu \in M_1^+(X)$ such that $f \in C(X)$ implies $L(f) = \int_X f \, \mathrm{d}\mu$.*

We note that the uniqueness follows from the fact that $A \in \mathscr{C}_0$ implies there is an $f \in C(X)$ with $f \equiv 1$ on A, $0 \le f < 1$ on $X\backslash A$. Hence $f^n \downarrow \chi_A$ so that $\mu(A) = \int \chi_A \, \mathrm{d}\mu = \lim L(f^n)$. Thus μ is determined by its values on $\mathscr{C}_0$ and in fact, for U open,

$$\mu(U) = \sup\{\mu(A) : A \subset U \text{ and } A \in \mathscr{C}_0\}.$$

We also have that for any $E \in \mathcal{B}$ there is a $B \in \mathcal{B}_0$ and a set N with

$$E = B \triangle N; \qquad N \subset D \in \mathcal{B} \quad \text{and} \quad \mu(D) = 0.$$

The essence of the Riesz Representation Theorem is that the correspondence between $M_1^+(X)$ and the state space, S, of $C(X)$ is affine and bi-unique. Henceforth we shall tacitly identify the elements of S as probability measures on X.

Theorem 5.1 leads to an identification of the space $M_{\mathbb{R}}(X)$ of regular signed Borel measures with the real linear span N of S in $C(X)^*$. Then Theorem 4.9 shows that, with $N_1 = \text{co}\,(S \cup -S)$, N and $C_{\mathbb{R}}(X)$ are isometric to $A(S)^*$ and $A(S)$ respectively. If $\mu \geq 0$ in $M_R(X)$ then Proposition 4.2 shows $\|\mu\| = \mu(1) = \mu(X)$. In general

$$\|\mu\| = p(N_1)\mu = \inf\{\alpha + \beta : \alpha, \beta \geq 0 \text{ and } \mu = \alpha\mu_1 - \beta\mu_2;\ \mu_1, \mu_2 \in S\}.$$

This in turn gives

$$\|\mu\| = \mu^+(1) + \mu^-(1)$$

where $\mu = \mu^+ - \mu^-$ is the Jordan decomposition of μ. Thus we will always identify $M_R(X)$ with $C_{\mathbb{R}}(X)^*$. In Chapter 4 we will extend this identification to the space $M(X)$ of regular complex-valued Borel measures as $C_{\mathbb{C}}(X)^*$.

If we have a *subspace* M (containing constants and separating points) of $C(X)$, with the state space S_M, then the elements of S_M can also be associated with probability measures, although the correspondence is no longer one-to-one. To see this, let $M^\perp$ $(= M^0)$ be the annihilator subspace of M in $C(X)^*$. We let

$$q : C(X)^* \to C(X)^*/M^\perp$$

be the *quotient* map and recall that $C(X)^*/M^\perp$ is isometric to M^* with $(qL)(f) = L(f)$. Because q is w^* to w^* continuous, the unit ball $C(X)_1^*$ is mapped onto M_1^*. In particular, from Proposition 4.2 q maps S onto S_M. Thus we have the following result.

5.2. THEOREM. *Let M be a separating subspace, containing the constants, in $C(X)$. Then $L \in S_M$ if and only if there is a probability measure μ on X with*

$$L(f) = \int_X f \, \mathrm{d}\mu.$$

In this case, we say μ *represents* L, meaning $q\mu = L$.

If $\phi : X \to S_M$ is the evaluation map, then $\phi(X)$ is a homeomorphic copy of X as a compact subset of S_M containing ext S_M (Theorem 4.3 and Corollary 4.4). Hence we can carry a $\mu \in M_1^+(X)$ to S_M by defining $\bar{\mu}(\phi(B)) = \mu(B)$

and $\bar{\mu}(S_M \backslash \phi(X)) = 0$. Since $f(x) = \phi(x)(f) = L_x(f)$, we can write

$$L(f) = \int_X f \, d\mu = \int_{S_M} f \, d\bar{\mu},$$

where, in the second integral, we evaluate f on elements of S_M via the duality $(M'M^*)$.

As we have seen, if $X = K$, a compact *convex* set, and $M = A(K)$ then ϕ identifies K with S_M and, in this case, each $L \in S_M$ can be represented by the point mass measure ε_x on K where $\phi(x) = L$.

On the other hand, if we start with K, embedded as the state space of $A(K)$, then we can take $X = (\text{ext } K)^-$.

5.3. THEOREM. *If K is a compact convex subset of a locally convex space and $X = (\text{ext } K)^-$ then for each $z \in K$ there is a probability measure μ with* supp $\mu \subset X \subset K$ *and $f(z) = \int_K f \, d\mu$ for all $f \in A(K)$.*

Proof. Let $\theta : A(K) \to M \subset C(X)$ be the restriction map. Since X is a boundary for $A(K)$, θ is an isometry onto M. Let $\theta^* : M^* \to A(K)^*$ be the adjoint. Clearly, θ^* maps S_M one-to-one onto K as the state space of $A(K)$. Hence, we can choose $\mu \in M_1^+(X)$ with $\theta^* q\mu = z$.

Then for $f \in A(K)$,

$$f(z) = (f, z) = (f, \theta^* q\mu) = (\theta f, q\mu) = \int_X f \, d\mu = \int_K f \, d\mu$$

where in the last integral we consider X as a subset of K with $\mu(K \backslash X) = 0$.

5.4. COROLLARY. *Let M be a separating subspace of $C(X)$ with constants. Then $L \in S_M$ if and only if there is a $\mu \in M_1^+(X)$ representing L and* supp $\mu \subset (\partial_M(X))^-$ $(= (\text{ext } S_M)^-)$.

Proof. By definition $\partial_M(X) = \text{ext } S_M$ and Theorem 5.3 applies with $K = S_M$.

Of course Theorem 5.3 and Corollary 5.4 are two sides of the same coin. This raises the question of whether we can further restrict the representing measure to live precisely on ext K. For example if K is a finite-dimensional polytope then each $z \in K$ is a convex combination of extreme points and the corresponding representing measure is the same convex combination of point-mass measures. The problem in the general setting is complicated by the fact that ext K need not be a Borel set, let alone compact. Thus we have gone as far as we can with the Riesz Representation Theorem. If, however, K is metrizable then ext K is in fact a G_δ set, and we *can* find a representing measure μ on K with $\mu(\text{ext } K) = 1$. This is the Choquet Theorem for K

metrizable and the corresponding result in the more general setting is called the Choquet–Bishop–de Leeuw Theorem.

6. MAXIMAL MEASURES AND THE CHOQUET–BISHOP–DE LEEUW THEOREM

To prove these facts requires some machinery involving the convex functions on K and their graphs in $K \times \mathbb{R}$. To begin with, if E is a locally convex space and f is an extended real-valued function on E we call

$$\{(a, r) \in E \times \mathbb{R}: f(a) = r\}$$

the *graph* of f, Since f is extended, the domain of the graph may be proper. If $F \in (E \times \mathbb{R})^*$ then F has the form

$$F(a, r) = (a, a^*) + rs$$

for some $a^* \in E^*$ and $s \in \mathbb{R}$. We say F is *vertical* if $F(0, 1) = 0$ (equivalently, $s = 0$). If H is the hyperplane $F \equiv c$, where F is *not* vertical then H is the graph of a unique $f = (\cdot, b^*) + t$ (translate of a linear functional) on E. In fact $(a, r) \in H$ if and only if $(a, -a^*/s) + c/s = r$ so that $f = (\cdot, -a^*/s) + c/s$.

Now consider K as the state space of $A(K)$. A real function f on K is *convex* if $f(\lambda x + (1-\lambda)y) \leq \lambda f(x) + (1-\lambda)f(y)$ whenever $x, y \in K$ and $0 \leq \lambda \leq 1$. We think of f as identically $+\infty$ off K so that the graph of f is a subset of $K \times R$ in $A(K)^* + \mathbb{R}$. Define the *epigraph* of f by

$$\operatorname{epi} f = \{(x, r) \in K \times \mathbb{R}: r \geq f(x)\}.$$

Then f is convex if and only if epi f is convex. Let $Q(K)$ denote the subcone of $C(K)$ consisting of the continuous convex functions. An important subset of $Q(K)$ consists of functions of the form $a_1 \vee \cdots \vee a_n$; $a_i \in A(K)$, $i = 1, \ldots, n$. We say f is *lower-semicontinuous* (l.s.c.) if

$$\{x \in K : f(x) \leq r\}$$

is closed for all real r. Then f is l.s.c. if and only if epi f is closed. We let $\check{Q}(K)$ denote the l.s.c. convex functions on K. Since K is compact, elements of $\check{Q}(K)$ are bounded below.

If f is any function on K, bounded below, let

$$A_f = \{a \in A(K) : a \leq f\}.$$

Let I denote the collection of finite subsets of A_f. Then I is a net directed by inclusion and for each $\alpha \in I$ we let

$$b_\alpha = \vee\{a : a \in \alpha\}.$$

Since $\alpha \subset \beta$ implies $b_\alpha \leq b_\beta$ we have $\lim_{\alpha \in I} b_\alpha = \sup\{a : a \in A_f\}$, which defines the *lower-envelope*, $\check{f}$, of f. The following theorem is a compilation of some of the important properties of $\check{f}$.

6.1. THEOREM.

(i) *If $f \in Q(K)$ then there is a sequence $(b_n) \subset Q(K)$ such that each b_n is the supremum of a finite family in A_f and*

$$b_n \uparrow f \text{ uniformly};$$

(ii) *if f is bounded below on K then $\check{f} \in \check{Q}(K)$;*

(iii) $\operatorname{epi} \check{f} = \overline{\operatorname{co}}\,(\operatorname{epi} f)$ *in $K \times R$;*

(iv) *if $g \in \check{Q}(K)$ and $g \le f$ then $g \le \check{f}$;*

(v) $(f_1 + f_2)\check{} \ge \check{f}_1 + \check{f}_2$ *(f_1, f_2 bounded below);*

(vi) $\alpha \ge 0$ *implies* $\alpha\check{f} = (\alpha f)\check{}$;

(vii) *if f is continuous,*

$$\check{f}(x) = \inf\left\{\sum_{i=1}^{n} \lambda_i f(x_i) : x = \sum_{i=1}^{n} \lambda_i x_i;\ x_i \in K, \sum \lambda_i = 1\right\};$$

(viii) *if $\mu \in M_1^+(K)$ then $\int_K \check{f}\,d\mu = \sup\{\int_K b_\alpha\,d\mu : \alpha \in I\}$ (f bounded below);*

(ix) *$f \in \check{Q}(K)$ is affine if and only if*

$$\{a \in A(K) : a < f\}$$

is directed. In this case $\mu(f) = f(x)$ whenever μ is a probability measure on K representing x;

(x) *if f is convex and u.s.c. then*

$$A^f = \{g \in Q(K) : g > f\}$$

is a net, directed down by $\ge$, with

$$f = \lim\{g : g \in A^f\} = \inf\{g : g \in A^f\}.$$

Proof. (i) Given $\varepsilon > 0$, for each $x_0 \in K$ and $r < f(x_0)$, $(x_0, r) \notin \operatorname{epi} f$ so that there is an $F \in A(K)^* \times \mathbb{R}$ with

$$F(x_0, r) < c \le \inf\{F(x, s) : (x, s) \in \operatorname{epi} f\}.$$

Thus F is not vertical (it increases along $\{x_0\} \times \mathbb{R}$) so there is an $a \in A_f$ with $a(x_0) > r$. Thus if $r = f(x_0) - \varepsilon/2$ we can find a neighbourhood of x_0 on which $a(x) > f(x) - \varepsilon$. Then a simple compactness argument shows there are $a_1, \ldots, a_n \in A_f$ with

$$f - \varepsilon < a_1 \vee \cdots \vee a_n \le f.$$

Since the supremum of a finite number of such functions is of the same type, (i) follows.

(ii) Since $\check{f}$ is an increasing limit of the net $(b_\alpha)_{\alpha\in I}$ in $Q(K)$, $\check{f}$ is l.s.c. convex, and hence in $\check{Q}(K)$.

(iii) Since, for each α, epi b_α is a closed convex set containing epi f we have

$$\operatorname{epi}\check{f}=\bigcap_{\alpha\in I}\operatorname{epi}b_\alpha\supset\overline{\operatorname{co}}\,(\operatorname{epi}f).$$

If $(x_0, r)\notin\overline{\operatorname{co}}\,(\operatorname{epi}f)$ then the above graph-separation argument produces $a\in A_f$ with $a(x_0)>r$. Hence $(x_0, r)\notin\operatorname{epi}\check{f}$ and (iii) is shown.

Part (iv) follows directly from (ii) and (iii). Parts (v) and (vi) are consequences of

$$A_{f_1}+A_{f_2}\subset A_{(f_1+f_2)}\quad\text{and}\quad \alpha A_f=A_{\alpha f}.$$

(vii) Let $\bar{f}(x)$ be given by the right-hand side. Then

$$\bar{f}(x)=\inf\{r:(x,r)\in\operatorname{co}(\operatorname{epi}f)\}$$

and hence $\bar{f}$ is the convex function defined by co (epi f). Thus epi $\bar{f}\supset$ epi $\check{f}$ so that $\check{f}\le\bar{f}$. Now, given $\varepsilon>0$, for each $x\in K$ choose a compact convex neighbourhood N_x of x in K on which $|f-f(x)|<\varepsilon/2$. Let $N_{x_1},\ldots,N_{x_n}$ be a finite subcover of K. Let $I_i=\{r\in\mathbb{R}:r\ge f(x_i)-\varepsilon/2\}$. Then epi $f\subset\bigcup_{i=1}^n N_{x_i}\times I_i$ so that epi $\check{f}\subset\overline{\operatorname{co}}\bigcup(N_{x_i}\times I_i)=\operatorname{co}\bigcup(N_{x_i}\times I_i)$ by virtue of the fact that each N_{x_i} is compact convex. Thus, for $x\in K$,

$$(x,\check{f}(x))=\sum_{i=1}^n\lambda_i(y_i,r_i);\qquad y_i\in N_{x_i},\qquad r_i\ge f(x_i)-\frac{\varepsilon}{2}.$$

Hence,

$$x=\sum_{i=1}^n\lambda_i y_i$$

and

$$\bar{f}(x)\le\sum_{i=1}^n\lambda_i f(y_i)\le\sum_{i=1}^n\lambda_i\left(f(x_i)+\frac{\varepsilon}{2}\right)\le\sum\lambda_i(r_i+\varepsilon)=\check{f}(x)+\varepsilon.$$

(viii) Choose an increasing *subsequence* $(b_n)_{n=1}^\infty$ such that

$$\sup\mu(b_n)=\sup\mu(b_\alpha)\qquad\left(\mu(b)\text{ means }\int_K b\,d\mu\right).$$

Let $f_0=\sup b_n=\lim b_n$. Then, by the Monotone Convergence Theorem

$$\mu(f_0)=\sup\mu(b_n).$$

If $\mu(\check{f})>\mu(f_0)$ then, since μ is regular, there is a compact set F and a rational r such that $\mu(F)>0$ and

$$\check{f}(x)>r\ge f_0(x)\quad\text{for all }x\in F.$$

Using the compactness of F and the lower-semicontinuity of $\check{f}$ we can find b_β ($\beta \in I$) and $\varepsilon > 0$ such that

$$b_\beta \geq r + \varepsilon \quad \text{on } F.$$

If $c_n = b_n \vee b_\beta$ then $c_n = b_\beta \geq b_n + \varepsilon$ on F and

$$\mu(c_n) \geq \mu(b_n) + \varepsilon\mu(F).$$

But then

$$\sup \mu(b_\alpha) \geq \sup \mu(c_n) \geq \sup \mu(b_n) + \varepsilon\mu(F) > \sup \mu(b_n).$$

Thus

$$\mu(\check{f}) = \mu(f_0) = \sup \mu(b_n) = \sup \mu(b_\alpha) \leq \mu(\check{f}).$$

(ix) If f is affine and a_i, $a_2 < f$ then

$$\text{co}\,(\text{graph } a_1 \cup \text{graph } a_2)$$

is compact, convex and disjoint from epi f so that graph separation yields an a such that $a_1 \vee a_2 < a < f$. The converse is clear. Using (viii) and the above, we have

$$\int_K f \, d\mu = \sup \mu(b_\alpha) = \sup \{\mu(a) : a \in A(K) \text{ and } a < f\} = \sup \{a(x) : a < f\}$$

$$= \check{f}(x) = f(x).$$

(x) If h is continuous and $h \geq f + \varepsilon$ then, given x with

$$x = \sum_{i=1}^{n} \lambda_i x_i, \qquad \sum_{i=1}^{n} \lambda_i = 1$$

and $x_i \in K$, we have

$$\sum_{i=1}^{n} \lambda_i h(x_i) \geq \sum_{i=1}^{n} \lambda_i f(x_i) + \varepsilon \geq f(x) + \varepsilon$$

so that, by part (vii), $h \geq f + \varepsilon$. Now for each $x \in K$ choose $a_x \in A(K)$ with $a_x \leq h$ and $a_x(x) > f(x) + \frac{1}{2}\varepsilon$. Then x belongs to a neighbourhood N_x with $a_x > f + \frac{1}{2}\varepsilon$ on N_x. Thus there is a finite set $a_1, \ldots, a_m$ in $A(K)$ such that if

$$g = a_1 \vee \cdots \vee a_m$$

then $g \in Q(K)$ and $f + \frac{1}{2}\varepsilon < g < h$. Since f is u.s.c., g_1, $g_2 \in A^f$ implies $h = g_1 \wedge g_2 \geq f + \varepsilon$ for some $\varepsilon > 0$. Hence the above shows that A^f is directed and, since f is the infimum of a family of continuous functions A^f converges to f.

If f is bounded above then we can define the *upper envelope* of f by

$$\hat{f}(x) = -(-f)\check{}(x).$$

Then

$$\hat{f}(x) = \inf\{a(x) : a \geq f, a \in A(K)\}$$

and

(xi) $$(f_1 + f_2)\hat{} \leq \hat{f}_1 + \hat{f}_2; \qquad (rf)\hat{} = r\hat{f} \quad (r \geq 0).$$

We note also that if $a \in A(K)$ then

(xii) $$(a + f)\hat{} = a + \hat{f}$$

and

(xiii) $$r\hat{f} \leq (rf)\hat{} \quad (r \leq 0).$$

(To obtain (xiii), apply (xi) to $0 = (rf - rf)\hat{}$.) Let $q : M_1^+(K) \to K$, where q is the restriction to $M_1^+(K)$ of the quotient map $C(K)^* \to C(K)^*/A(K)^\perp = A(K)^*$.

We let $S_x = q^{-1}(x)$. Thus $\mu \in S_x$ if and only if μ represents x; i.e. $\int_K a \, d\mu = a(x)$ for all $a \in A(K)$. We use the notation $\mu \sim x$. Note that the point mass $\varepsilon_x \in S_x$.

6.2. PROPOSITION. *If f is continuous then, $\hat{f}(x) = \sup\{\mu(f) : \mu \in S_x\}$ and $\check{f}(x) = \inf\{\mu(f) : \mu \in S_x\}$.*

Proof. We have $a \leq f$ implies $a(x) = \mu(a) \leq \mu(f)$ so that $\check{f}(x) \leq \mu(f)$. But $x = \sum \lambda_i x_i$ implies $\mu = \sum \lambda_i \varepsilon_{x_i} \in S_x$ and $\mu(f) = \sum \lambda_i f(x_i)$. Hence Theorem 6.1(vii) shows equality.

6.3. THEOREM. *The following are equivalent*:

(i) $x \in \operatorname{ext} K$;
(ii) $S_x = \{\varepsilon_x\}$;
(iii) $\check{f}(x) = f(x)$ *for all* $f \in C(K)$;
(iv) $\check{f}(x) = f(x) = \hat{f}(x)$ *for all* $f \in C(K)$.

Proof. If (i) holds then $S_x = q^{-1}(x)$ is a face of $M_1^+(K)$. Thus $\nu \in \operatorname{ext} S_x$ implies $\nu \in \operatorname{ext} M_1^+(K)$. But then, by Corollary 4.4 $\nu = \varepsilon_y$, $y \in K$. Then $a(y) = a(x)$ for all $a \in A(K)$, so that $y = x$. Hence $S_x = \{\varepsilon_x\}$. Since $\check{f}(x) = \inf\{\mu(f) : \mu \in S_x\}$, we have (ii) implies (iii). If (iii) holds for all $f \in C(K)$ then

$$\hat{f}(x) = -(-f)\check{}(x) = -(-f)(x) = f(x).$$

If (iv) holds at x and $x = \lambda y + (1 - \lambda)z$ with $x \notin \{y, z\}$, choose $f \in C(K)$ such that $f(x) = 1$ and $f(y) = 0 = f(z)$. Then $\check{f}(x) \leq 0$ and $\hat{f}(x) \geq 1$. Hence $x = y = z$ and so $x \in \operatorname{ext} K$.

Our object now is to find a mechanism for choosing measures in S_x where mass is tending to concentrate on ext K. Toward this end we define an ordering on the elements of $M_1^+(K)$ by $\mu > \nu$ if and only if $\mu(f) \geq \nu(f)$ for all $f \in Q(K)$. It follows from Theorem 6.1(viii) that $\mu > \nu$ implies $\mu(f) \geq \nu(f)$ for $f \in \check{Q}(K)$.

6.4. THEOREM.

(i) *$\mu > \nu$ implies $q\mu = q\nu$. If $\mu \in S_x$ then $\mu > \delta_x$.*

(ii) *For each ν there is a μ which is maximal with respect to $>$ and $\mu > \nu$.*

(iii) *If $f \in C(K)$ and $\nu \in M_1^+(K)$ then there is a $\mu > \nu$ such that $\mu(f) = \nu(\hat{f})$.*

Proof. If $a \in A(K)$ then $\pm a \in Q(K)$ so that $\mu(a) = \nu(a)$. The second part of (i) follows from Proposition 6.2.

To show (ii) we show every chain with respect to $>$ is bounded above. If $\mathcal{J}$ is such a chain then $\mathcal{J}$ is a net in $M_1^+(K)$ and hence there is a subnet, which we denote (ν_α) which converges w^* to ν_0. If $f \in Q(K)$ and $\nu \in \mathcal{J}$ then for each $\varepsilon > 0$ we can choose $\nu_\alpha > \nu$ and $|(\nu_0 - \nu_\alpha)(f)| < \varepsilon$. Then

$$\nu_0(f) \geq \nu_\alpha(f) - \varepsilon \geq \nu(f) - \varepsilon.$$

Hence $\nu_0 > \nu$. For (iii), we let $\phi(g) = \nu(\hat{g})$ define a sub-additive, positive homogeneous function on $C(K)$ such that

$$\phi(a+g) = \nu(a) + \phi(g) \quad \text{and} \quad r\phi(g) \leq \phi(rg)$$

(using (xi), (xii), and (xiii) above). Furthermore, if $A = \{g \in C(K) : \phi(g) \leq 1\}$ then

$$C(K)_1 - P \subset A \quad (P = \{g : g \geq 0 \text{ on } K\})$$

so that A^0 is a w^*-compact convex subset of *non-negative* measures in $C(K)_1^*$.

If $f \in A(K)$ then $f = \hat{f}$ so that we can take $\mu = \nu$. If $f \notin A(K)$ we can find $\eta \in A(K)^\perp$ and $\eta(f) = \phi(f) - \nu(f)$. Thus if $\bar{\nu} = \nu + \eta$ we have

$$\bar{\nu}(a + rf) = \nu(a) + r\phi(f) \leq \nu(a) + \phi(rf) = \phi(a + rf).$$

This shows, for $M = A(K) + \mathbb{R}f$,

$$\bar{\nu} \in (A \cap M)^0 = A^0 + M^\perp \quad (\text{since } A^0 \text{ is } w^*\text{-compact}).$$

Hence $\bar{\nu} = \mu + \zeta$ with $\mu \in A^0$ and $\zeta \in M^\perp$. It follows that

$$\mu = \nu \quad \text{on } A(K) \qquad \text{and} \qquad \mu(f) = \bar{\nu}(f) = \nu(\hat{f}).$$

Finally, if $g \in -Q(K)$ and $\nu(g) > 0$ then $g = \hat{g}$ and $g/\nu(g) \in A$. Since $\mu \in A^0$, $\mu(g) \leq \nu(g)$. But then $g \in -Q(K)$ implies $\nu(g + c) > 0$ for c a sufficiently large constant and $\mu(c) = \nu(c) = c$, since $c \in A(K)$. Thus $\mu \geq \nu$ on $Q(K)$ so that $\mu > \nu$.

We can now characterize the maximal measures by their preservation of integral values for envelopes of functions.

6.5. THEOREM. *Let $\mu \in M_1^+(K)$, then the following are equivalent:*
(i) *μ is maximal;*
(ii) *$\mu(\hat{f}) = \mu(f) = \mu(\check{f})$ for all $f \in C(K)$;*
(iii) *$\mu(\check{f}) = \mu(f)$ for all $f \in Q(K)$.*

Proof. That (i) implies the first equality in (ii) is immediate from Theorem 6.4(iii). The second equality follows from $\check{f} = -(-f)\hat{}$. If (iii) holds for μ and $\nu > \mu$ then $f \in Q(K)$ implies $\mu(f) \leq \nu(f)$. But $-\hat{f} \in \check{Q}(K)$ and Theorem 6.1(viii) shows $\mu(-\hat{f}) \leq \nu(-\hat{f})$, or

$$\nu(\hat{f}) \leq \mu(\hat{f}) = \mu(f).$$

Thus $\nu(f) \leq \nu(\hat{f}) \leq \mu(f)$, so that $\mu > \nu$. Hence μ is maximal.

If $f \in C(K)$ and

$$B_f = \{x : \hat{f}(x) = f(x)\} = \bigcap_{n=1}^{\infty} \{x : \hat{f}(x) - f(x) < 1/n\}$$

then B_f is a G_δ set containing ext K. On the other hand, if μ is maximal then Theorem 6.5(ii) shows $\mu(K \backslash B_f) = 0$ for all $f \in C(K)$. Moreover, ext $K = \bigcap \{B_f : f \in C(K)\}$ by Theorem 6.3. It would appear then that maximal measures live mainly on ext K. We next make some precise statements of this observation.

First, if K is metrizable then K is separable and we can find a sequence $(a_n)_{n=1}^{\infty}$ of non-negative functions in $A(K)_1$ which separates the points of K. Hence the function

$$h = \sum_{n=1}^{\infty} a_n^2/2^n$$

is continuous and *strictly convex*, meaning

$$h(\lambda x + (1-\lambda)y) < \lambda h(x) + (1-\lambda)h(y)$$

whenever $0 < \lambda < 1$ and $x \neq y$ in K. This follows since a_n^2 is strictly convex along any segment joining x and y where $a_n(x) \neq a_n(y)$.

Now, if $x \notin \text{ext}\, K$ then $h(x) < \hat{h}(x)$ so that

$$B_h = \text{ext}\, K \quad (\text{a } G_\delta \text{ set in } K).$$

Thus, we have the Choquet Theorem:

6.6 THEOREM (CHOQUET'S THEOREM). *If K is a metrizable compact convex set then for each $x \in K$ there is a $\mu \in M_1^+(K)$ such that $\mu \sim x$ and $\mu(\text{ext}\, K) = 1$.*

Let $X = (\text{ext } K)^-$. If G is compact in K and disjoint from X then there is an $f \in C(K)$ with

$$0 \leq f \leq 1, \qquad f|_G \equiv 0 \quad \text{and} \quad f|_X \equiv 1.$$

Hence $\hat{f} \equiv 1$ on K. Thus, if μ is maximal then $\mu(\hat{f}) = \mu(f)$ implies $\mu(G) = 0$. This shows supp $\mu \subset X$.

In the general case we show that μ maximal implies $\mu(B) = 0$ for every Baire set B disjoint from ext K. To do this, we need a lemma which amounts to a "maximum principle" for point-wise limits of convex functions.

6.7. LEMMA. *Let $(h_n)_{n=1}^\infty$ be a decreasing sequence of bounded functions in $\check{Q}(K)$ and let $h_n \downarrow h_0 \geq 0$ on K. If $h_0 = 0$ on* ext K *then $h_0 \equiv 0$ on K.*

Proof. Let $x_0 \in K$ be given and choose a sequence $(a_n)_{n=1}^\infty$ in $A(K)$ such that

$$a_n \leq h_n \quad \text{and} \quad h_n(x_0) - 1/n < a_n(x_0).$$

Let $\phi : K \to K_0 \subset \prod_{n=1}^\infty \mathbb{R}$; $\phi(x) = (a_n(x))_{n=1}^\infty$. Then $K_0 = \phi(K)$ is a compact convex (and hence metrizable) set in $\prod_n \mathbb{R}$. Let $p_n \in A(K_0)$ be the projection onto the nth coordinate so that

$$p_n \circ \phi = a_n.$$

If $y \in \text{ext } K_0$ then $y = \phi(x)$ for some $x \in \text{ext } K$. Hence

$$\limsup p_n(y) = \limsup a_n(x) \leq \lim h_n(x) = h_0(x) = 0.$$

Now, using Theorem 6.6, we choose $\mu_0 \in M_1^+(K_0)$ such that μ_0 represents $y_0 = \phi(x_0)$ and $\mu_0(\text{ext } K_0) = 1$. Thus $\mu_0(p_n) = p_n(y_0) = a_n(x_0) \to h_0(x_0)$. But $(p_n)_{n=1}^\infty$ bounded above in $A(K_0)$ implies we can apply Fatou's Lemma to $(-p_n)_{n=1}^\infty$ to conclude

$$0 \geq \int_{\text{ext } K_0} \limsup p_n \, d\mu_0 = \int_{K_0} \limsup p_n \, d\mu_0 \geq \limsup p_n(y_0) = h_0(x_0).$$

Since x_0 was arbitrary, $h_0 \equiv 0$ on K.

6.8. THEOREM (CHOQUET–BISHOP–DE LEEUW). *If K is a compact convex set then for each $x \in K$ there is a (maximal) $\mu \in M_1^+(K)$ which represents x and has measure zero on any Baire set disjoint from* ext K.

Proof. If $B \in \mathscr{B}_0$ then $\mu(B) = \sup\{\mu(A) : A \in \mathscr{C}_0 \text{ and } A \subset B\}$. Hence let A be a compact G_δ disjoint from ext K. Let $0 \leq f \leq 1$, $f \equiv 1$ on A and $f < 1$ on $K \backslash A$, $f \in C(K)$. Then $f^n \downarrow \chi_A$ and, since $(f^n)^\vee = f^n$ on ext K, $(f^n)^\vee \downarrow f_0 = 0$ on ext K. Thus $f_0 \equiv 0$ and $\mu(A) = \lim \mu(f^n) = \lim \mu(f^n)^\vee = \mu(f_0) = 0$.

Let $X = (\text{ext } K)^-$. In some applications we begin with functions defined only on X. Many of the same results apply—we summarize the situation below. If f is bounded on X then the envelopes for f can be defined as before:

$$\check{f}(y) = \sup \{a(y) : a \in A(K) \text{ and } a \leq f \text{ on } X\}$$

$$\hat{f}(y) = \inf \{a(y) : a \in A(K) \text{ and } a \geq f \text{ on } X\}.$$

Thus, for example, $\check{f} \in \check{Q}(K)$ and $(\check{f})^\vee = \check{f}$, where the second $^\vee$ is in the original sense of Theorem 6.1. The results of Proposition 6.2 and Theorem 6.5 still hold.

6.9. THEOREM. *Let $X = (\text{ext } K)^-$ and $f \in C(X)$.*
(i) $\check{f}(y) = \inf \{\mu(f) : \mu \in M_1^+(X)$ *and* $\mu \sim y \in K\}$;
(ii) *if μ is maximal on K then $\mu(\check{f}) = \mu(f)$.*

Proof. Assume, by translating f, that $1 \leq f$ on X. Let A denote the restriction $A(K)|_X$ and let $y \in K$ be given. Then, using Theorem 5.3, choose $\mu \in M_1^+(X)$ representing y with respect to $A(K)$. Let

$$C_f = \{g \in C(X) : g \leq f\}$$

so that

$$A_f = \{a \in A : a \leq f \text{ on } X\} = A \cap C_f.$$

Then

$$\check{f}(y) = \sup \{a(y) : a \in A_f\} = \sup \{\mu(a) : a \in A_f\} = \rho(A_f)\mu$$

where $\rho(A_f)$ is the support functional of A_f. But

$$\rho(A_f) = p(A_f^0) = p(C_f^0 + A^\perp)$$

where the latter set is w^*-closed since $C(X)_1 - P \subset C_f$, implying that C_f is a w^*-compact set in $M(X)$. Thus

$$\mu = \nu + \eta; \qquad \rho(C_f)\nu = \rho(A_f)\mu \quad \text{and} \quad \eta \in A^\perp.$$

Thus, $\nu \geq 0$, $\nu(1) = 1$, $\nu \sim y$ and

$$\nu(f) = \sup \{\nu(g) : g \leq f\} = \rho(C_f)\nu = \rho(A_f)\mu = \check{f}(y).$$

To show (ii), let F be a continuous extension of f to K. Then $\check{F} \leq \check{f}$. But if μ is maximal then supp $\mu \subset X$ and Theorem 6.5 shows

$$\mu(f) = \mu(F) = \mu(\check{F}) \leq \mu(\check{f}) \leq \mu(f).$$

7. BISHOP–PHELPS THEOREMS

Let A be a closed convex set in the Banach space E and assume $0 \in A$. Let A^* denote the domain of the support function ρ_A in E^*. Then A^* is a convex cone consisting of the linear functionals in E^* that are bounded above on A. We say (a, x) forms a *support pair* for A if $a \in A$ and

$$(a, x) = \rho_A(x).$$

We show here that the set of support pairs in $A \times A^*$ is plentiful. More precisely, the (product space) projection of support pairs to A is dense in the boundary of A and the projection to A^* is (norm) dense in A^*.

The proof rests heavily on the notion of *support cones*. Let $x \in A^*(\|x\| = 1)$ and $0 < \gamma < 1$. The *support cone* $P(x, \gamma)$ is the cone in E spanned by

$$\{a \in E_1 : (a, x) \geq \gamma\}.$$

Since $\|x\| = 1$ and $\gamma < 1$ this set has non-empty interior and

$$P(x, \gamma) = \{a : (a, x) \geq \gamma\|a\|\}.$$

If $a \in A$ we denote

$$A(a, x, \gamma) = A \cap (a + P(x, \gamma)).$$

If $A(a, x, \gamma)$ is the singleton $\{a\}$ then the Separation Theorem yields $y \in A^*$ with (a, y) a support pair and $y \geq 0$ on $P(x, \gamma)$.

7.1. THEOREM (BISHOP–PHELPS). *Let $a \in A$ and $x \in A^*$ with*

$$\rho_A(x) \leq (a, x) + \alpha \quad (\alpha > 0).$$

Then for each $0 < \gamma < 1$ there is a support pair (b, y) $(\|y\| = 1)$ with

$$\|b - a\| \leq \alpha/\gamma \quad \textit{and} \quad \|y - x\| \leq 2\gamma.$$

Proof. Let $P = P(x, \gamma)$ and let $A(\cdot)$ denote $A(\cdot, x, \gamma)$ on A. We note that if $b \in A(a)$ then $A(b) \subset A(a)$. Furthermore, if $c \in A(b)$ then

$$\gamma\|c - b\| \leq (c - b, x) \leq \sup\{(c, x) : c \in A(a)\} - (b, x)$$

so that

$$\text{diam}\,(A(b)) \leq (2/\gamma) \sup\,(A(a) - b, x).$$

We construct a sequence $(a_n)_{n=0}^{\infty}$ in A with $a_0 = a$ and

$$a_{n+1} \in A(a_n), \ \sup\,(A(a_n) - a_{n+1}, x) < 1/n.$$

Then $(A(a_n))_{n=0}^{\infty}$ is a nested sequence of sets with diameters decreasing to zero. Hence the completeness of E gives

$$\{b\} = \bigcap_{n=0}^{\infty} A(a_n).$$

Now $c \in A(b)$ implies $c \in A(a_n)$ for all n so that $c = b$. Thus $A(b) = \{b\}$. Since int $P \neq \phi$ this means the Separation Theorem can be applied to find $y \in E^*$ ($\|y\| = 1$) with

$$y \geq 0 \quad \text{on } P \qquad \text{and} \qquad (b, y) = \rho_A(y).$$

Thus (b, y) is a support pair and $b \in A(a)$ shows

$$\gamma\|b-a\| \leq (b-a, x) \leq \sup\,(A-a, x) \leq \alpha.$$

The fact that $y \geq 0$ on P shows

$$E_1 \cap \{c \in E : (c, y) \leq 0\} \subset \{c \in E : (c, x) \leq \gamma\}$$

The polar of this in E^* is

$$[0, 1/\gamma]x \subset E_1^* + \mathbb{R}^+ y.$$

Hence

$$(1/\gamma)x = \lambda y + w; \qquad \lambda \geq 0 \quad \text{and} \quad \|w\| \leq 1,$$

so that

$$x - \gamma\lambda y = \gamma w.$$

Thus

$$\|x - \gamma\lambda y\| \leq \gamma$$

and

$$|1 - \gamma\lambda| = |\,\|x\| - \|\gamma\lambda y\|\,| \leq \|x - \gamma\lambda y\| \leq \gamma.$$

Consequently,

$$\|x - y\| = \|x - \gamma\lambda y - (1 - \lambda\gamma)y\| \leq 2\gamma.$$

7.2. COROLLARY. *Let $\varepsilon > 0$ be given.*

(i) *For each $x \in A^*$ there is a support pair (a, y) with $\|y - x\| < \varepsilon$.*

(ii) *For each a in the boundary of A there is a support pair (b, x) with $\|b - a\| < \varepsilon$.*

Proof. Take $x \neq 0$ and apply Theorem 7.1 with any $a \in A$ to $x/\|x\|$ with any $\gamma < \varepsilon/2\|x\|$. For (ii) choose $c \notin A$ and $\|c - a\| \leq \varepsilon/2$. Then there is an $x \in A^*$ ($\|x\| = 1$) with

$$(c, x) > \rho_A(x).$$

Thus $\rho_A(x) \leq (a, x) + (c - a, x) \leq (a, x) + \varepsilon/2$ and Theorem 7.1 applies with $\gamma = \frac{1}{2}$.

Notes

Some general references for the background material discussed in this chapter are Day [**75**], Holmes [**131**] and Semadeni [**191**]. In particular Holmes [**131**; Section 17] has a discussion of completeness that incorporates some of the material in Section 3 concerning the gauge lemma and regularity of domain. The embedding of a compact convex set K as the state space of $A(K)$ (Theorem 4.7) is due to D. A. Edwards [**87**]. The standard text sources for the material in Sections 4–6 are Phelps [**169**] and Alfsen [**5**]. The latter has an excellent bibliography that traces the historical development of the Choquet theory. We merely note some of the highlights. The metrizable version, Theorem 6.6, was proved by Choquet [**60**] in 1956. The general case, Theorem 6.8, is in Bishop and de Leeuw [**40**]. The ordering $>$ or close variations of it are found in Bishop and de Leeuw [**40**] and Choquet [**61**]. Theorem 6.4(iii), the main development in showing the existence of maximal representing measure, occurs in Bonsall [**45**] and Choquet–Meyer [**63**]. The characterization of maximal measures in Theorem 6.5 is due to Mokobodski [**161**]. The Bishop–Phelps Theorem of Section 7 was proved by them in 1961 [**41**]. See also Bishop and Phelps [**42**]. An updated treatment with applications may be found in Phelps [**170**].

CHAPTER 2

Duality in Ordered Banach Spaces

We now take up the study of real Banach spaces that are partially ordered by means of a closed convex cone. We say the set P is a convex *cone* if

(i) $\lambda P \subset P$ for all $\lambda \geq 0$ and
(ii) $P + P \subset P$.

If P is a closed convex cone in the normed linear space E then (E, P) is partially ordered by

$$a \leq b \quad \text{if and only } b - a \in P.$$

We note that properties (i) and (ii) of P imply that

(i)′ $a \leq b$ and $c \in E$ implies $a + c \leq b + c$,
(ii)′ $a \leq b$ and $\lambda \geq 0$ implies $\lambda a \leq \lambda b$,
(iii)′ $a \geq 0$ and $b \geq 0$ implies $a + b \geq 0$.

Conversely, if (i)′, (ii)′, (iii)′ hold for a partial ordering "$\leq$" then $P = \{a : a \geq 0\}$ is the convex cone yielding "$\leq$" and P is referred to as the *positive cone* of "$\leq$".

Our principal source of examples for ordered Banach spaces consists of the $A(K)$ spaces introduced in Chapter 1 Section 4 (or equivalently, subspaces of $C_{\mathbb{R}}(X)$ containing constant functions). The ordering is the natural one, induced by taking P as the cone of non-negative functions. The cone P in $A(K)$ has the additional feature of having non-empty interior (indeed, $1 \in \operatorname{int} P$). This will not be the case in our general study.

Finally, observe that (E, P) induces a natural ordering on E^* with *dual* positive cone $P^* = -P^0 = \{x \in E^* : (a, x) \geq 0 \text{ for all } a \in P\}$. Thus, for example, $A(K)^*$ is ordered by P^*, where, using Proposition 1.4.2 and the fact that K is the state space of $A(K)$,

$$P^* = \bigcup \{\lambda K : \lambda \geq 0\}.$$

Notes

Some general reference books on ordered vector space theory are Jameson [**134**], Peressini [**168**], Vulikh [**213**] and Wong and Ng [**217**].

1. POSITIVE GENERATION AND NORMALITY

Every $a \in A(K)$ can be written as

$$a = a_1 - a_2; \qquad a_i \geq 0 \quad (i = 1, 2).$$

Indeed, if a_1 is any element $\geq a$, 0 then

$$a = a_1 - (a_1 - a).$$

The fact that a is bounded above on K guarantees the presence of many such a_1s.

Definition. The ordered space (E, P) is *positively generated* if each $a \in E$ can be written as

$$a = a_1 - a_2; \qquad a_i \geq 0 \quad (i = 1, 2).$$

Equivalently, $E = P - P$.

If E is a Banach space then we shall see that if $E = P - P$ then in fact one can find a bound (depending only on $\|a\|$) for the norms of the a_i $(i = 1, 2)$. Accordingly, we say E is *α-generated* $(\alpha \geq 0)$ if for each $a \in E$,

$$a = a_1 - a_2; \qquad a_i \geq 0 \quad (i = 1, 2)$$

and

$$\|a_1\| + \|a_2\| \leq \alpha \|a\|.$$

1.1. PROPOSITION. *The following are equivalent*:

(i) (E, P) *is α-generated*;

(ii) $E_1 \subset \alpha \operatorname{co}(P_1 \cup -P_1)$.

Proof. Let $0 \neq a = a_1 - a_2 \in E_1$ with $a_1, a_2 \geq 0$ and

$$0 < r = \|a_1\| + \|a_2\| \leq \alpha \|a\| \leq \alpha.$$

Then (with 0/0 taken to be 0)

$$a = r[(\|a_1\|/r)a_1/\|a_1\| + (\|a_2\|/r)(-a_2/\|a_2\|)] \in \alpha \operatorname{co}(P_1 \cup -P_1).$$

Conversely, given $b \neq 0$ let $a = b/\|b\|$. Then (ii) implies

$$a = \alpha(\lambda a_1 + (1-\lambda)(-a_2)) \quad (a_i \in P_1)$$

so that, with $b_1 = \|b\|\alpha\lambda a_1$, $b_2 = \|b\|\alpha(1-\lambda)a_2$,

$$b = b_1 - b_2; \qquad b_i \geq 0 \quad \text{and} \quad \|b_1\| + \|b_2\| \leq \alpha \|b\|.$$

If E is a normed space then we say a new norm p is *equivalent* to $\|\cdot\|$ if there are positive constants a and b such that

$$a\|\cdot\| \leq p \leq b\|\cdot\| \quad \text{on } E.$$

Of course equivalent norms induce the same topology on E.

1.2. THEOREM. *Let (E, P) be an ordered Banach space which is positively generated. Then there is an $\alpha > 0$ such that E is α-generated. Moreover there is a new norm p, equivalent to $\|\cdot\|$, such that*

(i) *$p = \|\cdot\|$ on P;*
(ii) *(E, p, P) is λ-generated for all $\lambda > 1$.*

Proof. Let $B = \text{co}\,(P_1 \cup -P_1)$. Then, by Proposition 1.3.1(iii), B is a pre-gauge set. But

$$E = P - P = \bigcup_{n=1}^{\infty} nB = \bigcup_{n=1}^{\infty} n\bar{B}$$

so Corollary 1.3.4 yields an α such that

$$E_1 \subset \alpha' B \quad \text{for all } \alpha' > \alpha.$$

Thus E is α'-generating for all $\alpha' > \alpha$. Since $B \subset E_1$ we have

$$B \subset E_1 \subset \alpha' B \quad \text{for all } \alpha' > \alpha.$$

Thus if $p = p(B)$ (the Minkowski functional of B) then p is a norm (since B is symmetric and absorbing) equivalent to $\|\cdot\|$. If $a \in P$ then $a \in B$ if and only if $a \in E_1$ so $p = \|\cdot\|$ on P. That (E, p, P) is α'-generating for any $\alpha' > 1$ now follows from Proposition 1.1 together with the fact that B is a regular pre-gauge set ($p(B) = p(\bar{B})$, 1.3.3).

Example 1.6 below shows that Theorem 1.2(ii) is *not* the same as being 1-generated.

Our next goal is to determine when an element $x \in (E, P)^*$ can be decomposed as the difference of two non-negative functionals, that is, when is (E^*, P^*) positively generated?

If we take the polar of the statement in Proposition 1.1(ii), using some polar calculus we get the property

$$(E_1^* - P^*) \cap (E_1^* + P^*) \subset E_\alpha^*$$

which is dual to α-generation in E.

This leads us to the following definition.

Definition. The ordered space (E, P) is *α-normal* if

$$(E_1 - P) \cap (E_1 + P) \subset E_\alpha.$$

The next proposition gives the usual characterization of normality. The proof is straightforward and is omitted.

1.3. PROPOSITION. *The following are equivalent:*

(i) *(E, P) is α-normal;*
(ii) *if $c \le x \le b$ then $\|x\| \le \alpha \max\{\|c\|, \|b\|\}$.*

1.4. THEOREM. *Let E be an ordered normed linear space. The following are equivalent:*

(i) *E^* is positively generated;*
(ii) *E^* is α-generated for some α;*
(iii) *E is α-normal.*

Proof. Since E^* is a Banach space, (i) implies (ii) by Proposition 1.2, and (iii) follows as the polar of (ii). If (iii) holds then, since $E_1 - P$ and $E_1 + P$ are regular pre-gauge sets, for any $\lambda > 1$

$$(E_1 - P)^- \cap (E_1 + P)^- \subset (E_\lambda - P) \cap (E_\lambda + P) \subset \alpha E_\lambda$$

and hence

$$(E_1 - P)^- \cap (E_1 + P)^- \subset \alpha E_1 = E_\alpha$$

which by taking polars, gives (ii).

Theorem 1.4 establishes the duality between α-normality and α-generation in one direction. To complete the picture we investigate the situation when E^* is α-normal. Theorem 1.4 yields that (E^{**}, P^{**}) is α-generated but (E, P) does not quite inherit the full strength of this conclusion.

Definition. The space (E, P) is said to be *approximately α-generated* if (E, P) is α'-generated for all $\alpha' > \alpha$. Thus, the statement in Theorem 1.2(ii) could be rephrased as saying (E, p, P) is approximately 1-generated.

1.5. THEOREM. *Let (E, P) be an ordered Banach space. The following are equivalent:*

(i) *E^* is α-normal;*
(ii) *E is approximately α-generated.*

Proof. The dual E^* is α-normal if and only if

$$(E_1^* - P^*) \cap (E_1^* + P^*) \subset E_\alpha^*.$$

Since the sets $E_1^* \pm P^*$ are already w^*-closed we can take the polar of the intersection and get the equivalent statement

(iii) $$E_1 \subset \alpha\, \overline{\mathrm{co}}\, (P_1 \cup -P_1)$$

But then $B = \text{co}(P_1 \cup -P_1)$ is a regular pre-gauge set and hence (iii) is equivalent to

$$E_1 \subset \alpha' \text{co}(P_1 \cup -P_1) \quad \text{for all } \alpha' > \alpha,$$

which is precisely (ii).

The following example of an approximately 1-generated space which is *not* 1-generated shows that Theorem 1.5 is the best conclusion.

1.6. EXAMPLE. Let c_0 denote the Banach space of sequences $x = (x_n)_{n=1}^{\infty}$ such that $\lim_{n\to\infty} x_n = 0$ with

$$\|x\| = \max\{|x_n|: n = 1, 2, \ldots\}.$$

Let

$$E = \left\{x \in c_0: x_1 + x_2 = \sum_{n=1}^{\infty} x_{n+2}/2^n\right\}.$$

Then E is a closed subspace (the zero set of the linear functional $(1, 1, -1/2, -1/2^2, \ldots) \in l^1 = (c_0)^*$). Moreover, E is positively generated, where $x \geq 0$ means each $x_n \geq 0$. Let

$$B = \text{co}(P_1 \cup -P_1) \quad \text{in } E.$$

We show $x = (1/2, -1/2, 0, 0, \ldots) \in \lambda B \backslash B$ for all $\lambda > 1$. If $x = \alpha y - \beta z$; $y, z \in P_1$ and $\alpha + \beta \leq 1$, then

$$1/2 = \alpha y_1 - \beta z_1$$

$$-1/2 = \alpha y_2 - \beta z_2.$$

Thus,

$$1 = \alpha(y_1 - y_2) + \beta(z_2 - z_1).$$

But $y_1 - y_2 \leq 1$, $z_2 - z_1 \leq 1$, and $y_i, z_i \geq 0$ $(i = 1, 2)$. It follows that $y_1 = z_2 = 1$, $y_2 = z_1 = 0$ and $\alpha = 1/2 = \beta$. But then

$$\sum_{n=1}^{\infty} y_{n+2}/2^n = 1 = \sum_{n=1}^{\infty} z_{n+2}/2^n \quad \text{and} \quad 0 \leq y_n, z_n \leq 1$$

implies $y_{n+2} = 1 = z_{n+2}$ for all n. Since $y, z \in c_0$ this is impossible. Hence $x \notin B$. If $0 < r < 1$ and $r' = \sum_{k=1}^{n} 1/2^k > r$ for some n then

$$rx = (1/2)(y - z)$$

where

$$y = (r, 0, r/r', \ldots, r/r', 0, \ldots,)$$

and

$$z = (0, r, r/r', \ldots, r/r', 0, \ldots,)$$

belong to P_1. Hence $x \in \lambda B$ for all $\lambda = 1/r > 1$.

It can be shown in similar fashion that $P_1 - P_1$ in the above example fails to be closed. In fact

$$(1, -1, 0, 0, \ldots,) \in \lambda(P_1 - P_1) \backslash (P_1 - P_1) \quad \text{for all } \lambda > 1.$$

2. ORDER UNIT AND BASE NORM SPACES

We observed in the introduction that if $E = A(K)$ then the function $1 \in \operatorname{int} P$. Indeed, we have

$$E_1 = (1 - P) \cap (1 + P)$$

which clearly implies 1-normality.

In general we say u is an *order unit* for the ordered norm space (E, P) if

$$E_1 = (u - P) \cap (u + P). \tag{1}$$

In this case we call (E, P) an *order unit space*. We shall see shortly that the presence of an order unit essentially characterizes $A(K)$ spaces.

We say the ordering in a linear space is *proper* if $P \cap -P = \{0\}$. This just means that $\pm a \geq 0$ if and only if $a = 0$.

2.1. Proposition. *Let (E, P) be an order unit space. Then*

(i) *E is 1-normal;*

(ii) *$u + E_1 \subset P$ (hence $u \in \operatorname{int} P$);*

(iii) *the ordering in (E, P) is proper;*

(iv) *if f is a non-negative (on P) linear functional on E then $f \in E^*$ and $\|f\| = f(u)$.*

Proof. Parts (i) and (ii) are immediate from equation (1) above. If $\pm a \geq 0$ then $-u \leq \lambda a \leq u$ for all $\lambda > 0$ so that $\lambda a \in E_1$ for all λ and hence, $a = 0$. For (iv) we have for any $a \in E_1$,

$$-u \leq a \leq u.$$

Hence $f \geq 0$ on P implies $|f(a)| \leq f(u)$. Therefore

$$\|f\| \leq f(u) \leq \|f\| \|u\| = \|f\|.$$

With the aid of an order unit we can define the *state space* of E (in close analogy to Chapter 1, Section 4) by

$$S = \{x \in E^*: x \geq 0 \text{ on } P \text{ and } (u, x) = 1\}.$$

From 2.1(iv) we have that $x \in S$ implies $\|x\| = 1$.

By an *order isometry* we mean an isometry between ordered normed spaces taking the positive cone of one onto that of the other.

2.2. THEOREM. *Let (E, P) be a normed order unit space. Then the state space S is w^*-compact, convex and E is order isometric to a dense subspace of $A(S)$. If E is a Banach space than E is order isometric to $A(S)$.*

Proof. From the definition of S we have that S is the intersection of a w^*-closed hyperplane with the dual cone and is contained in E_1^*. Thus S is w^*-compact and convex. Moreover $0 \neq y \in P^*$ implies $y/\|y\| \in S$. Since E is 1-normal, E^* is 1-generating and hence

$$E_1^* = \text{co}\,(P_1^* \cup -P_1^*) = \text{co}\,(S \cup -S).$$

If $\theta : E \to A(S)$ is the restriction map then

$$\|\theta a\| = \sup\,\{|(a, x)| : x \in S\} = \sup\,\{|(a, x)| : x \in E_1^*\} = \|a\|.$$

Since $\theta u = 1$ we have from Theorem 1.4.7 that θE is dense in, or if complete, equal to $A(S)$.

Definition. If P is a closed convex cone then a subset B of P is called a *base* for P if

(i) B is closed, convex and bounded, and
(ii) for each $0 \neq a \in P$ there is a *unique* $\lambda > 0$ such that $a/\lambda \in B$.

If (E, P) is an ordered normed space such that P has a base B and

$$E_1 = \overline{\text{co}}\,(B \cup -B)$$

then (E, P) is called a *base norm space.*

By Theorem 1.5 this implies E is approximately 1-generated so that in fact

$$E_1 = \bigcap_{\alpha > 1} \alpha\ \text{co}\,(B \cup -B).$$

We say (E^*, P^*) is a *dual base norm space* if the base B is w^*-closed (and hence compact). In this case the w^*-compactness gives

$$E_1^* = \text{co}\,(B \cup -B).$$

2.3. THEOREM. *Let E be a Banach space. Then the following are equivalent:*

(i) *(E, P) is an order unit space,*
(ii) *(E^*, P^*) is a dual base norm space.*

If these conditions hold then the base of P^* is the state space of E, E is order isometric to $A(S)$, and E^* is order isometric to $A(S)^*$.

Proof. If (i) holds then the state space S of E is w^*-compact, convex and in fact a base for P^* since

$$S = \{x \in P^* : (u, x) = 1\}.$$

Thus, the fact that

$$E_1^* = \text{co}\,(S \cup -S)$$

shows (ii). If (ii) holds and B is the base then the restriction map $\theta: E \to A(B)$ is an order isometry (into) since

$$E_1^* = \text{co}\,(B \cup -B).$$

Moreover, if u is defined to be identically 1 on B and extended (uniquely) to be linear on E^* then Corollary 1.2.5 shows $u \in E$. Thus θ is an order isometry onto $A(B)$.

We now show that a linear space with an appropriately defined order unit can be made into a (normed) order unit space. This will provide a strictly order-theoretic characterization of (dense subspaces of) $A(K)$ spaces.

Definition. Let P be a proper convex cone in the linear space E. We say the ordering on E induced by P is *Archimedean* if for $a, b \in E$,

$$ra \le b \quad \text{for all } r \ge 0 \text{ implies } a \le 0.$$

We say u is an *Archimedean order unit* if for $a \in E$

(a) $a \in E$ implies there is an $r \ge 0$ such that $-ru \le a \le ru$, and
(b) $ra \le u$ for all $r \ge 0$ implies $a \le 0$.

2.4. PROPOSITION. *Let u be an Archimedean order unit for (E, P), P a proper convex cone in E. Then*

(i) *E is Archimedean;*
(ii) *$u \in \text{core } P$;*
(iii) *$a \in P$ if and only if $u + ra \in P$ for all $r \ge 0$;*
(iv) *P is linearly closed.*

Proof. Part (i) is immediate. For (ii) let $a \in E$. Then for $0 < r$, r sufficiently small, we have from (a) above

$$-(1/r)u \le a.$$

Thus $u + ra \in P$. This shows $u \in \text{core } P$. For (iii), $u + ra \ge 0$ for all $r \ge 0$ if and only if $r(-a) \le u$ if and only if $-a \le 0$ if and only if $a \ge 0$. Now $u \in \text{core } P$, so (iv) holds if

$$u + a \in P \quad \text{whenever } u + \lambda a \in P \quad \text{for } 0 \le \lambda < 1.$$

But $u + r(u + a) = (r+1)[u + (r/(r+1))a]$ so that if $u + \lambda a \in P (\lambda < 1)$ then $u + r(u + a) \in P$ for all $r \ge 0$. Hence (iii) shows $u + a \in P$.

Now let

$$I = (u - P) \cap (u + P) = \{a \in E : -u \leqslant a \le u\}$$

and let $p = p(I)$, the Minkowski functional of I.

2.5. THEOREM. *Let u be an Archimedean order unit for (E, P) and let I and p be as above. Then p is a norm on E with unit ball I in which E is a normed order unit space with order unit u and closed convex positive cone P.*

Proof. The set I is symmetric and convex. Also Proposition 2.4(ii) and (iv) shows $0 \in \operatorname{core} I$ and I is linearly closed. Hence $p(I)$ is a semi-norm. If $p(a)=0$ then $\lambda a \in I$ for all $\lambda \geq 0$ and hence

$$-u \leq \lambda a \leq u \quad \text{for all } \lambda \geq 0.$$

Thus $\pm a \geq 0$ and, since P is proper, $a = 0$ and p is a norm. Since I is linearly closed, $I = \{a: p(a) \leq 1\}$ is the unit ball of (E, p). Moreover, $u + I \subset P$ so that $u \in \operatorname{int} P$. Hence the fact that P is linearly closed implies P is closed in (E, p). Thus the definition of I shows (E, p, P) is a normed order unit space.

We conclude this section with a characterization of cones with a compact base, S. This is useful since such cones can be identified with the dual cone in $A(S)^*$.

2.6. THEOREM. *Let P be a proper cone in a locally convex space E. The following are equivalent:*

(i) *P has a compact base S,*
(ii) *P is locally compact,*
(iii) *P is locally compact at* 0.

Proof. If (i) holds, choose $f \in E^*$ such that $\inf f(S) = 1$. If $U = \{a \in E: f(a) \leq 1\}$ then

$$U \cap P \subset \operatorname{co}(S \cup \{0\})$$

and hence is compact. Thus, for any $0 \leq \alpha < \beta$ we have

$$\{a \in P: \alpha \leq f(a) \leq \beta\}$$

is a compact neighbourhood in P. Since $P = \bigcup_{\alpha \geq 0} \alpha S$, (ii) is shown. If (iii) holds let U be a closed convex neighbourhood of 0 in E such that $U \cap P$ is compact. Let

$$S_0 = (U \backslash \operatorname{int} U) \cap P.$$

If $0 \neq a \in \operatorname{ext}(U \cap P)$ then $a \in S_0$. Since P is proper,

$$0 \in \operatorname{ext}(U \cap P) \backslash S_0.$$

Hence, by the Krein–Milman Theorem (1.1.3),

$$0 \notin \overline{\operatorname{co}}\,(S_0).$$

Thus there is an $f \in E^*$ with $f \geq 1$ on S_0. If $S = \{a \in P: f(a) = 1\}$ then S is a closed subset of $U \cap P$ and hence compact. Moreover, if $0 \neq a \in P$ then for some $\lambda > 0$, $a/\lambda \in S_0$ and hence $f(a) \geq \lambda > 0$ so that $a/f(a) \in S$. Thus S is a compact base for P.

2.7. COROLLARY. *If P is a locally compact cone in E (locally convex) then P is closed.*

Proof. Let $f \in E^*$ such that $f \geq 1$ on the base of P. Let $a \notin P$ and let U be a neighbourhood of a disjoint from the (possibly empty) compact set

$$A = \{b \in P: f(b) \leq f(a) + 1\}.$$

Then $U \cap \{b: f(b) < f(a) + 1\}$ is a neighbourhood of a disjoint from P.

3. DIRECTEDNESS AND ADDITIVITY

Another way of expressing the fact that an ordered normed space E is positively generated is to say E is *directed*, that is, for each $a, b \in E$ there is a $c \in E$ with $c \geq a, b$. As with positive generation it is possible to express the degree of directedness: we say E is (α, n)-*directed* (n a positive integer greater than one) if whenever $a_1, \ldots, a_n \in E_1$ there is an $a \in E_\alpha$ with

$$a \geq a_i \quad (i = 1, \ldots, n).$$

We also require an approximate version of directedness—thus we say E is *approximately* (α, n)-*directed* if E is (α', n)-directed for all $\alpha' > \alpha$. We give an example below of an approximately $(1, 2)$-directed space which is not $(1, 2)$-directed.

The dual order property has to do with the degree to which the triangle inequality can be reversed for positive elements. Specifically, we say E is (α, n)-*additive* if whenever $a_1, \ldots, a_n \in P$ then

$$\|a_1\| + \cdots + \|a_n\| \leq \alpha \|a_1 + \cdots + a_n\|.$$

A simple induction shows that a $(1, 2)$-additive (directed) space is in fact $(1, n)$-additive (directed) for all n. The same holds for approximately $(1, 2)$-additive spaces. In general $(\alpha, 2)$-additivity (directedness) implies (α', n)-additivity (directedness) for any n and some α' (depending on n and α). Finally, $(1, 2)$-additivity just amounts to $\|\cdot\|$ being additive on P and we just say E is additive or 1-additive in this case.

In general, if E is (α, n)-additive (directed) for all n and fixed α we will say simply E is α-*additive* (*directed*).

3.1. PROPOSITION. *The space E is* 1-*additive if and only if $B = \{a \in P: \|a\| = 1\}$ is a (convex) base for P.*

Proof. If $a, b \in P$ and $0 \neq r = \|a\| + \|b\|$ then

$$(a+b)/r = (\|a\|/r)(a/\|a\|) + (\|b\|/r)(b/\|b\|)$$

is a convex combination. The conclusions follow easily from this.

3.2. PROPOSITION. *The following are equivalent:*

(i) *The positive cone P has a base B;*

(ii) *the dual cone P^* has non-empty norm interior;*

(iii) *E is (α, n)-additive for all n (fixed α).*

If E is an ordered Banach space for which these conditions hold then every non-empty $\|\cdot\|$-bounded subset S which is directed by $\leq$ converges as a net to l.u.b. S.

Proof. If B is a base for P then by separating B from 0 we can find an element $x_0 \in P^*$ such that $x_0(b) \geq 1$ for all $b \in B$. If $B \subset E_\beta$ and $x \in (1/\beta)E_1^*$ then, for any $b \in B$,

$$(b, x_0 + x) \geq 1 - \|b\|\|x\| \geq 0.$$

Hence $x_0 + (1/\beta)E_1^* \subset P^*$ so (ii) holds.

If $x_0 + (1/\beta)E_1^* \subset P^*$ and $a \in P$ then for any $x \in E_1^*$

$$(a, -x) + \beta(a, x_0) \geq 0.$$

Hence $\beta(a, x_0) \geq (a, x)$ so that $(a, x_0) \geq (1/\beta)\|a\|$.

Now, given $a_1, \ldots, a_n \in P$,

$$\|x_0\|\|a_1 + \cdots + a_n\| \geq (a_1 + \cdots + a_n, x_0) \geq (1/\beta)(\|a_1\| + \cdots + \|a_n\|).$$

Hence E is (α, n)-additive with $\alpha = \beta\|x_0\|$.

If E is (α, n)-additive for all n take

$$S_0 = \{a \in P: \|a\| = 1\}.$$

Then if $S = \text{co}\, S_0$ we have S is disjoint from the interior of $(1/\alpha)E_1$. Then the Separation Theorem provides an $x_0 \in E^*$ such that for any $a \in P$,

$$\|a\| \leq (a, x_0) \leq \alpha\|a\|.$$

It follows that $B = \{a \in P: (a, x_0) = 1\}$ is a base for P.

Let E be α-additive. Suppose $(S, \leq)$ is not a Cauchy net. Then for some $\varepsilon > 0$ and each $a \in S$ there exists $b, c \geq a$ in S with $\|b - c\| \geq \varepsilon$. Hence, by the triangle inequality either $\|b - a\|$ or $\|c - a\| \geq \varepsilon/2$. Therefore we can construct an increasing sequence $(a_n) \subset S$ with $\|a_{n+1} - a_n\| \geq \varepsilon/2$. But, if S is bounded in norm by M,

$$2M \geq \|a_{n+1} - a_1\| = \left\| \sum_{k=1}^{n} (a_{k+1} - a_k) \right\| \geq \alpha^{-1} \sum_{k=1}^{n} \|a_{k+1} - a_k\| \geq n\varepsilon/2\alpha.$$

Hence S is Cauchy and so has a unique limit a_0 which clearly is l.u.b. S.

To establish the duality between directedness and additivity we need to formulate these properties as set inclusions. To do this we consider the n-fold Cartesian product

$$\Pi^n E \doteq \bar{E}$$

which we make into an ordered normed space by $\bar{a} \geq 0$ if and only if $a_i \geq 0$ $(i = 1, \ldots, n)$, $(\bar{a} = (a_1, \ldots, a_n))$ and

$$\|\bar{a}\| = \max\{\|a_i\|: i = 1, \ldots, n\}.$$

Then $(\bar{E})^* = \Pi^n E^*$ where

$$(\bar{a}, \bar{x}) = \sum_{i=1}^{n} (a_i, x_i).$$

Then the dual cone $\bar{P}^*$ in $\bar{E}^*$ is given by

$$\bar{x} \geq 0 \quad \text{if and only if each } x_i \geq 0$$

and

$$\|\bar{x}\| = \sum_{i=1}^{n} \|x_i\|.$$

The diagonal, $\bar{D}$, in $\bar{E}$ is given by

$$\bar{D} = \{\bar{a} \in \bar{E}: a_1 = \cdots = a_n\}.$$

Note that, since $\bar{D}$ is a subspace,

$$(\bar{D})^0 = \left\{\bar{x} \in \bar{E}^*: \sum_{i=1}^{n} (a, x_i) = 0 \text{ for all } a \in E\right\} = \{\bar{x} \in \bar{E}^*: \sum x_i = 0\}.$$

3.3. Proposition. *The ordered normed space E is (α, n)-directed if and only if*

$$\bar{E}_1 \subset \alpha(\bar{D}_1 - \bar{P}).$$

The proof is a simple verification and is omitted.

3.4. Proposition. *If $\bar{A} = \bar{D}_1 - \bar{P}$ then*

$$(\bar{A})^0 = \left\{\bar{x} \in \bar{P}^*: \left\|\sum_{i=1}^{n} x_i\right\| \leq 1, \bar{x} = (x_1, \ldots, x_n)\right\}.$$

Proof.

$$(\bar{A})^0 = (\bar{D}_1)^0 \cap P^*$$

and

$$(\bar{D}_1)^0 = \left\{\bar{x} \in \bar{E}^*: \sum_{i=1}^{n} (d, x_i) \leq 1 \text{ for all } d \in E_1\right\}$$

$$= \left\{\bar{x} \in \bar{E}^*: \left\|\sum_{i=1}^{n} x_i\right\| \leq 1\right\}.$$

3.5. THEOREM. *If E is an ordered Banach space then the following are equivalent*:

(i) *E is approximately (α, n)-directed,*
(ii) *E^* is (α, n)-additive.*

Proof. If (i) holds then

$$\text{(iii)} \qquad (\bar{A})^0 \subset \bigcap_{\alpha'>\alpha} \alpha' \bar{E}_1^* = \alpha \bar{E}_1^*.$$

But this says precisely that E^* is (α, n)-additive. If (ii) holds then taking the polar of (iii) gives

$$\bar{E}_1 \subset \alpha \operatorname{cl}(\bar{D}_1 - \bar{P}).$$

But then $\bar{D}_1 - \bar{P}$ is $\bar{E}$-regular so that Corollary 1.3.3 shows

$$\bar{E}_1 \subset \alpha'(\bar{D}_1 - \bar{P}) \quad \text{for all } \alpha' > \alpha$$

and hence (i) holds.

The most interesting case results when $\alpha = 1$.

3.6. COROLLARY. *If E is an ordered Banach space then the norm is additive on the dual cone P^* if and only if the open unit ball in E is directed.*

Theorem 2.3 characterizes the *dual* base norm spaces (those with a w^*-compact base). We can now characterize the dual spaces that happen to be base norm spaces (with base not necessarily w^*-compact.)

3.7. COROLLARY. *Let (E, P) be an ordered Banach space. The following are equivalent*:

(i) *E is 1-normal and approximately 1-directed;*
(ii) *E^* is a base norm space.*

Proof. If (i) holds, the dual properties give

$$E_1^* = \operatorname{co}(P_1^* \cup -P_1^*)$$

and $\|\cdot\|$ is additive on P^*. But then Proposition 3.1 says

$$B = \{x \in P^* : \|x\| = 1\}$$

is a base. Hence (ii) holds. Conversely, Proposition 3.1 in the other direction gives E^* is 1-additive. This together with the 1-generation of E^* gives (i) by Theorems 3.5 and 1.4.

To obtain the "dual" duality theorems the whole procedure is simply reversed. We take $\bar{E}$ as before but give $\bar{E}$ the l^1 product norm,

$$\|\bar{a}\| = \sum \|a_i\|$$

and then $\bar{E}^*$ has the l^∞ norm,

$$\|\bar{x}\| = \max \{\|x_i\|\}.$$

This direction is easier in that no approximate properties arise. We simply state the results.

3.8. THEOREM. *If E is an ordered norm space then the following are equivalent*:

(i) *E is (α, n)-additive*;
(ii) *E^* is (α, n)-directed.*

3.9. COROLLARY. *The following are equivalent*:

(i) *E is a base norm space*;
(ii) *E^* is an order unit space. If these conditions hold and B is the base for P and u the order unit for E^* then*

$$B = \{a \in P: (a, u) = 1\}.$$

In the case when $E = A(K)^*$ it is easy to check that the order unit space E^* coincides with the space $A^b(K)$ of all bounded real-valued affine functions on K with the supremum norm.

3.10. EXAMPLE. We can construct an approximately 1-directed space along the same lines as Example 1.6.

Let

$$E = \left\{\bar{x} \in c_0: x_1 + x_2 = \sum_{n=0}^{\infty} x_{n+3}\Big/2^n\right\}.$$

Let

$$a = (1, 0, 1, 0, 0, \ldots, 0, \ldots)$$
$$b = (0, 1, 1, 0, 0, \ldots, 0, \ldots)$$

If $c \in E_1$ and $c \geq a, b$ then $c_n \geq a_n \vee b_n$ forces $c_1 = c_2 = c_3 = 1$ and hence each $c_n = 1$, which is impossible if $c \in c_0$. Thus E fails to be 1-directed. It is not difficult to check, however, that E is α-directed for all $\alpha > 1$.

Notes

The pioneering work in abstract ordered vector spaces was conducted by a Russian school led by Kantorovich in the mid to late 1930s. The first duality result of the type considered herein is due to Grosberg and Krein [**122**] who proved Theorem 1.4 in 1939. The reverse duality (Theorem 1.5) was shown by Andô [**18**] (without regard to the constant α) and, in its present form, by Ellis [**100**]. The representations of order unit spaces and the results of Theorems 2.2 and 2.5 representing them as function spaces are essentially due to Kadison [**136**]. The notion of base-norm space and the duality of Theorem 2.3 is due to Ellis [**100**] and Edwards [**87**]. Theorem 2.6 on locally compact cones is shown by Klee [**142**]. The duality between additivity and directedness is established by Asimow [**23**]. The case $\alpha = 1$ (Corollary 3.6) is proved in Asimow [**22**] and Ng [**165**]. Corollary 3.9 is shown in Ellis [**100**]. Example 3.10 appears in [**22**].

4. HOMOGENEOUS FUNCTIONALS ON CONES AND THE DECOMPOSITION PROPERTY

In this section we develop some techniques involving homogeneous functionals on cones which will be used in characterizing Banach lattices and discussing their duality properties.

We take E to be an ordered norm space with closed convex proper positive cone P. We say P has the *decomposition property* if whenever

$$0 \leq a \leq b_1 + b_2 \quad (b_1, b_2 \geq 0)$$

there exist $a_1, a_2 \geq 0$ such that

$$a = a_1 + a_2 \quad \text{and} \quad a_1 \leq b_1, \qquad a_2 \leq b_2.$$

We say P is a *lattice cone* if for each $a, b \in P$ the least upper bound, $a \vee b$, and the greatest lower bound, $a \wedge b$, exist. Since P is proper, $a \vee b$, and $a \wedge b$ are unique. A lattice cone always has the decomposition property, as we can take

$$a_1 = a \wedge b_1 \quad \text{and} \quad a_2 = a - a_1.$$

Then $0 \leq a_1 \leq b_1$ and

$$a \leq (a + b_2) \wedge (b_1 + b_2) = a \wedge b_1 + b_2$$

so that $0 \leq a_2 \leq b_2$.

4.1. LEMMA. *If P has the decomposition property and*

$$a + b = \sum_{i=1}^{n} c_i \quad (a, b, c_1, \ldots, c_n \in P)$$

then $a = \sum_{i=1}^{n} a_i$, $b = \sum_{i=1}^{n} b_i$ with $c_i = a_i + b_i$ ($a_i, b_i \in P$).

Proof. The case $n=1$ is obvious. Assume the conclusion holds for n and let

$$a+b=\sum_{i=1}^{n+1} c_i.$$

Let $d=\sum_{i=1}^{n} c_i$. Then $d \leq a+b$ so that the decomposition property yields $d=a'+b'$; $a' \leq a$, $b' \leq b$. The induction hypothesis gives

$$a'=\sum_{i=1}^{n} a_i, \qquad b'=\sum_{i=1}^{n} b_i \quad \text{with } c_i=a_i+b_i.$$

Let $a_{n+1}=a-a'$ and $b_{n+1}=b-b'$.

Let p be a non-negative, positive homogeneous functional on P. We say p is *bounded* if there is a constant α such that

$$p(a) \leq \alpha\|a\| \quad \text{for all } a \in P.$$

If p is a non-negative, positive homogeneous functional then we can define the lower envelope of p by

$$\bar{p}(a)=\inf\left\{\sum_{i=1}^{n} p(a_i) \colon a=\sum_{i=1}^{n} a_i;\ a_i \geq 0\right\}.$$

The homogeneity of a functional p guarantees that p is convex on P if and only if $p(a+b) \leq p(a)+p(b)$; $a, b \in P$. In other words p is *sub-additive*. Similarly p is called *super-additive* if $p(a+b) \geq p(a)+p(b)$; $a, b \in P$. Thus p is super-additive if and only if p is concave.

4.2. THEOREM. *Let p be a non-negative, positive homogeneous functional on the closed proper cone P.*

(i) *$\bar{p}$ is a sub-additive, non-negative, positive homogeneous functional on P.*

(ii) *$\bar{p} \leq p$ on P and if q is any other sub-additive, positive homogeneous functional on P with $0 \leq q \leq p$ then $q \leq \bar{p}$.*

(iii) *If P has the decomposition property and p is super-additive then $\bar{p}$ is linear.*

Proof. If $a, b \in P$ and $\bar{p}(a)<r$, $\bar{p}(b)<s$ then let

$$a=\sum_{i=1}^{n} a_i \quad \text{and} \quad b=\sum_{j=1}^{m} b_j \quad (a_i, b_j \geqslant 0)$$

with

$$\sum p(a_i)<r \quad \text{and} \quad \sum p(b_j)<s.$$

Then

$$a+b=\sum a_i+\sum b_j$$

so that

$$\bar{p}(a+b) \le \sum p(a_i) + \sum p(b_j) < r+s.$$

Hence $\bar{p}$ is sub-additive. Since p is non-negative and positive homogeneous, it follows easily that $\bar{p}$ is as well. For (ii), it is clear that $\bar{p} \le p$ and, given $0 \le q \le p$ and $a = \sum_{i=1}^n a_i$, then

$$q(a) \le \sum q(a_i) \le \sum p(a_i)$$

so that $q \le \bar{p}$. For (iii) let $a, b \in P$ and $r > p(a+b)$. Then choose $a+b = \sum_{i=1}^n c_i$ with $\sum_{i=1}^n p(c_i) < r$. Then Lemma 4.1 gives

$$a = \sum a_i, \qquad b = \sum b_i \quad \text{and} \quad c_i = a_i + b_i.$$

Hence

$$\bar{p}(a) + \bar{p}(b) \le \sum p(a_i) + \sum p(b_i) \le \sum p(a_i + b_i) = \sum p(c_i) < r.$$

Of particular interest will be functionals p of the form

$$p(a) = (a, x_1) \wedge \cdots \wedge (a, x_n); \qquad x_1, \ldots, x_n \in P^*.$$

Then p is bounded, non-negative, positive homogeneous and super-additive. It is convenient to say p is the *point-wise infimum* of $x_1, \ldots, x_n$ and write $p = \inf\{x_i : i = 1, \ldots, n\}$. We will reserve the notation $x_1 \wedge \cdots \wedge x_n$ for the greatest lower bound of $x_1, \ldots, x_n$ in P^* when it exists.

If $p = \inf\{x_i\}_{i=1}^n$ then it is easy to see that $\bar{p}$ takes the form

$$\bar{p}(a) = \inf\left\{ \sum_{i=1}^n (a_i, x_i) : a = \sum_{i=1}^n a_i;\ a_i \ge 0 \right\}.$$

We now focus on functionals defined on a dual cone $P^* \subset E^*$ where (E, P) is an ordered normed space. Let p be a w^*-continuous non-negative positive homogeneous functional on P^*. We define $\check{p}$ by

$$\check{p}(x) = \sup\{(a, x) : a \in E \quad \text{and} \quad a \le p \text{ (point-wise) on } P^*\}.$$

We observe that $\check{p}$ is defined in close analogy with the $\check{f}$ of Chapter 1, Section 6. Indeed, if P^* has a w^*-compact base K then p is determined by its values on K (by positive homogeneity) and, if $f = p|_K$, then $\check{p}|_K = \check{f}$ and

$$\bar{p}|_K(x) = \inf\left\{ \sum_{i=1}^n \lambda_i f(x_i) : x = \sum_{i=1}^n \lambda_i x_i;\ x_i \in K,\ \sum_{i=1}^n \lambda_1 = 1 \right\}.$$

as in Theorem 1.6.1(vii).

4.3. THEOREM. *Let (E, P) be an ordered normed space and let p be a w^*-continuous non-negative positive homogeneous functional on P^*.*

(i) *$\bar{p}$ and $\check{p}$ are sub-additive positive homogeneous functionals with $0 \le \check{p} \le \bar{p} \le p$ on P^*.*

(ii) *Let* $B(p)=\{x\in P^*: p(x)\le 1\}$ *and* $A(p)=\{a\in E: a\le p$ *on* $P^*\}$. *Then* $B(p)^0=A(p)$, $B(\bar{p})$ *is convex and*

$$B(p)\subset B(\bar{p})\subset B(\check{p})=B(p)^{00}=w^*\ \overline{\mathrm{co}}\,(B(p)).$$

(iii) $\check{p}$ *is the Minkowski functional of* $B(p)^{00}$ *and the support functional of* $A(p)$.

Proof. Theorem 4.2(i) shows $\bar{p}$ is sub-additive homogeneous. The homogeneity of p and the definitions show $B(p)^0=A(p)$. By definition $\check{p}$ is the support functional of $A(p)$, a closed convex subset of E containing $-P$, so that $\check{p}$ is l.s.c., sub-additive and positive homogeneous on its domain of definition. But $a\in A(p)$ and $x\in P^*$ with $x=\sum_{i=1}^n x_i$ $(x_i\in P^*)$ imply

$$(a,x)=\sum (a,x_i)\le \sum p(x_i)$$

so that,

$$0\le \check{p}\le \bar{p}\le p.$$

In particular $B(p)\subset B(\bar{p})\subset B(\check{p})$ and $B(\bar{p})$ is convex since $\bar{p}$ is sub-additive. That $B(\check{p})=B(p)^{00}$ follows since $B(p)^0=A(p)$ and $\check{p}$ is the support functional of $A(p)$, hence the Minkowski functional of $A(p)^0=B(p)^{00}$. This completes (i), (ii) and (iii).

We conclude with some criteria for $\bar{p}$ to equal $\check{p}$.

4.4. THEOREM. *Let* (E,P) *be an ordered Banach space which is positively generated.*

(i) *If* $p=\inf\{a_i: a_i\in P, i=1,\ldots,n\}$ *then* $\bar{p}=\check{p}$, *and for each* $x\in P^*$ *there exist* $x_1,\ldots,x_n\in P^*$ *such that*

$$x=\sum x_i \quad \text{and} \quad \bar{p}(x)=\sum (a_i,x_i).$$

(ii) *If* $\|\cdot\|-\mathrm{int}\,P^{**}\ne\varnothing$ *in* E^{**} *then* $\bar{p}=\check{p}$ *whenever* p *is a non-negative* w^**-continuous positive homogeneous functional on* P^*.

Proof. We use the fact that E is in fact α-generated for some α to conclude that E is (β,n)-directed for some β and hence E^* is (β,n)-additive. Let $K=P_\beta^*$ and let $x\in P_1^*$. Then if

$$x=\sum_{i=1}^n x_i \quad (0\ne x_i\in P^*)$$

we have

$$x=\sum_{i=1}^n \lambda_i y_i;\quad 0\le\lambda_i \quad \text{and} \quad \sum\lambda_i=1;\quad y_i\in K$$

since if $r=\sum\|x_i\|$ then $r\le\beta\|x\|\le\beta$ and we can take $y_i=(r/\|x_i\|)x_i$, $\lambda_i=\|x_i\|/r$. Thus

$$\bar{p}(x)=\inf\left\{\sum_{i=1}^{n}(a_i,x_i)\colon x=\sum x_i\right\}$$

$$=\inf\left\{\sum_{i=1}^{n}\lambda_i a_i(y_i)\colon \sum\lambda_i=1,\ x=\sum\lambda_i y_i;\ y_i\in K\right\},$$

where we identify a_i with its restriction to K, an element of $A(K)$. We now invoke Theorem 1.6.1(vii) to conclude that

$$\bar{p}(x)=\sup\{a(x)\colon a\in A(K)\quad\text{and}\quad a\le p|_K\}.$$

Since each $a\in A(K)$ is w^*-continuous on P_1^* we have that $\bar{p}$ is w^*-l.s.c. on P_1^*. Since $\bar{p}$ is positive homogeneous, the Krein–Šmulyan Theorem (1.2.7) shows $\bar{p}$ is w-l.s.c. But then $\bar{p}\le\check{p}$ and hence equality holds.

Also, the fact that $\sum\|x_i\|\le\beta\|x\|$ whenever $x=\sum x_i$ in P^* implies that if $x=\sum x_i^k$ with

$$\sum_{i=1}^{n}(a_i,x_i^k)<\bar{p}(x)+\frac{1}{k}$$

then we can choose, by the Alaoglu Theorem, limit points $x_1,\ldots,x_n$ of $(x_1^k)_k,\ldots,(x_n^k)_k$ in P^* with $x=\sum_{i=1}^{n}x_i$ and $\bar{p}(x)=\sum_{i=1}^{n}(a_i,x_i)$.

For (ii) we use the hypothesis to conclude via Proposition 3.2 that E^* is β-additive. Thus the same argument applies to $x=\sum_{i=1}^{n}x_i$ for any positive integer n.

5. BANACH LATTICES AND THE RIESZ INTERPOLATION PROPERTY

Let (E,P) be an ordered normed space such that for each $a,b\in E$, $a\wedge b$ and $a\vee b$ exist and are unique. We then say (E,P) is a *lattice*. If (E,P) is a lattice then P must be proper, since $\pm a\in P$ implies both a and $-a$ equal $a\wedge(-a)$. The uniqueness of $a\wedge(-a)$ then forces $a=-a$ and hence $a=0$. In addition E is positively generated since $a=a\vee 0-(a\vee 0-a)\in P-P$.

5.1. PROPOSITION. *Let $a,b\in(E,P)$ be such that $a\wedge b$ exists.*

(i) *for any c, $(a+c)\wedge(b+c)$ exists and*

$$(a+c)\wedge(b+c)=a\wedge b+c;$$

(ii) *if $\lambda\ge 0$ then $(\lambda a)\wedge(\lambda b)=\lambda(a\wedge b)$;*

(iii) *$(-a)\vee(-b)$ exists and equals $-(a\wedge b)$.*

5.2. PROPOSITION. *Let (E,P) be an ordered normed space which is positively generated with P proper. The following are equivalent:*

(i) (E, P) *is a lattice;*
(ii) $a, b \in P$ *implies there is a unique* $a \wedge b \in P$.

Proof. Assume (ii) holds. Then given any $c, d \in E$ there is an $e \in E$ such that both $c+e, d+e \in P$. Then

$$c \wedge d = (c+e) \wedge (d+e) - e$$

shows $c \wedge d$ exists for any pair c, d. Then

$$c \vee d = -[(-c) \wedge (-d)].$$

Let (E, P) be a lattice. We define

$$a^+ = a \vee 0$$

$$a^- = -(a \wedge 0)$$

$$|a| = a \vee -a.$$

5.3. PROPOSITION. *Let* (E, P) *be a lattice. Then*
(i) $a+b = a \vee b + a \wedge b$;
(ii) $a = a^+ - a^-$;
(iii) *if* $a = b - c$; $b, c \geq 0$ *then* $b \geq a^+$, $c \geq a^-$;
(iv) $a^+ + a^- = |a| = a^+ \vee a^-$;
(v) $a^+ \wedge a^- = 0$.

Proof. For (i)

$$a - (a \vee b) = a + (-a \wedge -b) = 0 \wedge (a-b) = 0 \wedge (a-b) + b - b = b \wedge a - b.$$

Then (ii) follows with $b = 0$ in (i). (iii) is immediate from the definitions of a^+ and a^-. Now

$$|a| + a = a \vee -a + a = (2a) \vee 0 = 2a^+$$

so that

$$|a| = 2a^+ - (a^+ - a^-) = a^+ + a^-.$$

Thus $|a| \geq 0$ and so

$$|a| = |a| \vee 0 = (a \vee -a) \vee 0 = (a \vee 0) \vee (-a \vee 0) = a^+ \vee a^-.$$

Finally (v) follows from (i) applied to a^+ and a^-.

We have reserved the term *normed lattice* or, if E is complete, *Banach lattice* for lattices (E, P) with further relations between the order properties and the norm. Specifically we require that E be α-generated (or just positively generated if E is a Banach space) and β-normal for some constants α and β.

Our aim is to prove that (E^*, P^*) being a Banach lattice is a property that is essentially dual to (E, P) having the decomposition property (DP) of Section 4. It is convenient here to reformulate this as the *Riesz Interpolation Property* (RIP). To say (E, P) has the RIP means that whenever

$$c, d \leq a, b$$

there is an element e such that

$$c, d \leq e \leq a, b.$$

5.4. PROPOSITION. *Let (E, P) be an ordered normed space. The following are equivalent*:

(i) *P has the* DP;

(ii) *(E, P) has the* RIP.

Proof. Let (i) hold and apply Lemma 4.1 to the equation

$$(a-c)+(b-d)=(a-d)+(b-c)$$

to obtain

$$a_1, a_2, b_1, b_2 \geq 0 \text{ such that}$$

	$a-d$	$b-c$
$a-c$	a_1	a_2
$b-d$	b_1	b_2

with the rows and columns of the 2×2 array summing as indicated. Now let $e=a_2+c$ and note that $a-c=a_1+a_2$ implies

$$a_2+c=e=a-a_1.$$

Now

$$0 \leq a_2 \leq a-c$$

so that

$$c \leq e \leq a.$$

Now

$$e-d=(a-a_1)-d=(a-d)-a_1=b_1 \geq 0$$

and

$$b-e=b-(a_2+c)=(b-c)-a_2=b_2 \geq 0$$

so that

$$d \leq e \leq b$$

as well. If (ii) holds and

$$0 \le a \le b_1 + b_2 \quad (b_1, b_2 \ge 0)$$

then

$$a - b_2, \qquad 0 \le a, b_1$$

so that there is an a_1 such that

$$a - b_2, \qquad 0 \le a_1 \le a, b_1$$

and hence $0 \le a - a_1 \le b_2$ so that $a_2 = a - a_1$ decomposes a as required.

We can easily establish sufficient conditions for (E^*, P^*) to be a Banach lattice.

5.5. THEOREM. *Let (E, P) be a normed ordered space which is α-generated and β-normal for some α and β. If P has the* (DP) *then (E^*, P^*) is a Banach lattice. Furthermore, if $x, y \in P^*$ and $a \in P$ then*

$$(a, x \wedge y) = \inf \{(a_1, x) + (a_2, y) : a = a_1 + a_2;\ a_1, a_2 \ge 0\}.$$

If $x \in E^*$ and $a \in P$ then

$$(a, x \vee 0) = \sup \{(b, x) : 0 \le b \le a\}.$$

Proof. The dual properties show that (E^*, P^*) is β-generated and α-normal. Let $x, y \in P^*$ and define w on P by

$$w = [\inf \{x, y\}]^-$$

as in Section 4. Thus, for any $a \in P$,

$$w(a) = \inf \{(a_1, x) + (a_2, y) : a = a_1 + a_2;\ a_1, a_2 \ge 0\}.$$

Then Theorem 4.2(i) and (iii) show w is linear on P. In addition w is bounded on P since

$$0 \le w(a) \le (a, x) \wedge (a, y) \le \|a\|(\|x\| \wedge \|y\|).$$

Since E is α-generated w has a unique extension to $P - P = E$ such that if

$$c = a - b; \qquad a, b \in P \quad \text{and} \quad \|a\| + \|b\| \le \alpha\|c\|$$

then

$$|w(c)| \le w(a) + w(b) \le (\|x\| \wedge \|y\|)(\|a\| + \|b\|) \le \alpha(\|x\| \wedge \|y\|)\|c\|$$

so that $w \in P^*$. Also Theorem 4.2(ii) shows $w = x \wedge y$ in P^* and the formula for (a, w) holds. If $x \in E^*$ and y any element such that $y \ge x, 0$ then

$$y - (x \vee 0) = y + (-x) \wedge 0 = (y - x) \wedge y$$

and

$$
\begin{aligned}
(a, y)-(a, x \vee 0) &= \inf \{(a_1, y-x)+(a_2, y): a=a_1+a_2;\, a_1, a_2 \geq 0\} \\
&= \inf \{(a, y)-(a_1, x): 0 \leq a_1 \leq a\} \\
&= (a, y)-\sup \{(a_1, x): 0 \leq a_1 \leq a\}
\end{aligned}
$$

and the formula for $x \vee 0$ is shown.

If (E^*, P^*) is a Banach lattice then there are a number of interesting consequences (in fact, characterizations, as Theorem 5.7 below will show) that will be quite useful in later applications to $A(K)$ spaces.

To begin with, we say (E, P) is a *complete Banach lattice* if it is a Banach lattice such that every non-empty subset which is bounded above (in order) has a (unique) least upper bound.

Let F be a convex subset of the cone P. Then, using the definition of *face* in Chapter 1 Section 1 it is easily checked that F is a face of P if and only if F is a subcone of P for which

$$a \in F \quad \text{and} \quad 0 \leq b \leq a \quad \text{implies } b \in F.$$

We say the closed subcone F of P is *complemented* if there is a map σ_F, or just denoted σ if F is understood, such that $\sigma: P \to F$ and

(i) σ is positive homogeneous and additive,
(ii) $\sigma^2 a = \sigma a \leq a$ for all $a \in P$.

If F is complemented under σ then the *complementary subcone*

$$G = \{a \in P: \sigma a = 0\}$$

and every $a \in P$ has a *unique* representation

$$a = b + c; \qquad b = \sigma a \in F \quad \text{and} \quad c \in G.$$

In particular both F and G are faces of P. Conversely, if F and G are subcones of P such that every $a \in P$ can be represented uniquely as $b+c$, $b \in F$, $c \in G$, then F and G are complementary with map $\sigma_F a = b$.

5.6. PROPOSITION. *Let (E, P) be an ordered normed space which is α-generated and β-normal for some α and β.*

(i) *If F is a complemented subcone of P then σ_F extends uniquely to a bounded projection of E onto the subspace $F-F$ with null space $G-G$. In particular, $F-F$ and $G-G$ are $\|\cdot\|$-closed subspaces of E.*
(ii) *If F_1 and F_2 are complemented in P under σ_1 and σ_2 then $F_1 \cap F_2$ is complemented under $\sigma_1 \circ \sigma_2 = \sigma_{12} = \sigma_2 \circ \sigma_1$ and*

$$F_1 + F_2 \text{ is complemented under } \sigma_1 + \sigma_2 - \sigma_{12}.$$

Each $x \in F_1 + F_2$ has a unique representation

$$x = y + z_1 + z_2 \in F_1 \cap F_2 + F_1 \cap G_2 + F_2 \cap G_1$$

where G_1, G_2 are the faces complementary to F_1, F_2.

Proof. (i) Since E is positively generated the extension of σ_F to a projection onto $F - F$ with null space $G - G$ is automatic from the properties (i) and (ii) of σ_F. If $a \in P$ then $0 \leq \sigma_F a \leq a$ implies, by normality, that $\|\sigma_F a\| \leq \beta \|a\|$. If $c = a - b$; $a, b \in P$, with

$$\|a\| + \|b\| \leq \alpha \|c\|$$

then

$$\|\sigma_F c\| \leq \|\sigma_F a\| + \|\sigma_F b\| \leq \alpha\beta \|c\|.$$

(ii) For any $x \in P$ let $y = \sigma_2 x$ and write

$$y = \sigma_1 y + (y - \sigma_1 y).$$

Since F_2 is a face and $y \in F_2$, we have

$$\sigma_1 y \in F_1 \cap F_2 \quad \text{and} \quad y - \sigma_1 y \in G_1 \cap F_2.$$

It follows easily that $F_1 \cap F_2$ is complemented under $\sigma_1 \circ \sigma_2$ which, by symmetry, must equal $\sigma_2 \circ \sigma_1$. That $F_1 + F_2$ is complemented under $\sigma_1 + \sigma_2 - \sigma_{12}$ is easily verified using the fact that

$$\sigma_1|_{F_2} = \sigma_{12}|_{F_2} \quad \text{and} \quad \sigma_2|_{F_1} = \sigma_{12}|_{F_1}.$$

In particular, for $x = x_1 + x_2 \in F_1 + F_2$,

$$\begin{aligned} x &= (\sigma_1 + \sigma_2 - \sigma_{12})(x) \\ &= \sigma_{12}(x_1 + x_2) + (x_1 - \sigma_2 x_1) + (x_2 - \sigma_1 x_2) \in F_1 \cap F_2 + F_1 \cap G_2 + F_2 \cap G_1 \end{aligned}$$

which is easily seen to be unique.

Definition. We say a face F of P^* is *w^*-exposed* if for some $a \in P$

$$F = \{x \in P^* : (a, x) = 0\}.$$

5.7. THEOREM. *Let (E, P) be an ordered Banach space which is positively generated and normal. The following are equivalent:*

(i) *(E^*, P^*) is a Banach lattice;*

(ii) *P^* has the* DP;

(iii) (E^*, P^*) *is a complete Banach lattice*;
(iv) *each* w^**-closed exposed face of* P^* *is complemented*;
(v) *for any* $a_1, \ldots, a_n \in P$ *the functional* $[\inf\{a_i\}]^-$ *is linear on* P^*;
(vi) (E, P) *has the* RIP.

Proof. (i) implies (ii): as noted in Section 4, if P^* is a lattice then P^* has the DP. The rest is by definition. (ii) implies (iii): let S be a non-empty set bounded above by u. Let $S_0 = u - S$. Then S_0 is a non-empty subset of P^* and it suffices to show the greatest lower bound, w, of S_0 exists in P^*. Then $u - w$ is the least upper bound of S.

Let $T = \{x \in P^* : x \leq S_0\}$. Since $0 \in T$, T is not empty, and we let $D = \{(\alpha, \beta) : \alpha, \beta$ non-empty finite subsets of T, S_0 resp.$\}$. Since P^* has the DP we can use Proposition 5.4 (and a straightforward induction) to define a net

$$\{w(\alpha, \beta) : (\alpha, \beta) \in D\}$$

such that $w(\alpha, \beta)$ interpolates the families α and β:

$$x \leq w(\alpha, \beta) \leq y \quad \text{for all } x \in \alpha,\ y \in \beta.$$

Now, for any fixed $s_0 \in S_0$, the collection of $(\alpha, \beta) \in D$ such that $s_0 \in \beta$ is co-final in D so that, by normality and the Alaoglu Theorem we can find a subnet of $(w\ (\alpha, \beta))$ which converges w^* to $w \in P^*$.

Since P^* is w^*-closed it follows that for any $a \in P$, $x \in T$ and $y \in S_0$,

$$(a, x) \leq (a, w) \leq (a, y).$$

Thus $T \leq w \leq S_0$ so that $w = \text{g.l.b. } S_0$. Since E^* is positively generated this shows in particular (via Proposition 5.2) that E^* is a lattice, and hence, a complete Banach lattice.

(iii) implies (iv): let A be any w^*-closed face of P^*. For each $x_0 \in P^*$ the set

$$A(x_0) = \{y \in A : y \leq x_0\}$$

is bounded above and non-empty so that we can define

$$y_0 = \sigma x_0 = \text{l.u.b. } A(x_0).$$

Now $A(x_0)$ is directed by $\leq$ (since x, $y \in A(x_0)$ implies $x \vee y \in A(x_0)$) and hence is a net in A. Since $A(x_0)$ is bounded above and A is w^*-closed, there is a subnet converging to $z \in A$. Hence $A(x_0) \leq z$ so that $y_0 \leq z$. But, given any $a \in P$, $(a, x) \leq (a, y_0)$ for all $x \in A(x_0)$ so that $(a, z) \leq (a, y_0)$ and hence $z \leq y_0$. Thus $y_0 = z \in A$ and clearly $y_0 \leq x_0$.

If $x_1, x_2 \in P^*$ then

$$A(x_1) + A(x_2) = A(x_1 + x_2)$$

by virtue of the DP and the fact that A is a face. Also, for $\lambda \geq 0$,

$$\lambda A(x) = A(\lambda x)$$

so that σ is linear. Clearly $\sigma = \text{id}$ on A so that A is complemented in P^* under σ.

(iv) implies (v): Given $a, b \in P$ let $p = \inf\{a, b\}$ so that

$$\bar{p}(x) = \inf\{(a, x_1) + (b, x_2) : x = x_1 + x_2;\, x_1, x_2 \in P^*\}.$$

From Theorems 4.3 and 4.4 we have $\bar{p} = \check{p}$, the support functional of

$$A(p) = \{c \in E : c \leq a, b\}.$$

We note first that if F_1 and F_2 are w^*-closed exposed faces of P^* such that $\bar{p}|_{F_1}$ and $\bar{p}|_{F_2}$ are both linear then $\bar{p}|_{F_1+F_2}$ is also linear. This follows since F_1, F_2 are complemented (under σ_1, σ_2) and hence, by Proposition 5.6 $F = F_1 + F_2$ is complemented under $\sigma = \sigma_1 + \sigma_2 - \sigma_{12}$. But then the functional q defined by

$$q = \bar{p} \circ \sigma_1 + \bar{p} \circ \sigma_2 - \bar{p} \circ \sigma_{12}$$

is linear and equals $\bar{p}$ on $F_1 \cup F_2$. Moreover, if

$$x = y + z_1 + z_2 \in F_1 \cap F_2 + F_1 \cap G_2 + F_2 \cap G_1$$

then

$$qx = \bar{p}y + \bar{p}z_1 + \bar{p}z_2.$$

It follows from this that $\bar{p} \leq q \leq a, b$ on F. But if $x \in F$ and $x = x_1 + x_2$; $x_1, x_2 \in P^*$, then $x_1, x_2 \in F$ so that

$$qx \leq (a, x_1) + (b, x_2).$$

Thus $q \leq \bar{p}$ on F and therefore equality holds. We show that whenever (c_0, x_0) is a support pair for $A(p)$ then x_0 belongs to a w^*-closed face F_0 on which $\bar{p} = c_0$ is linear. Since $c_0 \leq a, b$ the sets

$$F_a = \{x \in P^* : (c_0, x) = (a, x)\} \quad \text{and} \quad F_b = \{x \in P^* : (c_0, x) = (b, x)\}$$

are in fact w^*-closed exposed faces. Now, by Theorem 4.4(i) we can choose

$$x_0 = x_1 + x_2 \quad (x_1, x_2 \in P^*)$$

with

$$\bar{p}(x_0) = (a, x_1) + (b, x_2).$$

But then

$$(c_0, x_1) + (c_0, x_2) = (c_0, x_0) = \bar{p}(x_0) = (a, x_1) + (b, x_2)$$

and hence $(c_0, x_1) = (a, x_1)$ and $(c_0, x_2) = (b, x_2)$. Thus $x_0 \in F_a + F_b$. Since $\bar{p} = c_0$ on $F_a \cup F_b$ the above remark shows $\bar{p} = c_0$ on $F_a + F_b$.

Finally, the Bishop–Phelps Theorem (1.7.1) shows that $\{x : (c, x)$ support pair for $A(p)\}$ is $\|\cdot\|$-dense in P^*. Thus

$$\sum (F : F\, w^*\text{-closed complemented face of } P^* \text{ and } \bar{p}|_F \text{ linear}\}$$

is a $\|\cdot\|$-dense sub-cone Q of P^*. But if $q = \bar{p}|_Q$ then q extends (uniquely) to a bounded linear functional (also denoted q) on the dense subspace $Q-Q$ and therefore an element of $P^{**} \subset E^{**}$. Since $q \leq a, b$ we have $q \leq \bar{p}$. Since $\bar{p}$ is l.s.c. on P^* we have that $\bar{p} = q$ on P^*. The general case of (v) is proved similarly.

(v) implies (vi): given $a, b \in P$ we wish to show that the set

$$A(p) = \{c \in E : c \leq a, b\}$$

is directed, where $p = \inf\{a, b\}$ on P^*. It follows from (v) that if

$$q = \inf\{a_i \in P : i = 1, \ldots, n\}$$

then $\bar{q}$ is linear. To show $A(p)$ is directed we can use the same technique as in Section 3. Thus let $\bar{E} = E \times E$ with

$$\|(a_1, a_2)\| = \max\{\|a_1\|, \|a_2\|\} \quad \text{and} \quad \bar{P} = P \times P$$

let $\bar{D}$ denote the diagonal and let $\bar{a} = (a, a)$ and $\bar{b} = (b, b)$. Let $q = \inf\{\bar{a}, \bar{b}\}$ on $\bar{P}^*$ so that by Theorems 4.3 and 4.4 $\bar{q}$ is the support functional of

$$\bar{A} \doteq \bar{A}(q) = \{(c, d) \in \bar{E} : (c, d) \leq (a, a), (b, b)\} = (\bar{a} - \bar{P}) \cap (\bar{b} - \bar{P}).$$

It is easy to see that $A(p)$ is directed in E if and only if

$$\bar{A} = \bar{A} \cap \bar{D} - \bar{P} \quad \text{in } \bar{E}.$$

Now, the support functional $\bar{q}$ of $\bar{A}$ satisfies for $(x, y) \in \bar{P}^*$,

$$\begin{aligned}\bar{q}(x, y) &= \inf\{((a, a), (x_1, y_1)) + ((b, b), (x_2, y_2)) : x = x_1 + x_2, y = y_1 + y_2\} \\ &= \inf\{((a, x_1) + (b, x_2)) + ((a, y_1) + (b, y_2)) : x = x_1 + x_2, y = y_1 + y_2\} \\ &= \bar{p}(x) + \bar{p}(y) = \bar{p}(x + y).\end{aligned}$$

On the other hand

$$(\bar{A} \cap \bar{D} - \bar{P})^0 = (\bar{A} \cap \bar{D})^0 \cap \bar{P}^*$$

so that if $\bar{\rho}$ is the support functional of $\bar{A} \cap \bar{D}$ and $(x, y) \in \bar{P}^*$,

$$\begin{aligned}\bar{\rho}(x, y) &= \sup\{((c, c), (x, y)) : (c, c) \in \bar{A} \cap \bar{D}\} \\ &= \sup\{(c, x + y) : c \leq a, b\} = \bar{p}(x + y).\end{aligned}$$

Thus $\bar{q} = \bar{\rho}|_{\bar{P}^*}$ so that

$$\bar{A}^0 = (\bar{A} \cap \bar{D} - \bar{P})^0.$$

We deduce that

$$\bar{A} = \|\cdot\| - \mathrm{cl}\,(\bar{A} \cap \bar{D} - \bar{P}).$$

We conclude by showing that $\bar{A} \cap \bar{D} - \bar{P}$ is closed.

Given $(c, d) \in \bar{A}$ choose $\bar{w} \in \bar{E}$ and

$$(c, d) = (z, z) - (p_1, p_2) + (w_1, w_2);$$

where

$$z \leq a, b; \qquad p_1, p_2 \geq 0 \quad \text{and} \quad w_1, w_2 \in E_1.$$

Since E is $(M, 2)$-directed (some M) there is a $q \in P$ with $q \geq w_1, w_2$ and $\|q\| \leq M$. Hence we have

$$c, d \leq z + q \leq a + q, b + q, \quad \text{or} \quad c, d, z \leq a + q, b + q, z + q.$$

The remark at the start shows that if $p' = \inf\{a, b, z + q\}$ then the procedure can be applied to p' (or a suitable translate to make $p' \geq 0$). Thus we can construct sequences $(z_n)_{n=1}^{\infty}$ and $(q_n)_{n=1}^{\infty}$ such that

$$\|q_{n+1}\| \leq \|q_1\|/2^n$$

and

$$c, d, z_n \leq z_{n+1} + q_{n+1} \leq a + q_{n+1}, b + q_{n+1}, z_n + q_n + q_{n+1}.$$

In particular, the normality of E shows, since

$$-q_{n+1} \leq z_{n+1} - z_n \leq q_n$$

that (z_n) is Cauchy and hence (z_n) converges to z interpolating c, d and a, b.

(vi) implies (i): This is immediate from Proposition 5.4 and Theorem 5.5.

6. ABSTRACT L AND M SPACES

A normed lattice (E, P) is *regular* if

$$|a| \leq |b| \quad \text{implies} \quad \|a\| \leq \|b\|.$$

This property implies that the norm is increasing on P and that $\|a\| = \|(|a|)\|$. Conversely, if the norm is increasing on P and $\|a\| = \|(|a|)\|$ then E is regular. We will see (Section 11) that in general normality and positive generation ensure that E can be re-normed to be regular.

In this section we will present versions of the Kakutani Representation Theorems for regular Banach lattices of type L and type M. The object is to give order-theoretic characterizations of two of the most common types of function spaces with lattice structure.

A regular Banach lattice is an *abstract M-space* if whenever $0 \leq a, b$ then

$$\|a \vee b\| = \|a\| \vee \|b\|.$$

In the absence of completeness we will use the term *pre-M*-space. Every M-space E can be realized as a closed Banach lattice-ordered subspace of $C_{\mathbb{R}}(X)$ (X compact Hausdorff). We content ourselves here with a proof of this fact in the event that E has an order unit, in which case $E = C_{\mathbb{R}}(X)$.

A regular Banach lattice for which the norm is additive on the positive cone is called an *abstract* L-space. The prototype for L-spaces is the space $L^1(\mu)$ of functions integrable with respect to a measure space $(X, \mathcal{B}, \mu)$.

We first give simple characterizations of L and M spaces in terms of their order properties and then demonstrate the duality between them.

6.1. THEOREM. *An ordered normed space (E, P) is a pre-M space if and only if*

(i) *$a, b \in P$ implies $a \wedge b$ exists in P;*

(ii) *E is 1-directed;*

(iii) *E is 1-normal.*

Proof. Let (E, P) be a pre-M space. Then clearly (i) holds and if $a, b \in E_1$, then $c = |a| \vee |b| \in E_1$ and $c \geq a, b$. If $c \leq a \leq b$ then

$$-(|b| \vee |c|) \leq a \leq |b| \vee |c|$$

and hence

$$|a| \leq |b| \vee |c|$$

and (iii) follows.

If (i), (ii) and (iii) hold then clearly (E, P) is a normed lattice. Then (iii) shows that $\|\cdot\|$ is increasing on P and, since

$$-|a| \leq a \leq |a|,$$

$\|a\| \leq \|(|a|)\|$. But (ii) shows that there is $b \geq a, -a$ with $\|b\| \leq \|a\|$. Thus

$$\|a\| \geq \|b\| \geq \|a \vee (-a)\| = \|(|a|)\|$$

and it follows that E is regular. Now, given $0 \leq a, b$, choose $c \geq a, b$ and $\|c\| = \|a\| \vee \|b\|$ by (ii). Then $0 \leq a \vee b \leq c$ and so, using (iii), $\|a \vee b\| \leq \|c\| = \|a\| \vee \|b\| \leq \|a \vee b\|$.

6.2. THEOREM. *An ordered Banach space (E, P) is an L-space if and only if*

(i) *$a, b \in P$ implies $a \wedge b$ exists in P,*

(ii) *E is 1-generated,*

(iii) *E is 1-additive.*

Proof. If E is an L-space then (i) and (iii) are automatic. Also,

$$\|a\| = \|(|a|)\| = \|a^+ + a^-\| = \|a^+\| + \|a^-\|$$

shows E is 1-generated. Let (i), (ii) and (iii) hold. Then E is a Banach lattice with norm additive on P. It remains only to check that E is regular. Property (iii) shows $\|\cdot\|$ is increasing on P. Also, given $a \in E$,

$$\|a\| = \|a^+ - a^-\| \leq \|a^+\| + \|a^-\| = \|(|a|)\|.$$

But, by (ii), $a = a_1 - a_2$ $(a_i \geq 0)$ with

$$\|a\| = \|a_1\| + \|a_2\|.$$

However, $0 \leq a^+ \leq a_1$ and $0 \leq a^- \leq a_2$ so that

$$\|a\| \geq \|(|a|)\|.$$

We can now apply the duality theorems of the previous sections to L and M spaces and immediately deduce the following theorem.

6.3. THEOREM.
(i) *If E is an L-space then E^* is an M-space.*
(ii) *If E is a pre-M-space then E^* is an L-space.*

We next *characterize* dual L and M spaces.

6.4. THEOREM. *Let (E, P) be an ordered Banach space. Then E^* is an M-space (if and) only if E is an L-space.*

Proof. If E^* is an M-space then we can conclude that
(i) P has the DP,
(ii) E is 1-additive,
(iii) E is approximately 1-generated.
In view of (i) and (ii), together with Proposition 3.2, we see that E is in fact a complete Banach lattice. In particular E is then 1-generated for, given $a = a_1 - a_2$ $(a_i \geq 0)$ we then have $0 \leq a^+ \leq a_1$ and $0 \leq a^- \leq a_2$. But (ii) implies $\|a^+\| \leq \|a_1\|$ and $\|a^-\| \leq \|a_2\|$. Thus (iii) shows $\|a\| = \|a^+\| + \|a^-\|$. Hence, by Theorem 6.2, E is an L-space.

6.5. THEOREM. *Let (E, P) be an ordered Banach space. Then E^* is an L-space if and only if*
(i) *P has the* DP,
(ii) *E is 1-directed,*
(iii) *E is 1-normal.*

Proof. If (i), (ii) and (iii) hold then their dual properties show, via 6.2, that E^* is an L-space. Conversely, if E^* is an L-space then (i) and (iii) hold but we conclude only that E is approximately 1-directed. However an iteration using (i) and (iii) will yield (ii). Thus let $a, b \in E_1$ and let $\lambda_n = (1 + 2^{-n})$ $(n \geq 1)$. Choose $c_1 \geq a, b, 0$; $\|c_1\| \leq \lambda_1$, and suppose $c_1, \ldots, c_n$ are given with

$c_n \geq a, b, 0$; $\|c_n\| \leq \lambda_n$, and $\|c_{k+1} - c_k\| \leq 1/2^k$ $(k = 1, \ldots, n-1)$. Then $c_n/\lambda_n \in E_1$ so that there exists c'_{n-1} such that

$$c'_{n+1} \geq a, b, c_n/\lambda_n \quad \text{and} \quad \|c'_{n+1}\| \leq \lambda_{n+1}$$

Now c_n, $c'_{n+1} \geq a, b, c_n/\lambda_n$ implies there exists c_{n+1} which interpolates. Thus

$$c_{n+1} \geq a, b, 0; \qquad \|c_{n+1}\| \leq \|c'_{n+1}\| \leq \lambda_{n+1} \quad \text{and} \quad c_n \geq c_{n+1} \geq c_n/\lambda_n.$$

Thus

$$0 \geq c_{n+1} - c_n \geq -(1 - 1/\lambda_n)c_n$$

and therefore

$$\|c_{n+1} - c_n\| \leq (1 - 1/\lambda_n)\lambda_n = \lambda_n - 1 = 2^{-n}.$$

Hence (c_n) is Cauchy and clearly has a limit $c \in E$ with $c \geq a, b$.

We next note some facts about normed lattices which will be useful in discussing the representation theorems below.

6.6. PROPOSITION. *Let (E, P) be an α-normal and β-generated ordered normed space. The following are equivalent:*

(i) *E is a normed lattice;*

(ii) *for each $a, b \in P$ and $p = \inf\{a, b\}$ on P^* the functional $\bar{p}$ is additive on P^* and extends uniquely to $a \wedge b \in P$.*

Proof. If E is a lattice then P^* has the DP so that $\bar{p}$ is additive and, given $x \in P^*$,

$$\bar{p}(x) = \sup\{(c, x) : c \leq a, b\} = (a \wedge b, x).$$

Conversely, if $\bar{p}$ is the restriction of an element $c \in E$ to P^* then by definition $c = a \wedge b$. Hence Proposition 5.2 shows E is a normed lattice.

If (E, P) is a normed lattice the elements $x \in P^*$ which preserves the lattice operations are called *lattice homomorphisms*. In view of Propositions 5.2 and 5.3 it suffices that

$$(a \wedge b, x) = (a, x) \wedge (b, x)$$

for all $a, b \in P$.

6.7. THEOREM. *Let (E, P) be a normed lattice. Then $x \in P^*$ is a lattice homomorphism if and only if the ray $\mathbb{R}^+ x$ is a face of P^*.*

Proof. If x is a homomorphism and $0 \leq y \leq x$ let $(a, x) = 0$. Then $(a^+, x) = (a, x) \vee 0 = 0 = (a^-, x)$ so that $(|a|, x) = 0$. Hence $(|a|, y) = 0 = (a^+, y) + (a^-, y)$ implies, since $y \geq 0$, that $(a, y) = 0$. Thus $y = \lambda x$ for some

$\lambda \geq 0$. If $\mathbb{R}^+x$ is a face of P^* and $a, b \in P$ then, by Proposition 6.6 and Theorem 4.4,

$$\begin{aligned}(a \wedge b, x) &= \inf \{(a, x_1) + (b, x_2) : x_1 + x_2 = x\} \\ &= \inf \{(a, \lambda x) + (b, \mu x) : \lambda + \mu = 1\} \\ &= (a, x) \wedge (b, x).\end{aligned}$$

6.8. PROPOSITION. *Let (E, P) be a normed lattice and let S be a non-empty subset for which $a = \text{l.u.b.}\,(S)$ exists. Then for any $b \in E$,*

$$b \wedge a = \text{l.u.b.}\,\{b \wedge s : s \in S\}.$$

Proof. Clearly $b \wedge a \geq b \wedge s$ for all $s \in S$. Let $c \geq b \wedge s$ for all $s \in S$. We use Proposition 5.3(i) to obtain $c \geq b + s - b \vee s$. Thus $c + b \vee a \geq c + b \vee s \geq b + s$ for all s so that

$$c + b \vee a \geq b + a$$

and hence $c \geq b \wedge a$. Thus $b \wedge a = \text{l.u.b.}\,\{b \wedge s : s \in S\}$

We next prove the Kakutani representation of an M-space E with order unit by showing $E = C_{\mathbb{R}}(X)$ where X is the set of extreme points of the state space S of E. This is tantamount to the Stone–Weierstrass Theorem (lattice version).

6.9. THEOREM (STONE–WEIERSTRASS). *Let X be a compact Hausdorff space and let the subspace E be a sub-lattice of $C_{\mathbb{R}}(X)$ which contains* 1 *and separates the points of X. Then E is dense in $C_{\mathbb{R}}(X)$.*

Proof. For each $x \neq y$ in X and $\alpha, \beta \in \mathbb{R}$ there is an $f \in E$ such that $f(x) = \alpha$ and $f(y) = \beta$. In fact if $r = f_0(x) - f_0(y) \neq 0$ for $f_0 \in E$ then

$$f = (\alpha/r)(f_0 - f_0(y) \cdot 1) - (\beta/r)(f_0 - f_0(x) \cdot 1).$$

Let $h \in C_{\mathbb{R}}(X)$ and $\varepsilon > 0$ be given. Let $x \in X$ be fixed and let $f_x = h(x) \cdot 1$. For each $y \neq x$ let $f_y \in E$ be such that $f_y(x) = h(x)$ and $f_y(y) = h(y)$. For each $y \in X$ let U_y be a neighbourhood of y on which $f_y > h - \varepsilon$. By compactness we can choose $y_1, \ldots, y_n$ such that if

$$g_x = f_{y_1} \vee \cdots \vee f_{y_n}$$

then $g_x(x) = h(x)$ and $g_x > h - \varepsilon$. For each x let V_x be a neighbourhood on which the g_x (as above) satisfies $g_x > h + \varepsilon$. Again, by compactness, we choose $x_1, \ldots, x_m$ such that if

$$g = g_{x_1} \wedge \cdots \wedge g_{x_m}$$

then $h - \varepsilon < g < h + \varepsilon$ on X.

Let (E, P) be a normed order unit space and recall from Section 2 that E is order-isometric to a dense subspace of $A(S)$, where S is the state space of E. If in addition E is a lattice then we can make a much stronger assertion. By a *lattice-isometry* we mean a linear isometry which preserves the lattice operations.

6.10. THEOREM. *Let (E, P) be an order unit space and a lattice. Let $X = \text{ext}\, S$, where S is the state space of E. Then X is weak*-compact and E is lattice-isometric to a dense sub-lattice of $C_{\mathbb{R}}(X)$. If E is a Banach space then E is lattice-isometric to $C_{\mathbb{R}}(X)$.*

Proof. Let e denote the order unit of (E, P). Since

$$S = \{x \in P^* : (e, x) = 1\}$$

it is immediate from Theorem 6.7 that $X = \text{ext}\, S$ is the set of lattice homomorphisms whose value at e is 1. Thus X is the intersection of the weak*-compact sets

$$S(a, b) = \{x \in S : (a \wedge b, x) = (a, x) \wedge (b, x)\} \quad (a, b \in P)$$

and hence is w^*-compact. Let $\theta : E \to C_{\mathbb{R}}(X)$ be the restriction map. Since the restriction of E to S is isometric to a dense subspace of $A(S)$ the Krein–Milman Theorem assures that θ is an isometry. Moreover each $x \in X$ being a lattice homomorphism shows

$$\theta(a \wedge b) = (\theta a) \wedge (\theta b) \quad \text{and} \quad \theta(a \vee b) = (\theta a) \vee (\theta b).$$

Since θE is a separating sub-lattice containing $1 = \theta e$ the rest follows from Theorem 6.9.

Observe that (E, P) being an order unit space ensures that properties (ii) and (iii) of Theorem 6.1 hold. Thus the hypotheses of Theorem 6.10 are equivalent to (E, P) being a pre-M-space with order unit, although this formulation is somewhat redundant, in view of Theorem 6.1.

6.11. COROLLARY. *Let E be a linear sub-lattice of $C_{\mathbb{R}}(Y)$, Y compact Hausdorff, which contains 1 and separates points. Then the evaluation map*

$$\theta : Y \to E^*$$

is a homeomorphism onto $X = \text{ext}\, S$.

Proof. We have from Corollary 1.4.4 that $X \subset \phi Y$. But each ϕy is a lattice homomorphism and hence belongs to X.

6.12. COROLLARY. *If X and Y are compact Hausdorff with $C_{\mathbb{R}}(X)$ order isometric to $C_{\mathbb{R}}(Y)$ then X is homeomorphic to Y.*

Proof. If $\psi: C_{\mathbb{R}}(Y) \to C_{\mathbb{R}}(X)$ is an order isometry then the adjoint ψ^* yields a homeomorphism between $\phi_X(X)$ and $\phi_Y(Y)$ where ϕ_X and ϕ_Y are the respective evaluation maps onto the extreme points of the respective state spaces. It follows from Corollary 6.11 that X and Y are homeomorphic.

An interesting and useful application of 6.10 can be made in connection with the Stone Representation Theorem for Boolean algebras. This will also be of value in the development leading to the characterization of L-spaces. A *Boolean algebra* is a partially ordered set $\mathscr{A}$ ($\leq$ is transitive, reflexive and anti-symmetric) such that

(i) $\mathscr{A}$ is a distributive lattice ($(A \wedge B) \vee C = (A \vee C) \wedge (B \vee C)$),
(ii) g.l.b. $\mathscr{A}$ and l.u.b. $\mathscr{A}$ (denoted 0 and I) exist in $\mathscr{A}$,
(iii) for each $A \in \mathscr{A}$ there exists a unique element A^c such that $A \wedge A^c = 0$ and $A \vee A^c = I$.

The axioms are, of course, modelled after set-theoretic operations $\cap$ and $\cup$. We shall see that each Boolean algebra $\mathscr{A}$ has a realization as the collection of *clopen* (both open and closed) subsets of a *totally disconnected* (TD) compact Hausdorff space. (We say X is *totally disconnected* if each point has a neighbourhood base of *clopen* sets.)

A subset A of X is clopen if and only if the characteristic function, denoted χ_A, is continuous. We observe that f is a characteristic function of a subset of X if and only if f has values of 0 and 1 only. This in turn is equivalent to

$$f \wedge (1-f) \equiv 0.$$

The subspace in $C_{\mathbb{R}}(X)$ (not closed in general) spanned by the continuous characteristic functions, denoted $\mathscr{S}(X)$, consists precisely of the continuous *simple functions*, that is, functions taking only finitely many values. Such a function f is associated with a *partition* of X into a disjoint union of non-empty clopen sets $A_1, \ldots, A_n$ with f constant on each A_i. The subspace $\mathscr{S}(X)$ is a sub-lattice of $C_{\mathbb{R}}(X)$ containing the constant functions.

6.13. PROPOSITION. *Let X be a compact Hausdorff space. Then X is* TD *if and only if $\mathscr{S}(X)$ is dense in $C_{\mathbb{R}}(X)$.*

Proof. If X is TD then $\mathscr{S}(X)$ separates the points of X and hence, by Theorem 6.9, is dense in $C_{\mathbb{R}}(X)$. Conversely, given $x \in U$, U open, there is a continuous function f such that $0 \leq f \leq 1$, $f(x) = 1$ and $f = 0$ on $X \backslash U$. Let a be an element of $\mathscr{S}(X)$ such that $\|f - a\| < \frac{1}{2}$. Then $\{x: a(x) \geq \frac{1}{2}\}$ is a clopen neighbourhood of x contained in U.

The collection of clopen subsets of X forms a Boolean algebra, $\mathscr{C}(X)$, under $\cup$ and $\cap$ which is lattice isomorphic to the subset of characteristic functions in $C_{\mathbb{R}}(X)$. The induced lattice operations on the characteristic functions are the natural ones inherited from $C_{\mathbb{R}}(X)$.

Now let $\mathscr{A}$ be a Boolean algebra. We wish to construct a formal replica of the space $\mathscr{S}(X)$ of simple functions on a compact Hausdorff space and use this to identify the elements of $\mathscr{A}$ with the continuous characteristic functions, denoted $\mathscr{C}(X)$. To do this we define a (formal) simple function $\bar{a}$ on $\mathscr{A}$ to be a finite collection of pairs $(A_i, a_i)_{i=1}^n$ with $0 \neq A_i \in \mathscr{A}$, $a_i \in \mathbb{R}$ and

$$A_1 \vee \cdots \vee A_n = I \quad \text{with} \quad A_i \wedge A_j = 0 \quad (i \neq j).$$

(We say such a collection of $(A_i)_{i=1}^n$ forms a *partition of* $\mathscr{A}$.)

We say $\bar{a} = (A_i, a_i)_{i=1}^n$ is *equivalent* to $\bar{b} = (B_j, b_j)_{j=1}^m$ if for each real number c

$$\bigvee \{A_i \colon a_i = c\} = \bigvee \{B_j \colon b_j = c\}.$$

Modulo this equivalence relation, the set of formal simple functions, denoted $\mathscr{S}(A)$, forms a normed order unit space which is a lattice, that is, a pre-M space with order unit. The operations are defined by

$$\lambda\bar{a} = (A_i, \lambda a_i)_{i=1}^n$$

$$\bar{a} + \bar{b} = (A_i \wedge B_j, a_i + b_j)_{i,j=1}^{n,m} \quad (A_i \wedge B_j \neq 0)$$

$$\|\bar{a}\| = \max \{|a_i| \colon i = 1, \ldots, n\}$$

$$\bar{a} \wedge \bar{b} = (A_i \wedge B_j, a_i \wedge b_j) \quad (A_i \wedge B_j \neq 0)$$

$$\bar{e} = (I, 1)$$

where $\bar{a}$, $\bar{b}$ are as above.

The simple function $\bar{a}$ is called a (formal) *characteristic function* if each a_i is zero or one: equivalently,

$$\bar{a} \wedge (\bar{e} - \bar{a}) = 0.$$

6.14. THEOREM (STONE). *Let $\mathscr{A}$ be a Boolean algebra and let $\mathscr{S}(\mathscr{A})$ be the associated space of (formal) simple functions. Let $X = \operatorname{ext} S$, S the state space of $\mathscr{S}(\mathscr{A})$. Then X (w^*-topology) is a* TD *compact Hausdorff space such that*

(i) *$\mathscr{S}(\mathscr{A})$ is lattice-isometric to $\mathscr{S}(X)$ and*

(ii) *$\mathscr{A}$ is (lattice) isomorphic to $\mathscr{C}(X)$.*

If Y is a TD *compact Hausdorff space with $\mathscr{A}$ isomorphic to $\mathscr{C}(Y)$ then Y is homeomorphic to X.*

Proof. From Theorem 6.10 we have $\mathscr{S}(\mathscr{A})$ is lattice-isometric to a dense sub-lattice of $C_{\mathbb{R}}(X)$. Let a denote the element of $C_{\mathbb{R}}(X)$ corresponding to $\bar{a} \in \mathscr{S}(\mathscr{A})$.

Now $\bar{a}$ is a characteristic function if and only if

$$\bar{a} \wedge (\bar{e} - \bar{a}) = 0,$$

in which case $a \wedge (1-a) = 0$ in $C_{\mathbb{R}}(X)$, showing that a is then a characteristic function on X. Hence each $\bar{a} \in \mathcal{S}(\mathcal{A})$ corresponds to $a \in \mathcal{S}(X)$ with the same values a_i. On the other hand, if f is a continuous characteristic function on X then, since $\{a: \bar{a} \in \mathcal{S}(\mathcal{A})\}$ is dense, there is an $\bar{a} \in \mathcal{S}(\mathcal{A})$ with $\|a-f\| < \frac{1}{2}$. Then either $|a_i - 1| < \frac{1}{2}$ or $|a_i| < \frac{1}{2}$. Let $A = \bigvee \{A_i: a_i \geq \frac{1}{2}\}$ and let $\bar{b}$ be the (formal) characteristic function of A. Then

$$\|b-f\| \leq \|b-a\| + \|a-f\| = \|\bar{b} - \bar{a}\| + \|a-f\| < 1.$$

Since both b and f are characteristic functions, $b = f$. This shows $\mathcal{S}(X) = \{a: \bar{a} \in \mathcal{S}(\mathcal{A})\}$. In particular, from Proposition 6.13, X is TD, and (i) is shown. The identification

$$A \to a, \qquad \bar{a} = \{(A, 1), (A^c, 0)\}$$

is a lattice isomorphism from $\mathcal{A}$ onto the characteristic functions of X, and hence $\mathscr{C}(X)$, showing (ii). The uniqueness assertion follows from the lattice-isometry of $\mathcal{S}(\mathcal{A})$ and $\mathcal{S}(Y)$ induced by $\mathcal{A} \leftrightarrow \mathscr{C}(Y)$. For then $C_{\mathbb{R}}(X)$ and $C_{\mathbb{R}}(Y)$ are order-isometric and Corollary 6.12 applies.

We now consider the representation theory for abstract L-spaces. We call an element u of the L-space (E, P) a *weak order unit* (also known as a *Freudenthal unit*) if $0 \neq u \in P$ and

$$a \wedge u = 0 \quad \text{implies } a = 0.$$

The first observation about L-spaces is that the conditions of Proposition 3.2 are met and this allows us to conclude a facial structure for P analogous to that of a dual lattice cone described in Section 5.

6.15. THEOREM. *Let (E, P) be an abstract L-space.*

(i) *Each closed face F of P is complemented with complementary face $G = \{b: b \wedge a = 0$ for all $a \in F\}$*

(ii) *If $a \in P$ and F_a is the smallest closed face of P containing a then the complementary face is*

$$G_a = \{b : b \wedge a = 0\}.$$

(iii) *The collection of closed faces of P forms a Boolean algebra*

$$\mathscr{B}(P) \quad \text{with} \quad F_1 \wedge F_2 = F_1 \cap F_2,\ F_1 \vee F_2 = F_1 + F_2$$

and complementation being the complementary face. The associated collection of projections is an isomorphic Boolean algebra with

$$\sigma_1 \wedge \sigma_2 = \sigma_1 \circ \sigma_2, \qquad \sigma_1 \vee \sigma_2 = \sigma_1 + \sigma_2 - \sigma_1 \circ \sigma_2, \qquad \sigma^c = \text{id} - \sigma.$$

Proof. Let F be a closed face and let $c \in P$. Define

$$\sigma_F c = \text{l.u.b. } A_c; \qquad A_c = \{a \in F: a \leq c\}.$$

Since A_c is directed in F and bounded in norm (by $\|c\|$), $\sigma_F c$ exists by Proposition 3.2. Then the DP shows σ_F has the required properties (i) and (ii) of Section 5 (see, for example, the proof of (iii) implies (iv) in 5.7). If $b \in G$, $a \in F$ and $0 \leq c \leq a, b$ then $c \in G \cap F = \{0\}$ so that $b \wedge a = 0$. Conversely, if $b \wedge a = 0$ for all $a \in F$ then in particular $b \wedge (\sigma_F b) = 0$. But $\sigma_F b \leq b$ so that $\sigma_F b = 0$, which shows $b \in G$.

For (ii), let $F_0 = \{b: 0 \leq b \leq \lambda a \text{ some } \lambda \geq 0\}$ and let

$$F = \{c: c = \text{l.u.b. } A, A \text{ a directed subset of } F_0\}.$$

Clearly $F \subset F_a$. Let $b \in P$ be given and let

$$A_b = \{c \in F: c \leq b\}.$$

Then each $c \in A_b$ is of the form $c = \text{l.u.b. } A_c$ where A_c is a directed subset of F_0. Let C be the set of suprema of finite subsets of $\bigcup \{A_c: c \in A_b\}$. Then $C \subset F_0$ and l.u.b. $C =$ l.u.b. $A_b \in F$. This shows that

$$\sigma_F b = \text{l.u.b. } A_b \in F$$

and hence F is a complemented (equivalently, closed) face of P. Thus $F = F_a$. If $b \wedge a = 0$ then clearly $b \wedge c = 0$ for all $c \in F_0$ and Proposition 6.8 applies to show $b \wedge c = 0$ for all $c \in F = F_a$. Finally, (iii) follows from (i) and Proposition 5.6.

6.16. THEOREM. *Let (E, P) be an abstract L-space and let $u \in P$ with $\|u\| = 1$. Let $E_u = \{a: -\lambda u \leq a \leq \lambda u \text{ some } \lambda \geq 0\}$. Then E_u is an abstract M-space with order unit u.*

Proof. The set E_u is a lattice ordered subspace of E. Furthermore, if $ra \leq u$ for all $r \geq 0$ then $(ra) \vee 0 = r(a \vee 0) = ra^+ \leq u$ for all $r \geq 0$. Hence $r\|a^+\| \leq 1$ for all r, so that $a^+ = 0$, showing $a \leq 0$. This implies that u is an Archimedean order unit for E_u so that, by Theorem 2.5, E_u is an order unit space with unit ball

$$I = \{a: -u \leq a \leq u\}.$$

Thus E_u is a pre-M-space with order unit u and it remains to show E_u is complete under p, the Minkowski functional of I.

Let $(a_n)_{n=1}^{\infty}$ be a Cauchy sequence in (E_u, p) and take a subsequence $(a_m)_{m=1}^{\infty}$ with

$$p(|a_{m+1} - a_m|) = p(a_{m+1} - a_m) < 1/2^m.$$

It follows that $|a_{m+1}-a_m|\le u/2^m$ and therefore

$$(a_{m+1}-a_m)^+ \quad \text{and} \quad (a_{m+1}-a_m)^- \quad \text{are both} \quad \le u/2^m.$$

Let

$$s_m=\sum_{k=1}^{m}(a_{k+1}-a_k)^+ \quad \text{and} \quad t_m=\sum_{k=1}^{m}(a_{k+1}-a_k)^-.$$

The sequences (s_m) and (t_m) are increasing and bounded above by u. Thus Proposition 3.2 applies again to show (s_m) and (t_m) are $\|\cdot\|$-convergent in E to s and t with $s=\text{l.u.b.}\,\{s_m\}$ and $t=\text{l.u.b.}\,\{t_m\}$. Since $0\le s, t\le u$ both s and t are in E_u with

$$(s-t)-(a_{m+1}-a_1)=(s-t)-\sum_{k=1}^{m}(a_{k+1}-a_k)=(s-s_m)-(t-t_m)$$
$$\le u/2^m+u/2^m$$

so that

$$p[(s-t)-(a_{m+1}-a_1)]\le p(s-s_m)+p(t-t_m)\le 1/2^{m-1}.$$

Thus $(a_n)_{n=1}^{\infty}$ converges to $a_1+(s-t)\in E_u$.

6.17. THEOREM (KAKUTANI). *Let (E, P) be an abstract L-space with weak order unit u, $\|u\|=1$, and let E_u be as in Theorem 6.16. Let $X=\text{ext}\ S$, where S is the state space of E_u.*

(i) *X(w^*-topology) is a* TD *compact Hausdorff space and the restriction map $\theta: E_u\to C_{\mathbb{R}}(X)$ is a lattice-isometry.*

(ii) *The continuous characteristic functions are lattice-isomorphic to the Boolean algebra $\mathcal{B}(P)$ of closed faces in P under the correspondence $F\to\theta(\sigma_F u)$.*

(iii) *There is a probability measure μ on X such that $\emptyset\ne U$, open, implies $\mu(U)>0$, and there exists a lattice-isometry ϕ of E onto $L^1(\mu)$ with $\phi|E_u=\theta$ (regarding $C_{\mathbb{R}}(X)$ as a subspace of $L^1(\mu)$).*

Proof. The Representation Theorem (6.10) for M-spaces with unit, together with Theorem 6.16, shows that X is compact Hausdorff and

$$\theta: E_u\to C_{\mathbb{R}}(X)$$

is a lattice-isometry. The characteristic functions in $C_{\mathbb{R}}(X)$ are identified by the property

$$f\wedge(1-f)\equiv 0$$

so that they are precisely the images under θ of

$$A=\{a\in E_u: a\wedge(u-a)=0\}.$$

Theorem 6.15(i) shows $\sigma_F u \in A$ for each $F \in \mathscr{B}(P)$ and clearly $F_1 \subset F_2$ implies $\sigma_{F_1} u \leq \sigma_{F_2} u$. Moreover, if $F_1, F_2 \in \mathscr{B}(P)$ with projections σ_1, σ_2 satisfying $\sigma_1 u \leq \sigma_2 u$, then $F_1 \subset F_2$. To see this let $b \in F_1 \cap G_2$ (G_2 complementary to F_2). Since $\sigma_1 u \in F_2$ we have $b \wedge (\sigma_1 u) = 0$. If $0 \leq c \leq b$, u then $c \in F_1$ so that $c = \sigma_1 c \leq \sigma_1 u$ and hence

$$0 \leq c \leq b \wedge (\sigma_1 u) = 0.$$

Thus $b \wedge u = 0$ and, since u is a weak order unit, $b = 0$. Then

$$F_1 + F_2 = F_1 \cap F_2 + F_1 \cap G_2 + F_2 \cap G_1 \subset F_2$$

so that $F_1 \subset F_2$. Next we observe that each $a \in A$ equals $\sigma_a u$, where σ_a is the projection to the closed face F_a spanned by a. This follows from 6.15(ii) since $a \wedge (u - a) = 0$ implies $u - a$ belongs to the complementary face G_a. Then

$$0 = \sigma_a (u - a)$$

gives $\sigma_a u = \sigma_a a = a$. This establishes (ii). We next show that X is TD. Let $x \in U$, open in X, and let D be a compact neighbourhood of x with $D \subset U$. Let $0 \leq f, g \leq 1$ in $C_{\mathbb{R}}(X)$ with

$$f(x) = 1 \quad \text{and} \quad f \equiv 0 \quad \text{on} \quad X \backslash D,$$

and

$$g \equiv 0 \quad \text{on} \quad D, \qquad g \equiv 1 \quad \text{on} \quad X \backslash U.$$

Let $\theta c = f$ and $\theta b = g$. Let σ be the projection to F_c and let $a = \sigma u$, $h = \theta a$. Now $c \leq u$ implies $c = \sigma c \leq a$ so that $h(x) = 1$. Since $g \wedge f = 0$, $b \wedge c = 0$. Thus b is in the complementary face G_c (by 6.15(ii)) and hence $b \wedge a = 0$. Thus $g \wedge h \equiv 0$ so that $h \equiv 0$ on $X \backslash U$. But then h is the characteristic function in $C_{\mathbb{R}}(X)$ of a clopen neighbourhood of x contained in U. This completes (i) and (ii). For (iii) we note first that E_u is $\|\cdot\|$-dense in E since, if F_u is the closed face spanned by u then $a \in G_u$ if and only if $a \wedge u = 0$ if and only if $a = 0$. Thus $F_u = P$.

But

$$F_u = \{c : c = \text{l.u.b. } A, A \text{ directed in } P_u\} \subset \|\cdot\| \text{ cl } P_u,$$

P_u the positive cone in E_u. Next, define $\mu(\theta a) = \|a\|$ for $a \geq 0$ in E_u and, since $\|\cdot\|$ is additive on P_u, extend μ uniquely to a linear functional on $C_{\mathbb{R}}(X)$. Clearly μ belongs to the state space of $C_{\mathbb{R}}(X)$ and hence is a probability measure, by the Riesz Representation Theorem. Define $\phi : E_u \to L^1(\mu)$ by $\phi(a) = \theta(a)$ and note that

$$\|\phi a\|_1 = \int_X |\theta(a)| \, \mathrm{d}\mu = \int_X \theta(|a|) \, \mathrm{d}\mu = \||a|\| = \|a\|.$$

Hence ϕ is a lattice-isometry of $(E_u, \|\cdot\|)$ with the subspace $C_{\mathbb{R}}(X) \subset L^1(\mu)$. Since $C_{\mathbb{R}}(X)$ is dense (in the L^1 norm) in $L^1(\mu)$ and E_u is $\|\cdot\|$-dense in E, ϕ extends to a lattice isometry from E onto $L^1(\mu)$.

We observe that (ii) says that the Stone Representation Space for $\mathscr{B}(P)$ is homeomorphic to the space X of (i) and (iii).

Notes

In the last three sections we have merely touched on the theory of vector lattices, a subject that literally fills volumes. Some standard references are Schaefer [**189**] and Luxemberg and Zaanen [**157**]. The decomposition property and its relation to vector lattices goes back to Riesz [**178, 179**]. The idea of complemented order ideals in lattices appears in [**178**] and the result of Theorem 5.5 in [**179**]. The reverse direction ((i) implies (iv) in Theorem 5.7) is due to Andô [**18**]. The fact that (iv) implies (i) in Theorem 5.7 was shown by Ellis [**108**] for the case $E = A(K)$. We will have more to say about this in Section 7 on simplexes. The machinery set up in Section 4 appears in Asimow and Ellis [**31**].

The representation theorems for spaces of type L and type M are given by Kakutani [**139, 140**]. Theorem 6.4 is due to Ellis [**100**]. The various characterizations of ordered Banach spaces whose duals are L-spaces was of some interest in the 1960s when Effros [**93**] termed them *simplex spaces.* A somewhat different order-theoretic characterization from the one in Theorem 6.5 is due to Davies [**72**]. The directedness of E_1 is the essential feature. This is developed somewhat differently in [**72**] and we discuss his approach in Section 11. The results of Theorem 6.10 and Corollary 6.11 go back to 1940–41 in papers by Stone [**203**], Kakutani [**140**], Krein and Krein [**143**], Yosida [**218**]. The representation of Boolean algebras in Theorem 6.14 is the well-known theorem of Stone [**202**].

7. CHOQUET SIMPLEXES

An ordered Banach space (E, P) is order-isometric to an $A(K)$ space if and only if E is an order unit Banach space, in which case K is the state space of E. As we saw in Section 6 if E is an order unit space and a lattice then E is order-isometric to $C_{\mathbb{R}}(X)$, where $X = \operatorname{ext} K$ (and is compact). In this section we study in more detail an important situation which is intermediate to these, namely, E is an order unit space with the RIP. Theorem 6.5 shows E^* is then an abstract L-space and the state space is then a w^*-compact base for the lattice ordered cone P^*.

Definition. A compact convex set K (of a locally convex space) is called a *Choquet simplex* if $A(K)^*$ is a lattice.

In view of Proposition 5.2 this is equivalent to the cone P^* spanned by K being a lattice. To summarize we state the following.

7.1. THEOREM. *Let K be a compact convex subset of a locally convex space. The following are equivalent*:

(i) *$A(K)$ has the* RIP;
(ii) *$A(K)^*$ is an abstract L-space*;
(iii) *K is a Choquet simplex.*

Theorem 5.7 shows that K inherits a rich facial-geometric structure from the general properties of dual lattice cones. In this regard we can simply identify the elements of $A(K)$ with the w^*-continuous positive homogeneous additive functionals on P^*. In addition each w^*-closed face of K spans a w^*-closed (by Theorem 2.6, Corollary 2.7) face of P^* and conversely, the intersection of faces of P^* with K are faces of K. It is conventional to call a face of K *split*, or a *split face*, if it is the intersection of a complemented face of P^* with K. Thus, from Theorem 5.7, we have the following.

7.2. THEOREM. *The compact convex set K is a simplex if and only if each closed exposed face of K is split.*

The notion of an infinite-dimensional simplex was originally formulated in connection with uniqueness of representing measures concentrated on the extreme points, as formulated in Chapter 1, Section 6. This is clearly a property that finite-dimensional simplexes possess; in this case one says a point of the n-dimensional simplex has a unique set of $n+1$ barycentre coordinates with respect to the vertices.

We recall the set-up of Chapter 1, Section 6, where $M_1^+(K)$ is the state space of $C_{\mathbb{R}}(K)$ consisting of the probability measures on K, and

$$q : C_{\mathbb{R}}(K)^* \to C_{\mathbb{R}}(K)^*/A(K)^{\perp}$$

is the quotient map and the quotient space is identified with $A(K)^*$. Then q maps $M_1^+(K)$ onto K (as the state space of $A(K)$) with $q\mu = x$, where

$$\int_K f \, d\mu = (f, x) = f(x); \qquad f \in A(K).$$

Let $\partial M_1^+(K)$ denote the subset of $M_1^+(K)$ consisting of maximal measures. ($\partial M_1^+(K)$ is never used to denote ext $M_1^+(K)$.) Then Theorem 1.6.4 shows the restriction of q to $\partial M_1^+(K)$ is onto K and Theorem 1.6.5 shows $\partial M_1^+(K)$ is convex in $M_1^+(K)$. Moreover, $\partial M_1^+(K)$ is a (not necessarily closed) face of $M_1^+(K)$, for, if $\mu \in \partial M_1^+(K)$ and $\mu = (\nu + \eta)/2$; ν, $\eta \in M_1^+(K)$ then we can choose a maximal $\nu_1 > \nu$. Then

$$\mu_1 = (\nu_1 + \eta)/2 > (\nu + \eta)/2 = \mu$$

so that $\mu_1 = \mu$ and hence $\nu_1 = \nu$. Similarly, η is maximal.

7.3. THEOREM (CHOQUET–MEYER). *Let K be a compact convex set. The following are equivalent*:

(i) *K is a simplex*;
(ii) *for each continuous concave function f on K, $\check{f}$ is affine*;
(iii) *for each $x \in K$ there is a unique element $\mu \in \partial M_1^+(K)$ with $q\mu = x$.*

Proof. We identify K as the state space of $A(K)$ so that (i) implies the cone P^* spanned by K is a lattice and in particular has the DP. Given f continuous and concave we can assume, by adding an appropriate multiple of 1, that $f \geq 0$. If we consider f as a homogeneous super-additive function on P^* (see the discussion preceding Theorem 4.3) then by Theorems 4.2(iii), 4.3 and 4.4(ii), (in conjunction with Proposition 3.2) we see that $\check{f}$ is additive on P^*, hence affine on K. If (ii) holds then, by Theorem 1.6.1(ix), for $-f \in Q(K)$

$$\{a \in A(K)\colon a < \check{f}\}.$$

is directed. Hence, using Theorem 1.6.1(viii), for any $\mu \in M_1^+(K)$ with $q\mu = x$

$$\mu(f) = \sup\{\mu(a)\colon a \in A(K) \quad \text{and} \quad a < \check{f}\}$$
$$= \sup\{a(x)\colon a < \check{f} = \check{f}(x)\}.$$

Now if $\mu, \nu \in \partial M_1^+(K)$ and $q\mu = x = q\nu$ then, by Theorem 1.6.5,

$$\mu(f) = \mu(\check{f}) = \check{f}(x) = \nu(\check{f}) = \nu(f).$$

Thus $\mu = \nu$ on the linear sub-lattice $Q(K) - Q(K)$ of $C_{\mathbb{R}}(K)$. But the Stone–Weierstrass Theorem (6.9) implies this subspace is dense so that $\mu = \nu$, completing (ii) implies (iii). If (iii) holds and $\partial M(K)^+$ denotes the subcone spanned by $\partial M_1^+(K)$ in $C_{\mathbb{R}}(K)^*$, the fact that $\partial M_1^+(K)$ is a face of $M_1^+(K)$ shows $\partial M(K)^+$ is a face of the (lattice ordered) cone in $C_{\mathbb{R}}(K)^*$. But then $\partial M(K)^+$ is a sublattice and q is a one-to-one map of $\partial M(K)^+$ onto P^* in $A(K)^*$. It follows from this that P^* is lattice ordered, establishing (i).

The comments at the beginning of this section indicate that the order unit spaces $A(K)$ with K a simplex come very close to being abstract M-spaces with order unit and hence, $C_{\mathbb{R}}(X)$ spaces. If $A(K)$ is a *lattice* then $X = \operatorname{ext} K$ (as in Theorem 6.10) is w^*-compact and $A(K) = C_{\mathbb{R}}(X)$. We next show that the converse is true.

7.4. PROPOSITION. *Let K be compact convex and let $X = \operatorname{ext} K$ be a closed subset of K. Then $\mu \in M_1^+(K)$ is maximal if and only if $\mu(X) = 1$.*

Proof. If $\mu(X) = 1$ then for $f \in C_{\mathbb{R}}(K)$ we have, by Theorem 1.6.3 that $f = \check{f}$ on X so that $\mu(f) = \mu(\check{f})$, implying by Theorem 1.6.5, that μ is maximal. If μ is maximal then Theorem 1.6.8 shows $\mu(F_0) = 0$ for all compact G_δ sets disjoint from X. If F is compact and $F \cap X = \varnothing$ then there is an $f \in C_{\mathbb{R}}(K)$ with $0 \le f \le 1$ and $f \equiv 1$ on F, $f \equiv 0$ on X. Then $F_0 = \{x : f(x) = 1\}$ is a compact G_δ with

$$F \subset F_0 \subset K \backslash X.$$

Thus, the regularity of μ shows $\mu(K \backslash X) = 0$.

7.5. THEOREM. *Let K be a Choquet simplex with $X = \operatorname{ext} K$. Then $A(K)$ is a lattice if and only if X is compact. In this case $A(K)$ is lattice isometric to $C_{\mathbb{R}}(X)$.*

Proof. If $A(K)$ is a lattice then it is an abstract M-space with an order unit and state space K. But then Theorem 6.10 shows X is compact. Conversely, let X be compact and let $\theta : A(K) \to C_{\mathbb{R}}(X)$ be the restriction map. Then Proposition 7.4 shows $\mu \in \partial M_1^+(K)$ if and only if μ belongs to the state space of $C_{\mathbb{R}}(X)$. Thus the probability measures, $M_1^+(X)$, on X as a subset of $M_1^+(K)$ are precisely $\partial M_1^+(K)$. Thus $(f, \theta^*\mu) = \int_K f \, d\mu = \int_X f \, d\mu = (f, q\mu)$ so that θ^* is one-to-one on $M_1^+(X)$ and hence on $C_{\mathbb{R}}(X^*)$. It follows that $\theta(A(K))$ is dense and therefore (since θ is an isometry) onto $C_{\mathbb{R}}(X)$. Since θ is order-preserving, one-to-one and onto it is a lattice-isometry.

Simplexes with the set of extreme points closed are called *Bauer simplexes* and Theorem 7.5 shows they are precisely the state spaces (probability measures) on a compact Hausdorff space.

We will take up the theory of Choquet simplexes in greater detail in Chapter 3. We conclude here with a result that allows for a significant strengthening of the RIP in $A(K)$.

7.6. THEOREM (EDWARDS' SEPARATION THEOREM). *Let K be a Choquet simplex and let $-f$, g be convex* u.s.c. *functions on K with $g \le f$. Then there is an $a \in A(K)$ with $g \le a \le f$ on K.*

Proof. Assume first that strict inequality holds between g and f so that there is an $\varepsilon > 0$ with $g < f - \varepsilon$. Now Theorem 1.6.1(x) shows

$$\{b \in Q(K) : b > g\}$$

is directed down with infimum equal to g. Hence by compactness and the fact that f is l.s.c. we can choose $b_g \in Q(K)$ with

$$g < b_g < f - \varepsilon/2.$$

Furthermore, the set of b in $Q(K)$ that are the suprema of a finite subset of $A(K)$ is dense in $Q(K)$ so we can assume

$$b_g = \sup \{a_i \in A(K) : i = 1, \ldots, n\}.$$

Similarly we can find $b_f = \inf \{c_j \in A(K): j = 1, \ldots, n\}$ with

$$g < b_g < b_f < f \text{ on } K.$$

Then the RIP in $A(K)$ shows there is an $a \in A(K)$ with

$$g < a < f.$$

Now, if $g \leq f$, we can iterate the above to construct a Cauchy sequence $(a_n)_{n=1}^{\infty}$ in $A(K)$ with

$$g - 1/2^n < a_n < f + 1/2^n$$

and

$$a_n - 1/2^n < a_{n+1} < a_n + 1/2^{n+1}.$$

For, if $a_1, \ldots, a_n$ are chosen then

$$g \vee a_n - 1/2^{n+1} < f \wedge a_n + 1/2^{n+1}$$

so that a_{n+1} can be chosen as required. Thus $a = \lim a_n \geqslant A(K)$ and interpolates g and f.

7.7. Corollary. *Let F be a closed face of the Choquet simplex K with $-f$, g convex* u.s.c. *functions on K with*

$$g \leq f.$$

If $a \in A(F)$ and

$$g|_F \leq a \leq f|_F$$

then there is a $b \in A(K)$ such that

$$b|_F = a \quad \textit{and} \quad g \leq b \leq f.$$

Proof. Let $r > s$ be such that

$$r \geq \sup \{g(y), a(x): y \in K, x \in F\}$$

and

$$s \leq \inf \{f(y), a(x): y \in K, x \in F\}.$$

Extend a to a_r, a_s on K by taking $a_r \equiv r$ on $K \backslash F$ and $a_s \equiv s$ on $K \backslash F$. Then $-a_r$, a_s are convex and u.s.c. on K and hence so are

$$(-a_r) \vee (-f) = -(a_r \wedge f) \quad \text{and} \quad a_s \vee g.$$

Since

$$a_s \vee g \leq a_r \wedge f$$

Theorem 7.6 yields

$$a_s \vee g \leq b \leq a_r \wedge f.$$

But

$$(a_s \vee g)|_F = a = (a_r \wedge f)|_F$$

so that

$$b|_F = a.$$

Notes

Theorem 7.1 (the direction (iii) implies (i)) was established by work of Lindenstrauss [**153**], Semadeni [**190**] and Edwards [**88**] in 1964–65, although there seems to have been a lapse of some time before the relevance of the 1962 paper of Andô [**18**] establishing the DP in the pre-dual of a lattice was recognized. The fact that the faces of a Choquet simplex are *split* was observed by Alfsen [**2**] in 1965. The terminology was introduced by Alfsen and Andersen [**6**]. Theorem 7.2 is due to Ellis [**108**]. In this connection it is interesting to note that the property of each extreme *point* being split does *not* characterize infinite-dimensional simplexes. Indeed it is shown in Asimow [**23**] that the state space of a function algebra always has this property; however the uniqueness of maximal representing measures need not hold. Theorem 7.3 is found in Choquet and Meyer [**63**]. Theorem 7.5 is due to Bauer [**34**]. Theorem 7.6 and its corollary are due to Edwards [**88**].

8. DECOMPOSITION OF DUAL CONES

In this section and the next we will initiate a general study of order-geometric properties which allow for various types of extension theorems. For example, if F is a closed face of a simplex K then the decomposition of K by F and F^c, together with the Edwards Separation Theorem, allows for extensions of continuous affine functions on F to elements of $A(K)$ satisfying certain prescribed conditions.

Typical questions concerning a compact convex set K and a closed convex subset F are the following:

1. When does each element of $A(F)$ extend to an element of $A(K)$?
2. When can we find $a \in A(K)$ such that a is non-negative and $F = a^{-1}(0)$?
3. If $a \in A(K)$ is non-negative on F can we find a non-negative $b \in A(K)$ such that $b = a$ on F?

The first question is the key to various interpolation and approximation results in function spaces, dealt with in Chapter 4 and the second question relates to the problem of identifying peak sets for function spaces. The third question, concerning the positive extension property, turns out to be relevant to the problem of finding norm-preserving extensions. We return to these questions in detail in Chapter 4.

We first formulate these matters in terms of homogeneous functions on dual cones and then, in Section 10 apply the general results to $A(K)$ spaces.

We shall assume throughout that (E, P) is an ordered Banach space that is α-normal and β-generated for some constants α and β. Let Q^* denote a w^*-closed subcone of P^*. We note that, by the Krein–Šmulyan Theorem, Q^* is w^*-closed if and only if Q_1^* is w^*-compact. We let N denote the w^*-closed linear span of Q^* in E^*. Thus

$$N = w^* - \mathrm{cl}\,(Q^* - Q^*).$$

Let $Q = \{a \in E\colon (a, x) \geq 0 \text{ for all } x \in Q^*\}$. Thus

$$Q = -(Q^*)^0$$

and is a closed cone in E containing P. If $M = Q \cap -Q$ then

$$M = \{a \in E\colon (a, x) = 0 \text{ for all } x \in Q^*\}$$

(sometimes denoted $(Q^*)^\perp$) and clearly

$$M = N^0.$$

Our point of view is to consider elements of E as functions, $A_0(P^*)$ (w^*-continuous positive homogeneous and additive) on P^*. In fact we can represent (E, P) as the space $A_0(P_1^*)$, the subspace of $A(P_1^*)$ of functions vanishing at 0. We have from Corollary 1.4.8 that the restriction of E to $A(P_1^*)$ is dense, and, since E^* is positively generated, we have for some α,

$$\|a\| \leq \alpha \|a|_{P_1^*}\|_\infty \leq \alpha \|a\|.$$

Thus the restriction map is onto and the restriction norm is equivalent to the original. In similar fashion (as in Theorem 1.4.7) we can identify the linear space $Q^* - Q^*$ with unit ball co $(Q_1^* \cup -Q_1^*)$ as the dual space of $A_0(Q_1^*)$, with dual cone Q^*.

The analogue to the first question above concerns the restriction map of E to $A_0(Q_1^*)$. The answer is provided by Theorem 1.3.5, since E is dense in $A_0(Q_1^*)$. Thus let

$$\theta : E \to A_0(Q_1^*)$$

be the restriction map and note that θ^* is the (w^*-continuous) inclusion map taking $Q^* - Q^*$ with unit ball co $(Q_1^* \cup -Q_1^*)$ to itself with unit ball $(Q^* - Q^*)_1$.

8.1. THEOREM. *The following are equivalent*:

(i) *The restriction map θ is onto*,

(ii) *there is a number r such that for each $s > r$ and $a_0 \in A_0(Q_1^*)$ with $\|a_0\| \leq 1$ there is an $a \in E$ with $\theta a = a_0$ and $\|a\| < s$*,

(iii) *$Q^* - Q^*$ is w^*-closed*,

(iv) Q^*-Q^* *is* $\|\cdot\|$*-closed,*
(v) $(Q^*-Q^*)_1 \subset r\,\mathrm{co}\,(Q_1^* \cup -Q_1^*)$.

To proceed we must be more precise about the relation between Q^* and $M=(Q^*)^\perp$. We say the w^*-closed cone Q^* is *self-determined* if $N \cap P^* = Q^*$ (N the w^*-closed span of Q^*). Thus Q^* is self-determined if and only if it is the intersection of a w^*-closed subspace with P^*. By polarizing we can characterize this property in E.

8.2. PROPOSITION. *The w^*-closed subcone Q^* of P^* is self-determined if and only if*

$$Q=(P+M)^-.$$

Given $Q^* \subset P^*$ and a closed subspace J in E we say Q^* is a *section of P^* by J* if

(i) $Q^*=J^0 \cap P^*$,
(ii) $N=w^*\text{-cl}\,(Q^*-Q^*)=J^0$.

Property (i) is sometimes written as $Q^*=J^\perp$ so that (ii) says (by polarizing) $J=J^{\perp\perp}$.

8.3. PROPOSITION. *Let J be a closed subspace of E and let $Q^*=J^0 \cap P^*$. Then Q^* is a section of P^* by J if and only if*

$$J=(J-P)^- \cap (J+P)^-.$$

Proof. Q^* is a section by J if and only if (ii) holds, which means

$$w^*\text{-cl}\,(J^0 \cap P^* - J^0 \cap P^*)=J^0.$$

By polarizing, this in turn is equivalent to

$$(J-P)^- \cap (J+P)^- = J.$$

Such subspaces are automatically *order ideals* ($c \le a \le b$ with $c, b \in J$ implies $a \in J$) since they are of the form $(Q^*)^\perp$. They are sometimes called *almost Archimedean* (the reason for the terminology will become apparent shortly) and characterize the subspaces $J=J^{\perp\perp}$.

If Q^* is a section by J then for each $x \in P^*\backslash Q^*$ there is an $a \in J$ such that $(a, x) \neq 0$. If Q^* is a subset of P^* such that for each $x \in P^*\backslash Q^*$ there is an $a \in P \cap (Q^*)^\perp$ with $(a, x)>0$, we say Q^* is *semi-exposed.* This implies that Q^* is a self-determined *face* of P^*.

8.4. PROPOSITION. *The set Q^* is semi-exposed in P^* if and only if*

$$Q=(P-M\cap P)^-$$

where $Q=\{a \in E\colon (a, x) \ge 0$ for all $x \in Q^*\}$ and $M=Q \cap -Q$.

Proof. Q^* is semi-exposed if and only if

$$\begin{aligned} Q^* &= \{x \in P^*: (a, x) = 0 \text{ for all } a \in M \cap P\} \\ &= \{x \in P^*: (a, x) \leq 0 \text{ for all } a \in M \cap P\} \\ &= (M \cap P)^0 \cap P^*. \end{aligned}$$

By taking polars, Q^* is semi-exposed if and only if

$$Q = (P - M \cap P)^-.$$

If J is a closed subspace of E then J is an ordered Banach space with positive cone $J \cap P$. Thus we say the subspace J of E is *positively generated* if $J = J \cap P - J \cap P$.

8.5. COROLLARY. *If Q^* is a section of P^* by J and J is positively generated then Q^* is semi-exposed.*

Proof. We have $Q^* = J^0 \cap P^*$ and $J = M$ so that

$$Q = (P - J)^- = (P - J \cap P + J \cap P)^- = (P - J \cap P)^-.$$

8.6. PROPOSITION. *If Q^* is semi-exposed in P^* and X is a w^*-compact subset of $P^* \backslash Q^*$ then there is an $a \in E$ with $a \equiv 0$ on Q^*, $a \geq 0$ on P^* and $a > 0$ on X.*

Proof. For each $x \in X$ there is an $a_x \in M \cap P$ and a neighbourhood N_x of x in X such that $a_x > 0$ on N_x. Then, by compactness, we can choose $x_1, \ldots, x_n$ such that $a = a_{x_1} + \cdots + a_{x_n}$ is the required functional.

8.7. COROLLARY. *If Q^* is a semi-exposed w^*-G_δ in P^* then there is an $a \in E$ with*

$$a \geq 0 \quad \textit{on} \quad P^* \quad \textit{and} \quad Q^* = \{x \in P^*: (a, x) = 0\}.$$

Proof. Let $Q^* = \bigcap_{n=1}^{\infty} U_n$, U_n open in P^*, and let $X_n = P_1^* \backslash U_n$. Then

$$P_1^* \backslash Q^* = \bigcup_{n=1}^{\infty} X_n$$

and if a_n is chosen in $(M \cap P)_1$ with $a_n > 0$ on X_n then

$$a = \sum_{n=1}^{\infty} a_n / 2^n$$

is the required function in $(M \cap P)_1$.

Such a Q^* is called *exposed* and Corollary 8.7 shows that Q^* is exposed if and only if Q^* is a semi-exposed G_δ. Moreover, Q^* is semi-exposed if and only if Q^* is the intersection of exposed sets of P^*.

Next, let Q^* be a section by J in P^* and consider the property in question (3). We say Q^* has the *positive extension property* (PEP) if for each $a \in E$ with $a \geq 0$ on Q^* there is a $b \in E$ with

$$b \geq 0 \quad \text{on} \quad P^* \quad \text{and} \quad b = a \quad \text{on} \quad Q^*.$$

In terms of Q and J we can characterize this quite easily.

8.8. PROPOSITION. *If Q^* is a section by J in P^* then Q^* has the* PEP *if and only if*

$$Q = J + P.$$

Combining this with Proposition 8.3 and Corollary 8.5 we have the following.

8.9. COROLLARY. *Let J be a closed subspace of E. Then J gives rise to a section $Q^* = J^0 \cap P^*$ with the* PEP *if and only if*

(i) *$J + P$ is closed in E,*
(ii) $J = (J - P) \cap (J + P)$.

If in addition

(iii) $J = J \cap P - J \cap P$,

then Q^* is a semi-exposed face of P^* with the PEP.

If $q: E \to E/J$ is the quotient map then the adjoint $q^*: (E/J)^* \to E^*$ (using Theorem 1.1.2(v)) identifies $(E/J)^*$ with J^0 and is an isometry (where J^0 has unit ball J_1^0). Moreover, properties (i) and (ii) in Corollary 8.9 are equivalent to the cone qP being closed and proper in E/J. Thus, the dual cone

$$-(qP)^0 = (q^*)^{-1}(P^*) = P^* \cap J^0 = Q^*$$

so that $(E/J)^*$ is order-isometric to (J^0, Q^*). In particular, the quotient space $(E/J, qP)$ is Archimedean if (i) and (ii) hold since (ii) implies qP is proper and

$$rq(a) \leq q(b) \quad \text{if and only if } b - ra \in P + J.$$

Thus $rq(a) \leq q(b)$ for all $r \geq 0$ if and only if

$$b/r - a \in P + J \quad \text{for all} \quad r \geq 0.$$

Then, letting $r \to \infty$, we get $-a \in P + J$ so that $qa \leq 0$.

A closed subspace J of E satisfying (i), (ii) and (iii) is called an *Archimedean order ideal.* If in addition

$$\text{(iv)} \qquad J^0 = Q^* - Q^*$$

(so that the statements of 8.1 also hold) then J is called *strongly Archimedean.*

8.10. COROLLARY. *Let J be a closed subspace of E and let $Q^* = J^0 \cap P^*$. The following are equivalent:*

(i) *J is Archimedean;*

(ii) *Q^* is a section of P^* by J such that for each $a \in Q$ there is an $m \in J$ such that*

$$a + m \geqslant a, 0.$$

Proof. If J is Archimedean then $a \in Q$ implies there is a $b \in P$ with $b|_{Q^*} = a|_{Q^*}$; i.e. $b - a = m_1 \in J$. Since J is positively generated, choose $m_2 \in J$ with $m_2 \geq a - b, 0$. Then

$$m_2 + b = m_2 + m_1 + a \geq a, b \geq a, 0.$$

Conversely, (ii) clearly implies J is positively generated with the PEP.

Verifying these conditions can, in practice, be quite difficult. What we seek next are sufficient conditions of a geometric nature (involving Q^* in P^*) for various of these properties to hold.

To motivate the next definition we consider the case where Q^* is a complementary face of P^* with projection π. Then π is a bounded (by normality and positive generation) projection of E^* onto $N = Q^* - Q^*$.

Let σ denote the complementary projection with range L. Then both N and L are norm-closed subspaces. Let

$$q : E^* \to E^*/N \quad \text{(quotient map)}$$

and

$$\rho : E^*/N \to L, \quad \text{with } \rho \circ q = \sigma.$$

Then ρ is an isomorphism onto L with

$$\|\rho(qx)\| = \|\sigma x\| = \|\sigma(x+N)\| \leq \|\sigma\|\,\|qx\|$$

so that ρ is a Banach space isomorphism. The point here is that the induced quotient norm on L is equivalent to the restriction norm. The quotient norm, $\|qz\|$, is the distance $d(N, E_1^*)(z)$ (Chapter 1, Section 3) of z from N, and if $z \in L$,

$$\|qz\| \leq \|z\| \leq \|\sigma\|\,\|qz\|.$$

In particular if $x \in P^*$ then

$$x = y + z; \qquad y \in Q^*, \quad z \in P^*$$

with

$$\|z\| \leq \|\sigma\|\,\|qz\| = \|\sigma\|\,\|qx\|.$$

Thus x can be decomposed into an element of Q^* and an element of P^* far (relative to $\|x\|$) from N.

Now let Q^* be a section of P^* by J, so that $N = J^0$. We can measure the distance from N in E^* by taking any norm-closed convex neighbourhood W of 0 in E^* and letting

$$d(N, W)(x) = p(N + W)(x) = \inf \{r \geq 0 \colon x \in N + rW\}.$$

For convenience we let the N be understood and denote this by $d_W(x)$.

We define Q^* to be a *decomposable* section of P^* by J if there is a positive homogeneous map $\pi : P^* \to Q^*$ satisfying

(i) $\pi = 1$ (the identity map) on Q^*,

(ii) for each $x \in P^*$, $(1-\pi)(x) \in P^*$,

(iii) $\|(1-\pi)x\| \leq d_W(x)$ for some (norm) neighbourhood W of 0 in E^*.

Thus each $x \in P^*$ can be decomposed, as in the case where Q^* is complemented, but the set of elements $z = x - \pi x$ need not, in general, be convex. Indeed Q^* may be "surrounded" by $(1-\pi)P^*$, and in particular, need not be a face of P^*.

8.11. THEOREM. *Let Q^* be a decomposable section of P^* by J. Then*

(i) *$P^* + J^0$ is w^*-closed in E^*,*

(ii) *$P + J$ is norm-closed in E.*

Proof. We can assume that the neighbourhood W in (iii) is $E^*_{1/r}$ for some $r > 0$, so that $x \in P^*_1$ implies

$$d_W(x) \leq r \quad \text{if and only if} \quad x \in J^0 + E^*_1.$$

(i) If $u \in (P^* + J^0)_1$ then

$$u = y + z + n; \qquad y \in Q^*, \qquad n \in J^0, \qquad z \in P^*$$

and

$$\|z\| \leq d_W(y+z) \leq r\|u\| \leq r.$$

Thus

$$(P^* + J^0)_1 \subset (Q^* + J^0 + P^*_r) = J^0 + P^*_r.$$

Hence

$$(P^* + J^0)_1 = (J^0 + P^*_r)_1$$

which is w^*-closed. Hence, by the Krein–Šmulyan Theorem, $P^* + J^0$ is w^*-closed.

(ii) The decomposability is equivalent to

$$P^* \cap (J^0 + E^*_1) \subset Q^* + P^*_r$$

which yields

$$Q \cap (P+E_1)^- \subset (P+J_r)^-$$

in E. We now apply the Gauge Lemma (1.3.2) with

$$A \leftrightarrow P, \qquad B \leftrightarrow J_r, \qquad D \leftrightarrow Q.$$

If $a \in Q$ then $\bar{d}(a)\text{-} = \bar{d}(P, J_r)$ and

$$\bar{d}(a) \leq d(P, E_1) < \infty$$

so that $d(P, J_r)(a) = \bar{d}(a) < \infty$. Hence $a \in P + J_s$ for all $s > r$. This shows $(J+P)^- = Q = J + P$.

8.12. COROLLARY. *If Q^* is a decomposable section of P^* by J and*

$$q: E^* \to E^*/J^0$$

then qP^* is w^*-closed in E^*/J^0 and

$$(E^*/J^0, qP^*) \cong (J, J \cap P)^* \quad (\textit{order-isometry}).$$

Proof. From Theorem 8.11(i) we have qP^* is w^*-closed in E^*/J^0 with the usual identification of E^*/J^0 and J^* (q is the adjoint of the inclusion map $i: J \subset E$). Moreover,

$$qP^* = (qP^*)^{00} = (i^{-1}(-P))^0 = (J \cap P)^*.$$

If $(J, J \cap P)$ is positively generated then (Corollary 8.5) Q^* is semi-exposed. This leads us to the notion of positive decomposability: we say Q^* is a *positively decomposable* section of P^* by J if there is an $h \in E^{**}$ with $h \equiv 0$ on J^0 and the set W in (iii) can be taken as

$$W = \{z \in E^*: h(z) \leq 1\}.$$

Then

$$d_W(x) = \inf\{r \geq 0: x \in J^0 + rW\} = \max\{h(x), 0\}.$$

Thus on P^* we have

$$x = y + z; \qquad y \in Q^* \quad \text{and} \quad \|z\| \leq h(x).$$

8.13. THEOREM. *Let Q^* be a positively decomposable section of P^* by J. Then $(J, J \cap P)$ is an approximately α-directed Banach space ($\alpha = \|h\|$).*

Proof. From Corollary 8.12 and Theorem 3.5 it suffices to show the quotient norm on qP^* is α-additive. Let $(x_i)_{i=1}^n \subset P^*$ be given with $x = \sum x_i$. Let $z_i = x_i - \pi x_i$ and $z = x - x$. Then

$$\sum \|qx_i\| \leq \sum \|z_i\| \leq \sum h(x_i) = h(x) = h(x+n) \quad (\text{any } n \in J^0).$$

Thus

$$\sum \|qx_i\| \leqslant \|h\| \|qx\|.$$

8.14. COROLLARY. *If Q^* is a positively decomposable section of P^* by J then J is an Archimedean order ideal and Q^* is a semi-exposed face of P^* with the* PEP.

8.15. COROLLARY. *Let (E, P) have the* DP *and let J be a closed subspace of E with $Q^* = J^0 \cap P^*$. The following are equivalent:*

(i) *Q^* is a face of P^* and a section of J;*
(ii) *J^0 is a positively generated order ideal in E^*;*
(iii) *J is a strongly Archimedean order ideal in E;*
(iv) *J is a positively generated order ideal in E.*

Proof. Since (E, P) has the DP Q^* is a face of P^* if and only if Q^* is complemented. In this case $Q^* - Q^*$ is w^*-closed and Q^* is a positively decomposable section of P^*. Thus, since Q^* is a section by J, (i) implies that $J^0 = Q^* - Q^*$ and, since Q^* is a face, that J^0 is an order ideal.

If (ii) holds it follows that Q^* is a facial section by J so that Corollary 8.14 shows J is an Archimedean (hence strongly Archimedean) order ideal in E. Then (iii) implies (iv) by definition.

If (iv) holds then we observe first that Q^* is a face since $Q^* = \{x \in P^*: (a, x) = 0 \text{ for all } a \in J\} = \{x \in P^*: (a, x) = 0 \text{ for all } a \in J \cap P\}$ and hence is an intersection of faces. To conclude (i) then, we must show

$$J = (Q^* - Q^*)^0$$

or equivalently

$$J^0 = Q^* - Q^*$$

since $Q^* - Q^*$ is w^*-closed. Clearly $Q^* - Q^* \subset J^0$. If $x \in J^0$ and $a \in J \cap P$ then, using Theorem 5.5,

$$(a, x \vee 0) = \sup \{(b, x): 0 \leq b \leq a\} = 0$$

since J is an order ideal. Since J is positively generated, we have

$$x \vee 0 \in J^0 \cap P^* = Q^*.$$

Hence $J^0 = Q^* - Q^*$.

Notes

Theorem 8.1, in the context of $A(K)$ spaces, was shown by Edwards [**87**] and Alfsen [**3**]. In the latter the number r is related to the bound on the norm of the extension function. This is discussed in greater detail in Chapter 4 in connection with interpolation in complex function spaces. The notion of self-determinacy is introduced in Alfsen [**1**]. An example of a non-self-determining face is given by Asimow [**25**]. Proposition 8.3 (the $A(K)$ case) is found in Ellis [**101**]. Other related results characterizing order ideals whose annihilators determine faces of the dual cone are found in Jameson [**134**], Ellis [**100**], Bonsall [**44**] and Asimow [**26**].

In the years 1968–70 much work on $A(K)$ spaces centred on the various extension properties of functions in $A(F)$, F a face of K. The two main properties are incorporated in the condition of Corollary 8.10(ii). These are the positive extension property (PEP) and the directedness in $A(K)$ of the null functions on F. As Corollary 8.10 shows these are tantamount to the conditions of Corollary 8.9 and hence the usefulness of designating the Archimedean ideals. This terminology and the applicability to $A(K)$ is found in Størmer [**206**] and, in the case of strongly Archimedean, Alfsen [**3**]. Geometric conditions guaranteeing these extension properties were first noticed for simplexes where each closed face is split (Alfsen [**3**]) and also discussed in Effros [**93**] and Lazar [**147**]. Asimow [**23**] gives geometric conditions on K and F (in terms of gauges) that assure the directedness of J. In Asimow [**25**] the more restrictive but geometrically more appealing notion of positive decomposability is introduced for compact convex sets as a sufficient condition for the Archimedean extension properties of Corollary 8.10.

9. STABILITY OF SPLIT SETS AND THE CHARACTERIZATION OF COMPLEMENTED SUBCONES

Let E be a Banach space with bounded complementary projections π and σ whose ranges are the closed complementary subspaces N and L.

Let S be a closed convex set containing zero with Minkowski functional $p = p_S$. We say S is *split with respect to* π if

$$p = p \circ \pi + p \circ \sigma.$$

The next two results will clarify the geometric content of this definition.

9.1. PROPOSITION. *Let A and B be closed convex subsets of N and L respectively, each containing zero. Let*

$$S = \overline{\text{co}}\,(A, B).$$

Then

(i) $S = \text{co}\,(A, B) + P_A + P_B$ *where*

$$P_A = \{a \in N : p_A(a) = 0\} \quad \textit{and} \quad P_B = \{a \in L : p_B(a) = 0\},$$

(ii) $p_S = p_A \circ \pi + p_B \circ \sigma.$

Proof. Let $c_n \to c$ with

$$c_n = \lambda_n a_n + \mu_n b_n; \qquad a_n \in A, b_n \in B \quad \text{and} \quad \lambda_n + \mu_n \leq 1.$$

By passing to a subsequence we can assume $\lambda_n \to \lambda$ and $\mu_n \to \mu$. Now

$$\pi c = \lim \pi c_n = \lim \lambda_n a_n$$

so that

$$p_A \circ \pi(c) \leq \liminf p_A(\lambda_n a_n) \leq \lambda.$$

Hence $\pi c \in \lambda A$ and similarly $\sigma c \in \mu B$. Therefore

$$c = \pi c + \sigma c = \lambda(\pi c/\lambda) + \mu(\sigma c/\mu),$$

where we mean 1 for 0/0. Then both (i) and (ii) follow, since $\lambda + \mu \leq 1$.

9.2. PROPOSITION. *The following are equivalent*:

(i) *S is split with respect to π,*
(ii) *$S = \overline{\text{co}}\,(S \cap N, S \cap L)$,*
(iii) *$S = \overline{\text{co}}\,(\pi S, \sigma S)$.*

Proof. If $p_S = p_S \circ \pi + p_S \circ \sigma$ then

$$p_S = p_A \circ \pi + p_B \circ \sigma,$$

where $A = S \cap N$, $B = S \cap L$ so, using Proposition 9.1, (ii) holds. If (ii) holds then Proposition 9.1(i) shows $\pi S \subset S \cap N$, $\sigma S \subset S \cap L$. Thus equality holds and (iii) follows. Clearly (iii) implies (ii) and hence, by Proposition 9.1(ii), (i).

9.3. COROLLARY. *If S and T are split with respect to π then so is*

$$R = \overline{\text{co}}\,(S, T) \quad \textit{and} \quad \pi R = \overline{\text{co}}\,(\pi S, \pi T).$$

We now consider a Banach space E with a closed subspace M for which the polar N in E^* is the range of a bounded projection π. We let L and σ denoted the complementary subspace and projection. We take A, B as closed convex subsets containing zero in E such that both A^0 and B^0 are split. The next result shows the preservation of the "splitness" of polar sets under the formation of intersections in E. We formulate this without reference to order properties in E for application in Chapter 4, although our immediate goal is to determine the order properties of subspaces in (E, P) giving rise to complemented subcones of P^*.

9.4. THEOREM (ANDÔ). *Let A, B be closed convex sets containing zero in the Banach space E, such that A^0, B^0 are split with respect to π, whose range, N, is the polar of the closed subspace $M \subset E$. Let $C = A \cap B$. Then*

(i) *$A+M$ and $B+M$ are closed in E,*
(ii) *if $C^0 = \|\cdot\|\text{-}\overline{\mathrm{co}}\,(A^0, B^0)$ in E^* then C^0 is split and $(A+M)\cap(B+M) = C+M$,*
(iii) *in general, if $\|\sigma\| \le 1$, where $\sigma = 1-\pi$, then $(A+M)\cap(B+M) \subset A \cap (B+E_s)+M$ for all $s>0$.*

Proof. We use the Gauge Lemma to show $A+M$ is closed, with $A \leftrightarrow A$, $B \leftrightarrow M_1$, $D \leftrightarrow (A+M)^-$. We need to show

$$(A+M)^- \cap (A+E_r) \subset (A+M_{kr})^-, \qquad (*)$$

for then $x \in (A+M)^-$ implies (in the notation of 1 Section 3)

$$\bar{d}(A, M_1)(x) \le kd(A, E_1)(x) < \infty$$

so that $d(A, M_1)(x) = \bar{d}(A, M_1)(x) < \infty$ and hence $x \in A+M$. It suffices to show the polar of $(*)$ holds. Thus consider the support functional $\rho = \rho(A+M_{kr})$, where $k > \|\sigma\|$. Then

$$\rho = \rho(A) + kr\rho(M_1) = p(A^0) + krp(N+E_1^*).$$

Now, for any $n \in N$,

$$\|\sigma x\| = \|\sigma x + \sigma n\| = \|\sigma(x+n)\| \le k\|x+n\|.$$

Hence

$$\|\cdot\| \circ \sigma \le kp(N+E_1^*).$$

Since A^0 is split we then have

$$\begin{aligned}\rho &\ge p(A^0)\circ\pi + p(A^0)\circ\sigma + r\|\cdot\|\circ\sigma \\ &= p(A^0)\circ\pi + (p(A^0)+r\|\cdot\|)\circ\sigma\end{aligned}$$

which says, by Proposition 9.1,

$$(A+M_{kr})^\circ \subset \|\cdot\|\text{-}\overline{\mathrm{co}}\,(A^0 \cap N, (A+E_r)^0),$$

from which $(*)$ follows by taking polars. Hence $A+M$ is closed. If

$$C^0 = \|\cdot\|\text{-}\overline{\mathrm{co}}\,(A^0, B^0)$$

then, by Corollary 9.3, C^0 is split and

$$C^0 \cap N = \|\cdot\|\text{-}\overline{\mathrm{co}}\,(A^0 \cap N, B^0 \cap N)$$

so that

$$(C+M)^- = (C+M) = (A+M)\cap(B+M)$$

and hence (ii) is shown.

Assume now that the complementary projection σ with range L in E^* has $\|\sigma\| \le 1$. Let $B^s = B+E_s$ and $C^s = A \cap B^s$. We shall apply the Gauge Lemma with

$$A \leftrightarrow C^s, \qquad B \leftrightarrow M_1, \qquad D \leftrightarrow (A+M) \cap (B+M).$$

As above, we require

$$(A+M) \cap (B+M) \cap (C^s + E_r) \subset (C^s + M_r)^-. \tag{**}$$

To prove (**) we first evaluate the support functional, $\rho(C^s + M_r)$, of $C^s + M_r$. Now

$$\rho(M_r) = rp(N + E_1^*) \geq r\|\cdot\| \circ \sigma \quad (\text{since } \|\sigma\| \leq 1).$$

Next

$$\rho(C^s) = p((C^s)^0) = p[\|\cdot\|\text{-}\overline{\text{co}}\,(A^0, (B^s)^0)]$$

since $(B^s)^0$ is w^*-compact (Theorem 1.1.5). Thus,

$$\rho(B^s) = \rho(B) + s\|\cdot\| \geq \rho(B) \circ \pi + (\rho(B) + s\|\cdot\|) \circ \sigma$$

since B^0 is split and $\|\sigma\| \leq 1$. Hence

$$(B^s)^0 = \|\cdot\|\text{-}\overline{\text{co}}\,(B^0 \cap N, (B^s)^0 \cap L)$$

so that

$$\begin{aligned} \|\cdot\|\text{-}\overline{\text{co}}\,(A^0, (B^s)^0) &= \overline{\text{co}}\,(A^0 \cap N, B^0 \cap N, A^0 \cap L, (B^s)^0 \cap L) \\ &\subset \overline{\text{co}}\,(D^0, (C^s)^0 \cap L), \end{aligned}$$

where

$$D = (A+M) \cap (B+M), D^0 = w^*\text{-}\overline{\text{co}}\,(A^0 \cap N, B^0 \cap N) \subset N.$$

Hence

$$\rho(C^s) \geq \rho(D) \circ \pi + \rho(C^s) \circ \sigma$$

and

$$\rho(C^s + M_r) = \rho(C^s) + \rho(M_r) \geq \rho(D) \circ \pi + \rho(C^s + r\|\cdot\|) \circ \sigma.$$

Therefore,

$$(C^s + M^r)^0 \subset \overline{\text{co}}\,(D^0, (C^s + E_r)^0)$$

so that (**) holds, and hence (iii) is shown.

Assume now that Q^* is a w^*-closed complemented subcone of P^* under the projection π with $N = Q^* - Q^*$. Since N is the range of a projection N is $\|\cdot\|$-closed and hence, using Theorem 8.1, N is w^*-closed. Combining this with Corollary 8.9, we have that in particular Q^* is a section of P^* by the strongly Archimedean ideal $J = N^0$ in E. If $Q = (-Q^*)^0$ then $J = Q \cap -Q$, and $Q = J + P$.

We have $(E/J, qP)$ is order isomorphic to $A_0(Q^*)$ and

$$a_0 \le b_0 \quad \text{in } A_0(Q^*)$$

if and only if there are $a, b \in E$ with

$$qa = a_0, \qquad qb = b_0 \quad \text{and} \quad b - a \in P + J = Q,$$

and $a_0 = b_0$ in $A_0(Q^*)$ if and only if $a - b \in J$. Thus we say $a \le b(Q)$ if $qa \le qb$, or equivalently

$$a|_{Q^*} \le b|_{Q^*}.$$

We write $a \le b + \varepsilon$ if there is a $z \in E_\varepsilon$ with $a \le b + z$.

Since E is positively generated we can, by renorming E, assume E^* is 1-normal, so that the norm in E^* is monotonic on P^*. This guarantees that $\|\pi\|$ and $\|\sigma\|$ are less than or equal to 1 so the hypothesis of Theorem 9.4(iii) is satisfied.

9.5. THEOREM. *Let Q^* be a w^*-closed complemented subcone of P^* and let a_i, $b_j (i = 1, \ldots, n; j = 1, \ldots,. m)$ c, c_0 be elements of E such that*
(i) $b_j \le c \le a_i$ *for all* $1 \le i \le n$, $1 \le j \le m$,
(ii) $b_j \le c_0 \le a_1(Q)$, $1 \le i \le n$, $1 \le j \le m$.
Then for each $\varepsilon > 0$ there exists $m \in J$ such that

$$b_j - \varepsilon \le c_0 + m \le a_i, \qquad 1 \le i \le n; \qquad 1 \le j \le m.$$

(*In short, if the b's and a's can be interpolated in E and c_0 interpolates the b's and a's on Q^* then c_0 has an extension in $E = A_0(P_1^*)$ that interpolates in E.*)

Proof. By translating we can assume $c = 0$ in E. let

$$A_i = \{c \in E : c \le a_i\} = a_i - P.$$

Then the support functional $\rho(A_i)$ just equals a_i on P^* (since $0 \le a_i$) and clearly

$$a_i = a_i \circ \pi + a_i \circ \sigma \quad \text{on } P^*$$

so that A_i^0 is split. If

$$A = \bigcap_{i=1}^{n} A_i = \{c : c \le a_1, \ldots, a_n\}$$

then the support functional $\rho(A)$ equals

$$\inf\{a_1, \ldots, a_n\}^-$$

which we shall denote as $\bar{p}$. But the definition of $\bar{p}$ (Section 4) shows that $\bar{p}$ is also the support functional of

$$\{c \in E^{**} : c \le a_1, \ldots, a_n \quad \text{in } (E^{**}, P^{**})\}$$

and hence the Minkowski functional of

$$\|\cdot\|\text{-}\overline{\text{co}}\,(A_1^0, \ldots, A_n^0).$$

This shows the w^*-and $\|\cdot\|$-closed convex hulls of $A_1^0, \ldots, A_n^0$ coincide. Hence

$$(A_1+M)\cap\cdots\cap(A_n+M)\subset A_1\cap\cdots\cap A_n+M.$$

Thus

$$c_0\leq a_1, \ldots, a_n(Q)$$

if and only if c_0 "extends" to

$$c+m\leq a_1, \ldots, a_m(P); \qquad m\in J.$$

Similarly

$$b_1, \ldots, b_m\leq c_0(Q)$$

if and only if for some $m'\in J$

$$b_1, \ldots, b_m\leq c_0+m'.$$

Thus hypothesis (ii) says

$$c_0\in(A+M)\cap(B+M)$$

which, by Theorem 9.4(iii) gives

$$c_0\in A\cap(B+E_\varepsilon)+M.$$

We now present properties in the ordered Banach space (E, P) that characterize the w^*-closed complemented subcones of P^*. Let Q be a closed convex cone containing P in E and let $M=Q\cap-Q$ be the annihilator of the dual subcone Q^* of P^*. As above we write $a\leq b(Q)$ for the ordering induced by Q. In the next theorem, part (ii) characterizes complemented subcones in terms of the polar cones in E and part (iii) characterizes the annihilators of complemented subcones.

9.6. THEOREM. *The following are equivalent*:

(i) *Q^* is a w^*-closed complemented subcone of P^*,*

(ii) *if a, b_1, b_2 are given in E with*

$$0\leq b_1, b_2$$

$$0\leq a\leq b_1, b_2(Q)$$

then for each $\varepsilon>0$ there is an $m\in M$ such that

$$-\varepsilon\leq a+m\leq b_1, b_2,$$

(iii) (a) *if* $a_1, a_2 \in M$, $b \in E$ *with* $a_1, a_2 \leq b$ *then for each* $\varepsilon > 0$ *there is an* $a \in M$ *such that*

$$a_1, a_2 \leq a \leq b + \varepsilon,$$

(b) *if* $a \in M$, $b_1, b_2 \in P$ *with* $a \leq b_1 + b_2$ *then for each* $\varepsilon > 0$ *there exist* $a_1, a_2 \in M$ *such that*

$$a = a_1 + a_2 \quad \text{and} \quad a_i \leq b_i + \varepsilon \quad (i = 1, 2).$$

If J *is a closed subspace of* E *satisfying* (iii)(a), (b) *then* $J^0 \cap P^*$ *is a complemented subcone of* P^* *with annihilator* J.

Proof. (i) implies (ii) is immediate from 9.5. To show (iii)(a) we have

$$0 \leq b - a_1, \qquad b - a_2 \quad \text{and} \quad 0 \leq b \leq b - a_1, \qquad b - a_2(Q)$$

so that (i) gives $a \in M$ with

$$-\varepsilon \leq b - a \leq b - a_1, \qquad b - a_2$$

or

$$-b - \varepsilon < -a < -a_1, -a_2$$

which, multiplying by -1, gives (iii)(a). For (b) we have

$$0 \leq b_1 + b_2 - a \quad \text{and} \quad 0 \leq b_2 - a \leq b_1 + b_2 - a(Q)$$

so that there is an $a_1 \in M$ such that

$$-\varepsilon \leq b_2 - a + a_1 \leq b_1 + b_2 - a.$$

Thus

$$-\varepsilon - b_2 + a \leq a_1 \leq b_1$$

so that, if $a_2 = a - a_1$ then $a = a_1 + a_2$ with

$$a_1 \leq b_1 \quad \text{and} \quad a_2 \leq b_2 + \varepsilon.$$

This shows (iii) holds if M is the annihilator of Q^*. Next let J be any subspace for which (iii) holds and let $J^\perp = J^0 \cap P^*$. We show $J^\perp$ is complemented with annihilator $J = J^{\perp\perp}$. First, if $x \in P^* \backslash J^\perp$ then there is an $a_1 \in J$ with $(a_1, x) > 0$. Then (iii)(a) shows there is an $a \in J$ with $a \geq a_1, 0$, so that $a \in P \cap J$ and $(a, x) > 0$. This shows $J^\perp$ is a semi-exposed face. We show next that $J = J^{\perp\perp}$. Thus let

$$a \in (J + P)^- \cap (J - P)^- \quad \text{and} \quad \varepsilon > 0$$

be given. Then

$$a = m_1 + p_1 + z_1 = m_2 - p_2 + z_2; \qquad m_i \in J, \qquad p_i \in P \quad \text{and} \quad \|z_i\| < \varepsilon.$$

Hence

$$m = m_2 - m_1 = p_1 + p_2 + w, \qquad \|w\| \leq 2\varepsilon.$$

Since E is positively generated, for some constant α, we have

$$m \leqslant p_1 + p_2 + p; \qquad p \geqslant 0 \quad \text{and} \quad \|p\| \leq \alpha\varepsilon.$$

Hence

$$m = a_1 + a_2; \qquad a_i \in J \quad \text{and} \quad a_i \leq p_i + 2\alpha\varepsilon \quad (i = 1, 2).$$

Now

$$m - a_1 = a_2 \leq p_2 + 2\alpha\varepsilon = (m - p_1 - w) + 2\alpha\varepsilon$$

so that

$$-a_1 \leq -p_1 - w + 2\alpha\varepsilon.$$

Thus

$$p_1 + w - 2\alpha\varepsilon \leq a_1 \leq p_1 + 2\alpha\varepsilon$$

and the normality of E gives

$$\|a_1 - p_1\| \leq \beta\varepsilon \quad \text{for some constant } \beta.$$

Therefore

$$a = m_1 + p_1 + z_1 \in J + E_{\gamma\varepsilon}$$

for some constant γ depending only on α and β, so that

$$(J + P)^- \cap (J - P)^- = J.$$

Thus we have that J is an Archimedean order ideal and we conclude by constructing the projection π of P^* onto $J^\perp$.

Given $a \in P$, $x \in P^*$ let

$$p(a, x) = \lim_{\varepsilon \to 0} \sup \{(m, x): m \in J \cap A_\varepsilon(a)\},$$

where $A_\varepsilon(a) = (a - P) + E_\varepsilon$. Then $0 \leq p(a, x) \leq (a, x)$ and we show first that for fixed x, $p_x \doteq p(\cdot, x)$ is positive homogeneous and additive on P: if $p_x(a_i) > \lambda_i (i = 1, 2)$ and $\varepsilon > 0$ then we have $m_i \leq a_i + \varepsilon/2$, $(i = 1, 2)$ and $(m_i, x) > \lambda_i$. Thus

$$m_1 + m_2 \in A_\varepsilon(a_1 + a_2)$$

and

$$(m_1 + m_2, x) > \lambda_1 + \lambda_2.$$

This shows

$$p_x(a_1 + a_2) \geq p_x(a_1) + p_x(a_2).$$

If $p_x(a_i) < \lambda_i (i=1,2)$ then choose $\varepsilon > 0$ such that

$$(m_i, x) \le \lambda_i \quad \text{for all } m_i \in J \cap A_\varepsilon(a_i) \quad (i=1,2).$$

For an appropriate constant $\alpha > 0$,

$$m \in A_{\alpha\varepsilon}(a_1 + a_2)$$

implies, by (iii)(b),

$$m = m_1 + m_2, \qquad m_i \in J \cap A_\varepsilon(a_i) \quad (i=1,2).$$

Hence $(m, x) \le \lambda_1 + \lambda_2$. This shows

$$p_x(a_1 + a_2) \le p_x(a_1) + p_x(a_2).$$

The fact that $p_x(ra) = rp_x(a)(r \ge 0)$ is straightforward.

Next, if a is fixed then $p_a = p(a, \cdot)$ is positively homogeneous and sub-additive from the definition of p. The super-additivity follows from the property (iii)(a) in similar fashion to the above arguments. Thus we have that for each $x \in P^*$, p_x extends to an element of E^* such that

$$0 \le p_x \le x$$

and $x \in J^\perp$ implies $p_x = 0$. Conversely, if $p_x = 0$ then $(m, x) = 0$ for all $m \in J$ so that $x \in J^0 \cap P^* = Q^*$. The additivity of each p_a shows the operator $x \to p_x$ is additive on P^* and

$$0 \le p \le id, \qquad Q^* = \{x \in P^* : p_x = 0\}$$

so that $id - p = \pi$, the complementing projection.

Notes

The general definition of split sets and (the stability) Theorem 9.4 are due to Andô [**19**]. Theorem 9.5 is a generalized version of the extension theorem for a split face of a compact convex set found in Alfsen and Andersen [**6**] and Asimow [**25**]. Various characterizations of the annihilators of complemented subcones have been given. Alfsen and Andersen [**6**] describe them in terms of "near-lattice ideals" in the case of split faces of $A(K)$ and Perdrizet [**167**] gives a general condition which he calls hypostrict.

10. APPLICATION TO $A(K)$

We now focus on the case where (E, P) is an order unit space with order unit u. Then $E = A(K)$ where K is the state space of E and so K is a w^*-compact base for the dual cone P^*. This allows us to describe properties of P^* by restricting to K (and vice versa). Thus F is a closed face of K if and only if the

(w^*-closed) cone

$$Q^* = \bigcup_{\lambda \geq 0} \lambda F$$

is a face of P^*, and so forth. In particular, F is a *section* of K by the closed subspace $J \subset A(K)$ providing

$$F = J^0 \cap K \quad \text{and} \quad J = F^\perp = N^0$$

where N is the w^*-closed linear span of F given by

$$N = w^*\text{-cl}\,(Q^* - Q^*), \quad \text{and} \quad Q^* = \bigcup_{\lambda \geq 0} \lambda F,$$

is the w^*-closed linear span of F.

We first identify the Archimedean ideals in $A(K)$.

10.1. THEOREM. *Let J be a closed subspace of $A(K)$ and let $F = J^0 \cap K$. The following are equivalent*:

(i) *$A(K)/J$ is Archimedean ordered with proper cone $\hat{P} = qP$ (q the quotient map)*,

(ii) *$(A(K)/J, \hat{u})$ is a normed order unit space, where $\hat{u} = q1$*,

(iii) *$J + P$ is closed and $J = (J + P) \cap (J - P)$*,

(iv) *F is a section of K by J with the* PEP.

Proof. If (i) holds then, since $a \in A(K)$ implies $\pm a \leq \lambda 1$ for some $\lambda > 0$, $\pm \hat{a} \leq \lambda \hat{u}$ and hence $\hat{u}$ is an Archimedean order unit for $A(K)/J$ (as defined in Section 2). Thus $(A(K)/J, \hat{u})$ is a normed order unit space with unit ball

$$\hat{I} = \{\hat{a} : \pm \hat{a} \leq \hat{u}\}$$

and positive cone $\hat{P} = qP$, by Theorem 2.5. Hence we have

$$qE_1 \subset \hat{I}$$

and so

$$\hat{u} + qE_1 \subset \hat{u} + \hat{I} \subset \hat{P}$$

so that int $\hat{P}$ (quotient norm) is non-empty. Since $\hat{P}$ is linearly closed with non-empty interior, $\hat{P}$ is closed in the quotient norm, and hence $J + P$ is closed. Then

$$J = (J + P) \cap (J - P)$$

follows as well, since $\hat{P}$ is proper. This shows (iii). Now, the equivalence of (iii) and (iv) follows from Corollary 8.9. If (iii) holds and $r\hat{a} \leq \hat{b}$ for all $r \geq 0$ then

$$r\hat{a} \leq \hat{u} \quad \text{for all } r \geq 0.$$

Thus

$$ra \in 1 + J - P \quad \text{for all } r \geq 0$$

so that

$$a \in (J-P)^{-} = J - P$$

and hence $\hat{a} \in -\hat{P}$. Then

$$J = (J+P) \cap (J-P)$$

shows $\hat{P}$ is proper so (i) holds.

This allows us to characterize the Archimedean and strongly Archimedean order ideals of $A(K)$, using Corollary 8.10.

10.2. COROLLARY. *Let J be a closed subspace of $A(K)$ with $F = J^0 \cap K$.*
(1) *The following are equivalent:*
 (i) *$A(K)/J$ is Archimedean ordered with proper cone $\hat{P}$ and J is positively generated,*
 (ii) *F is a section of K by J satisfying: for each $a \in A(K)$, $a \geq 0$ on F, there is a $b \in A(K)$ such that*

$$b|_F = a|_F \quad \text{and} \quad b \geq a, 0.$$

 (iii) *J is an Archimedean order ideal.*
(2) *The following are equivalent:*
 (i) *J is a strongly Archimedean order ideal,*
 (ii) *property* 10.2(1)(ii) *is satisfied and $a \in A(F)$ implies there is $b \in A(K)$ with $b|_F = a|_F$.*

We say F is a *decomposable section* of K by J if the cone Q^* spanned by F is a decomposable section of P^* by J. The notion of positive decomposability is defined similarly.

10.3. THEOREM. *Let F be a section of K by J.*
(1) *The following are equivalent:*
 (i) *F is a decomposable section of K,*
 (ii) *there is a $\delta > 0$ such that each $x \in K$ can be written as a convex combination of y and z where*

$$y \in F \quad \text{and} \quad z \in K \quad \text{with } \|z - J^0\| \geq \delta.$$

(2) *The following are equivalent:*
 (i) *F is a positively decomposable section of K,*
 (ii) *there is an $h \in A(K)^{**}$ such that $h \equiv 0$ on J^0 and each $x \in K$ can be written as a convex combination of y and z where*

$$y \in F \quad \text{and} \quad z \in K \quad \text{with } h(z) \geq 1.$$

Proof. If (1)(i) holds then for some $r \geqslant 1$ each $x \in P^*$ can be written

$$x = y + z; \qquad y \in Q^* \quad \text{and} \quad \|z\| \leq r\|x - J^0\|.$$

Then, using the additivity of the norm, (ii) follows easily (with $\delta = 1/r$). The converse is also straightforward, as is the proof of (2), using the remark preceding Theorem 8.13.

10.4. COROLLARY. *Let F be a positively decomposable section of K by J. Then J is an Archimedean order ideal. Thus the extension property of Corollary* 10.2(1)(ii) *holds.*

We now consider the case where F is a split face of K. Since the norm is additive on P^* we can use 4.4(ii) to obtain a strengthening of 9.5.

10.5. THEOREM. *Let F be a split face of K and let $-f$, g be continuous convex functions on K.*

(i) *If*

$$g \leq 0 \leq f$$

and $a \in A(F)$ with

$$g|_F \leq a \leq f|_F$$

then for each $\varepsilon > 0$ there is a $b \in A(K)$ such that

$$b|_F = a \quad \text{and} \quad g \leq b \leq f + \varepsilon.$$

(ii) *If*

$$g \leq 0 < f$$

and $a \in A(F)$ with

$$g|_F \leq a \leq f|_F$$

then there is a $b \in A(K)$ such that

$$b|_F = a \quad \text{and} \quad g \leq b \leq f.$$

Proof. Since F is split the restriction map from $A(K)$ to $A(F)$ is onto. Furthermore, it is readily verified, using Theorem 4.4(ii) that the sets

$$A_f = \{a \in A(K) \colon a \leq f \quad \text{on} \quad K\}$$

$$A_g = \{a \in A(K) \colon a \geq g \quad \text{on} \quad K\}$$

have polars which split in $A(K)^*$ as in Theorem 9.5. It follows that $A_f + J$ is closed. If

$$\hat{A}_f = \{a \in A(K) \colon a \leq f \quad \text{on} \quad F\}$$

then

$$(A_f)^0 = \{x \in P^*: \bar{f}(x) \le 1\}$$

and

$$(\hat{A}_f)^0 = \{x \in Q^*: \bar{f}|_{Q^*}(x) \le 1\}, \qquad Q^* = \bigcup_{\lambda > 0} \lambda F.$$

Since Q^* is a face of P^*, $x \in Q^*$ implies

$$\bar{f}(x) = (f|_{Q^*})^-$$

so that

$$(A_f)^0 \cap J^0 = (\hat{A}_f)^0.$$

By taking polars of this we have

$$A_f + J = \hat{A}_f.$$

Thus (i) follows from Theorem 9.4(iii) and (ii) follows from Theorem 9.4(ii) since in the case $0 < f$ we have $(A_f)^0$ is w^*-compact.

Of course much more can be said if the complementary face of F is also closed. In this case we have a complete duality between the complementing projections and their adjoints in $A(K)$.

10.6. THEOREM. *Let $K = \text{co}\,(F_1 \cup F_2)$, F_i closed and convex in K. Let $A_i = \{a \in A(K): a|_{F_i} \equiv 0\}$. Then the following are equivalent:*

(i) *The F_i are complementary split faces of K,*

(ii) $A(K) = A_1 \oplus A_2$.

Proof. In general if $E = M_1 \oplus M_2$ with projections π_i onto M_i, then $E^* = M_1^\perp \oplus M_2^\perp$ with adjoint projections $\sigma_i = \pi_i^*$ such that range $\sigma_i = (\ker \pi_i)^\perp = M_j^\perp$. Thus if (i) holds and π_i is the corresponding projection in $A(K)^*$ then π_i is w^*-continuous so that $\sigma_i = \pi_i^*$ is well-defined on $A(K)$. Then (ii) is immediate. Conversely, (ii) yields the dual fact that

$$A(K)^* = N_1 \oplus N_2$$

where $N_i = (Q_i^* - Q_i^*)^-$ and Q_i^* the w^*-closed cone spanned by F_i. But then $K = \text{co}\,(F_1 \cup F_2)$ so that if $x \in K$,

$$x = x_1 + x_2; \qquad x_i \in Q_i^*.$$

Since $Q_i^* \subset N_i$ this representation is unique and the Q_i^* are complemented subcones of P^* in $A(K)^*$.

It is convenient to have a characterization of split faces in terms of maximal measures. If $0 \ne \mu \in C_{\mathbb{R}}(K)^*$ and $|\mu|/\|\mu\|$ is maximal ($|\mu| = \mu^+ + \mu^-$) then we say μ is a *boundary measure*. Let $\partial M(K)$ denote the set of boundary measures on K. Let

$$q: C_{\mathbb{R}}(K)^* \to C_{\mathbb{R}}(K)^*/A(K)^{\perp} = A(K)^*$$

be the quotient map, as usual, and $M_1^+(K)$ the probability measures on K.

10.7. LEMMA. *Let F be a closed face of K and let G be a compact subset of K disjoint from F. Then given ε $(0<\varepsilon<1)$ and $k \geq 1$, there exists $a \in A(K)$ such that*

$$a \geq 0 \quad \textit{on } K$$

$$a < \varepsilon \quad \textit{on } F$$

$$a \geq k \quad \textit{on } G.$$

Proof. The set G is covered by a finite collection $(N_i)_{i=1}^n$ of compact convex neighbourhoods in K, each disjoint from F. Then, since F is a face,

$$F \cap \mathrm{co}\,(\bigcup N_i) = \varnothing$$

so that we can assume, by replacing G with $\mathrm{co}\,(\bigcup N_i)$, that G is convex. Now consider the set H in $A(K)^* \times \mathbb{R}$ given by

$$H = \mathrm{co}\,(G \times \{k\}, K \times \{0\}).$$

Again, since F is a face, H can be *strictly* separated from $F \times \{\varepsilon\}$ $(0<\varepsilon<1)$ by a hyperplane in $A(K)^* \times \mathbb{R}$ which is the graph of an $a \in A(K)$ with

$$a \geq 0 \quad \text{on } K, \qquad a < \varepsilon \quad \text{on } F, \qquad a \geq k \quad \text{on } G.$$

10.8. PROPOSITION.

(i) *If F is a closed face of K and $\mu \in M_1^+(K)$ with $q\mu = x \in F$ then* $\operatorname{supp} \mu \subset F$.

(ii) *If F is a closed face of K and $\mu \in \partial M_1^+(K)$ with $q\mu \in F$ then $\mu \in \partial M_1^+(F)$.*

(iii) *If f is a bounded* l.s.c. *function on K and μ is maximal on K then $\mu(\check{f}) = \mu(f)$.*

Proof. (i) Given G compact in $K \backslash F$ we can apply Lemma 10.7 with $k=1$ to obtain $a \in A(K)$ such that

$$\varepsilon > a(x) = \int_K a \, \mathrm{d}\mu \geq \mu(G)$$

so that $\operatorname{supp} \mu \subset F$.

(ii) We have from (i) that $\operatorname{supp} \mu \subset F$ and it suffices to show (by Theorem 1.6.5) that $\mu(\check{f}) = \mu(f)$ for all $f \in C_{\mathbb{R}}(F)$. If $g \in C_{\mathbb{R}}(K)$ and $f = g|_F$ then

Theorem 1.6.1(vii) shows (since F is a face)

$$\check{f}(x)=\check{g}(x)$$

for each $x \in F$. Hence

$$\mu(\check{f})=\mu(\check{g})=\mu(g)=\mu(f).$$

(iii) We have, as in Theorem 1.6.1(viii), for example,

$$\mu(f)=\sup\{\mu(g): g \in C_{\mathbb{R}}(K) \text{ and } g \leq f\}.$$

Thus given $\varepsilon>0$ there is a $g \in C_{\mathbb{R}}(K)$ with

$$g \leq f \quad \text{and} \quad \mu(f)<\mu(g)+\varepsilon.$$

Then, using Theorem 1.6.5,

$$\mu(f) \geq \mu(\check{f}) \geq \mu(\check{g})=\mu(g)>\mu(f)-\varepsilon.$$

If F is a closed subset of K then $\mu \rightarrow \mu|_F$ is a complementing projection of $C_{\mathbb{R}}(K)^*$ onto $C_{\mathbb{R}}(F)^*$. We let $A(K)^{\perp}$ denote the measures in $A(K)$ representing 0 in $A(K)^*$ and let $\partial A(K)^{\perp}$ denote the boundary annihilating measures. Then P^* and Q^* denote the cones in $A(K)$ spanned by K and F respectively.

10.9. THEOREM. *Let F be a closed face of K and let*

$$\sigma_F \mu=\mu|_F \quad \textit{for } \mu \in C_{\mathbb{R}}(K)^*.$$

The following are equivalent:

(i) *F is a split face of K,*

(ii) *$\mu \in \partial A(K)^{\perp}$ implies $\sigma_F \mu \in A(K)^{\perp}$.*

(iii) *$\sigma \circ q=q \circ \sigma_F$ on $\partial M(K)$ well-defines σ which makes Q^* complemented in P^*.*

Proof. We show (i) implies (iii) first.

Let F be split and let σ be the complementing projection on $A(K)^*$ to $A(F)^*$.

(a) If $\mu \in \partial M_1^{+}(K)$ then $q\mu \in F$ if and only if $\mu(F)=1$ (by Proposition 10.8(i)).

Let $F'=(\ker \sigma) \cap K$ be the complementary face.

(b) If $\mu \in \partial M_1^{+}(K)$ then $q\mu \in F'$ if and only if $\mu(F)=0$. To show (b) let $f=1-\chi_F$. Now, using Theorem 1.6.1(iii),

$$\operatorname{epi} \check{f}=\operatorname{co}(K \times\{1\}, F \times\{0\})+\{0\} \times[0, \infty).$$

Now $\check{f}$ is l.s.c. and, since F is split, $\check{f}$ is affine. Hence, using Proposition

10.8(iii) and Theorem 1.6.1(ix), $q\mu \in F'$ if and only if

$$1 = \check{f}(q\mu) = \int \check{f}\, d\mu = \int f\, d\mu = \mu(K \backslash F)$$

if and only if $\mu(F) = 0$. Thus (a) shows that

$$(\sigma \circ q)\mu = (q \circ \sigma_F)\mu \quad \text{for } \mu \in \partial M(K) \cap (\text{range } \sigma_F)$$

and (b) shows the same for $\partial M(K) \cap (\ker \sigma_F)$. Hence (iii) holds. That (ii) and (iii) are equivalent is straightforward and (iii) implies (i) since q maps the non-negative boundary measures onto P^*.

Another geometric property, intermediate to "split face" and "decomposable face", is the notion of a *parallel face*. The closed convex subset F or K is called a *parallel face* if there is a convex subset F' of K such that

$$K = \text{co}\,(F \cup F')$$

and the *coefficients* in the convex combination are uniquely determined. A trapezoid in $\mathbb{R}^2$ forms the most immediate model. Clearly a split face is parallel, and the fact that the function identically one on F and identically zero on F' has a unique *affine* extension to K shows that a parallel face is positively decomposable. In fact, if $a = \chi_F$ then $\hat{a}$ is the (u.s.c.) affine function just described.

There is a measure theoretic characterization of parallel faces similar to the one for split faces.

10.10. THEOREM. *Let F be a closed face of the compact convex set K. Then F is a parallel face if and only if*

$$\mu(F) = 0 \quad \textit{for all } \mu \in \partial A(K)^{\perp}.$$

Proof. Let F be a parallel face and let $\mu \in \partial A(K)^{\perp}$ be given. Then $\mu = \mu_1 - \mu_2$ where we can assume the μ_i are maximal probability measures representing the same $x \in K$. We write

$$\mu_i = \lambda_i \nu_i + (1 - \lambda_i)\eta_i$$

where

$$\text{supp } \nu_i \subset F \quad \text{and} \quad \eta_i(F) = 0.$$

As in the split case, it is easy to see, using the a and $\hat{a}$ described above, that $\eta_i(F) = 0$ if and only if $q\eta_i \in F'$. Hence the coefficients λ_i in the representation of μ_i are the same for $i = 1, 2$. Thus $\mu_1(F) = \mu_2(F)$ so that $\mu(F) = 0$. Conversely, the measure condition guarantees that $\lambda_1 = \lambda_2$ for any $\mu = \mu_1 - \mu_2 \in \partial A(K)^{\perp}$. Hence, if we let

$$F' = \{z \in K : \mu(F) = 0 \quad \text{for some} \quad \mu \in \partial M_1^+(K) \quad \text{with} \quad q\mu = z\}$$

then $\mu(F)=0$ for *all* $\mu \in \partial M_1^+(K)$ representing z. It follows that

$$K = \mathrm{co}\,(F \cup F')$$

and the coefficients of the convex combinations are uniquely determined for each $x \in K$.

Notes

Theorem 10.5 with the stronger extension property (than that of Theorem 9.5) is due to Andersen [**17**]. The existence of an order unit in the pre-dual is essential and he uses a $1/2^n$ argument which has been incorporated into Theorem 9.4(iii) in our present proof. The measure theoretic characterization of split faces is due to Alfsen and Andersen [**6**] and is also discussed for complex function spaces in Alfsen and Hirsberg [**9**]. This will be dealt with in detail in Chapter 4. The idea of parallel faces appears in Hirsberg [**125**] and has been used by Briem [**54**] to give a characterization of simplexes analogous to that of Theorem 7.2.

11. THE RIESZ NORM

Another way of employing the basic properties of normality and positive generation in ordered Banach spaces is through the notion of a Riesz norm.

Definition. An ordered Banach space is said to have a *Riesz norm* if the following two properties are satisfied:

(a) if $-x \leq y \leq x$ then $\|y\| \leq \|x\|$;

(b) if $\|y\| < 1$ then there exists x such that $\|x\| < 1$ and $-x \leq y \leq x$.

It is easy to check that these properties imply 2-normality and approximately 2-generation.

Our aim is to demonstrate that properties (a) and (b) are dual and that as a consequence (E, P) has a Riesz norm if and only if (E^*, P^*) does.

Let S be a convex subset of $E(E^*)$ containing the origin.

Definition. We say S is

(a)′ *absolutely order convex* if $x \in S$ and $-x \leq y \leq x$ implies $y \in S$,

(b)′ *absolutely dominated* if $y \in S$ implies there exists $x \in S$ with $-x \leq y \leq x$.

Thus (E, P) has a Riesz norm if and only if E_1 satisfies (a)′ and $\mathrm{int}\, E_1$ satisfies (b)′.

As previously we give set theoretic versions of these properties and position our polar calculus artillery. Let $\bar{E} = E \times E$ and $\bar{P} = P \times P$.

Define

$$S_D = \{(x, x) \in \bar{E} : x \in S\}$$

$$S_G = \{(x, -x) \in \bar{E} : x \in S\}$$

$$S_A = \{(x, y) \in \bar{E} : x + y \in S\}$$

$$S_B = \{(x, y) \in \bar{E} : x - y \in S\}.$$

Note that $(\bar{S})_D = (S_D)^-$, $(\bar{S})_A = (S_A)^-$, etc.

11.1. PROPOSITION.

(1) *The following are equivalent:*
 (i) *S is absolutely dominated;*
 (ii) $S_G \subset S_D - \bar{P}$;
 (iii) $S_B \subset S_A \cap \bar{P} + E_D$.

(2) *The following are equivalent:*
 (i) *S is absolutely order convex;*
 (ii) $(S_D - \bar{P}) \cap E_G \subset S_G$;
 (iii) $S_A \cap \bar{P} \subset S_B$.

Proof. These are routine verifications. We simply note that, for example, (1)(ii) says for $y \in S$, $\pm y \leq x$ for some $x \in S$ which is equivalent to $-x \leq y \leq x$ so that (1)(i) and (1)(ii) are equivalent. If (1)(ii) holds and $(x, y) \in S_B$ then there exists $w \in S$ with

$$-w \leq x - y \leq w$$

so that

$$(x, y) = (w + x - y, w - x + y)/2 + (x + y - w, x + y - w)/2$$

gives (1)(iii). We omit the rest.

11.2. PROPOSITION.
 (i) $(S_D)^0 = (S^0)_A$;
 (ii) $(S_G)^0 = (S^0)_B$:
 (iii) $(S_A)^0 = (S^0)_D$;
 (iv) $(S_B)^0 = (S^0)_G$.

Proof. $(a, b) \in (S_D)^0$ if and only if $(x, a) + (x, b) = (x, a + b) \leq 1$ for all $x \in S$ if and only if $a + b \in S^0$ if and only if $(a, b) \in (S^0)_A$. For (iii), $(S_A)^0 = ((S^{00})_A)^0 = ((S^0)_D)^{00} = (S^0)_D$. Similarly for the rest.

It is immediate from these two propositions that (a)′ and (b)′ are dual order properties, except for the usual interjection of a *closure* in the pre-dual, leading to an approximate version of absolute domination. This is, of course, the reason for using int E_1 in property (b).

11.3. THEOREM. *Let (E, P) be an ordered Banach space. Then E has a Riesz norm if and only if E^* has a Riesz norm.*

Proof. If E has a Riesz norm then so does E^* by direct application of Propositions 11.1 and 11.2 with $S = \operatorname{int} E_1$. The converse follows similarly except that we only deduce from the order convexity of E_1^* that

$$(E_1)_B = [(E_1)_A \cap \bar{P} + E_D]^-.$$

But then E_B is a closed subspace of $\bar{E}$ so that we can conclude the E_B regularity of the restriction of the right-hand side to E_B (from Corollary 1.3.3) to obtain

$$(E_1)_B = (E_\alpha)_A \cap \bar{P} + E_D$$

for all $\alpha > 1$. This gives (b)$'$.

We recall from Section 6 that if E is a Banach lattice then E is *regular* if the norm is increasing on P and $\|x\| = \||x|\|$.

11.4. PROPOSITION. *Let (E, P) be a Banach lattice. Then E is regular if and only if E has a Riesz norm.*

Proof. If E is regular and $y \in E_1$ then $-|y| \le y \le |y|$ with $\|y\| = \||y|\|$ so that E_1 is absolutely dominated. The monotonicity of the norm on P shows that if $-x \le y \le x$ then $|y| \le |x|$ and $\|y\| \le \|x\|$ so that absolute convexity holds. If E has a Riesz norm and $0 \le y \le x$ then $-x \le y \le x$ so that $\|y\| \le \|x\|$ gives monotonicity. If $\|y\| < 1$ and $-x \le y \le x$ with $\|x\| < 1$ then $|x| \ge |y|$ so that $\| |y| \| < 1$. Regularity follows.

11.5. COROLLARY. *Let (E, P) be a Banach lattice that is positively generated and normal. Then E can be given an equivalent Riesz norm.*

Proof. Since E is normal we can renorm E with the unit ball

$$(E_1 - P)^- \cap (E_1 + P)^-$$

to obtain a norm $\|\cdot\|$ satisfying 1-normality. Then define

$$p(x) = \||x|\|.$$

It is easily verified that p is an equivalent Riesz norm.

Notes

The material in this section is developed in detail in the book of Wong and Ng [**217**]. We use their terminology here. Theorem 11.3 is due to Davies [**73**].

CHAPTER 3

Simplex Spaces

1. THE FACIAL TOPOLOGY FOR K, AND THE CENTRE FOR $A(K)$.

If K is any compact convex set then we will call a subset E of ext $K = \partial K$ *facially closed* if $E = \partial F$ for some closed split face F of K. Recall that F is split if and only if the subcone $A(F)^*_+$ is complemented in $A(K)^*_+$.

1.1. THEOREM. *The facially closed subsets of ∂K form the closed sets for a topology for ∂K, under which ∂K is a compact space.*

Proof. Proposition 2.5.6 shows that if F_1 and F_2 are split faces then so are co $(F_1 \cup F_2)$ and $F_1 \cap F_2$. Thus, if $E_j = \partial F_j$ $(j = 1, 2)$ are facially closed then so are

$$E_1 \cap E_2 = \partial(F_1 \cap F_2)$$

and

$$E_1 \cup E_2 = \partial(\text{co } F_1 \cup F_2).$$

Next consider an arbitrary intersection, $\bigcap_\alpha E_\alpha$, where

$$E_\alpha = \partial F_\alpha; \quad F_\alpha \text{ split}.$$

Let σ_α be the complementing projection of $A(F_\alpha)^*$ in $A(K)^*$. Since the finite intersections of split faces are split we can assume the F_α's are directed down by inclusion. Thus, for each $x \in A(K)^*_+$, $(\sigma_\alpha(x))_\alpha$ is a directed set, bounded below, and hence 2.3.2 shows $\{\sigma_\alpha x\}$ converges in norm to

$$\sigma x = \text{g.l.b. } \{\sigma_\alpha(x)\}.$$

Since

$$0 \leq \sigma x \leq \sigma_\alpha x \leq x \quad \text{for all } \alpha,$$

$$0 \leq \sigma x \leq x$$

and

$$\sigma x \in \bigcap_{\alpha} A(F_\alpha)_+^* = A(F)_+^*; \qquad F = \bigcap_{\alpha} F_\alpha.$$

Moreover, if $x \in F$ then $\sigma_\alpha x = x$ for all α, so that $\sigma x = x$. Hence σ is a complementing projection of $A(K)_+^*$ onto $A(F)_+^*$ and so F is split and

$$\bigcap_{\alpha} E_\alpha = \partial F.$$

We have therefore shown that the facially closed sets form the closed sets for the facial topology of K. If $\{E_\alpha\}$ is a family of facially closed sets satisfying the finite intersection property, and with $E_\alpha = \partial F_\alpha$ for some closed split faces F_α, then $\{F_\alpha\}$ has the finite intersection property. Since each F_α is compact, $\bigcap_\alpha F_\alpha$ is a non-empty closed split face with extreme boundary $\bigcap_\alpha E_\alpha$. Therefore ∂K is compact for the facial topology.

Recall that a simplex K is a Bauer simplex if $X = \partial K$ is compact, in which case $A(K) \cong C_{\mathbb{R}}(X)$ and $K = M_1^+(X)$ (2.7.5). If F is a closed face then F is split and $F = \overline{\text{co}}\, E$, where $E = F \cap X$ is a closed subset of X. Conversely, if E is a closed subset of X then using the Separation Theorem,

$$\begin{aligned} F = \overline{\text{co}}\, E &= \{\mu \in M_1^+(X): \mu(f) \leq 1 \text{ whenever } f \in C_{\mathbb{R}}(X) \text{ and } f \leq 1 \text{ on } E\} \\ &= \{\mu \in M_1^+(X): \text{supp}\, \mu \subset E\}. \end{aligned}$$

The latter set is clearly split under the complementing map, $\mu \to \mu|_E$.

1.2. COROLLARY. *A compact convex set K is a Bauer simplex if and only if the facial topology of ∂K is Hausdorff.*

Proof. If K is a Bauer simplex then ∂K is compact in the relative topology and, by the above remark, every closed subset of ∂K is facially closed. Therefore the relative and facial topologies of ∂K coincide.

Conversely, suppose that the facial topology is Hausdorff. Assume that there exists a point $x \in \overline{\partial K} \backslash \partial K$, and hence distinct points $y, z \in \partial F$, where F is the smallest closed split face of K which contains x. Then, because the facial topology is Hausdorff, there exist closed split faces G, H of K such that $K = \text{co}\,(G \cup H)$ and $y \notin G$, $z \notin H$. Since $x \in \overline{\partial K}$, either $x \in G$ or $x \in H$. But this contradicts the minimality property of F. Consequently ∂K is closed for the relative topology and hence the two topologies for ∂K coincide. If C is a closed face of K it now follows that ∂C is facially closed, and hence C is a split face of K. Therefore, by Theorem 2.7.2, K is a simplex, and hence a Bauer simplex.

It will be of importance to determine the nature of the real-valued functions on ∂K which are facially continuous.

We say that a function $f \in A(K)$ belongs to the *centre* of $A(K)$ if given $g \in A(K)$ there exists an $h \in A(K)$ such that $h(x) = f(x) \cdot g(x)$ for all $x \in \partial K$. Since functions in $A(K)$ are determined by their values on ∂K the function h will be uniquely determined by f and g. Moreover we will have $h(x) = f(x) \cdot g(x)$ for all $x \in \overline{\partial K}$.

It is evident that, when restricted to $\overline{\partial K}$, the central functions form a uniformly closed algebra of continuous real-valued functions on $\overline{\partial K}$ containing the constants. If K is a Bauer simplex then $A(K)|\partial K = C_{\mathbb{R}}(\partial K)$ and hence every $f \in A(K)$ is central. On the other hand if, for example, K is a closed plane disc (or square) then it is easy to check that the centre of $A(K)$ consists of precisely the constants.

Our aim is to prove that the facially continuous functions are precisely the central functions restricted to ∂K.

We denote $\{f \in A(K): f \equiv 0 \text{ on } F \text{ and } f \geq 0 \text{ on } K\}$ by $F_{\perp}^{+}$.

1.3. Lemma.

(i) *Let F and G be closed split faces of K with $f \in A(F)$, $g \in A(G)$ such that $f = g$ on $F \cap G$. Then there exists $h \in A(K)$ such that $h = f$ on F and $h = g$ on G.*

(ii) *Let F_j, G_j $(j = 1, \ldots, n)$ be closed split faces of K such that $F_j \cap G_j = \phi$ and $K = \text{co}\,(\bigcup_{j=1}^{n} G_j)$. Then, if $f \in A(K)^+$ and $\varepsilon > 0$, there exist $f_j \in (F_j)_{\perp}^{+}$ with*

$$f \leq \sum_{j=1}^{n} f_j \leq f + \varepsilon.$$

Proof. (i) Consider graph $f \subset F \times \mathbb{R}$ and graph $g \subset G \times \mathbb{R}$. Then 2.5.6 (ii) shows each $x \in \text{co}\,(F \cup G)$ has a unique representation in $\text{co}\,(F \cap G, F' \cap G, F \cap G')$ where F', G' are the complementary faces. It follows that

$$\text{co}\,(\text{graph}\, f \cup \text{graph}\, g)$$

is a compact convex subset of $K \times \mathbb{R}$ which equals graph h_0, h_0 a uniquely determined affine function on $\text{co}\,(F \cup G)$. Since graph h_0 is compact, $h_0 \in A\,(\text{co}(F \cup G))$. Then, by 2.10.5, h_0 extends to $h \in A(K)$.

(ii) Let $h_1 \in A(\text{co}\,(F_1 \cup G_1))$ satisfy $h_1 = 0$ on F_1, while $h_1 = f$ on G_1. Then, since $\text{co}\,(F_1 \cup G_1)$ is a split face of K, we can find an extension f_1 of h_1, $f_1 \in A(K)^+$ and $f_1 \leq f + \varepsilon/n$ (2.10.5). In particular we have $f_1 \in (F_1)_{\perp}^{+}$ and $f_1 = f$ on G_1.

Now suppose that we have constructed $f_j \in (F_j)_{\perp}^{+}$ for $j = 1, \ldots, r$, such that

$$\sum_{j=1}^{r} f_j \leq f + \frac{r\varepsilon}{n} \quad \text{on } K$$

while

$$f \leq \sum_{j=1}^{r} f_j \quad \text{on co}\left(\bigcup_{j=1}^{r} G_j\right).$$

Put

$$g = f + \frac{r\varepsilon}{n} - \sum_{j=1}^{r} f_j \in A(K)^+,$$

and let $h_{r+1} \in A(\text{co}\,(F_{r+1} \cup G_{r+1}))$ satisfy $h_{r+1} = 0$ on F_{r+1} and $h_{r+1} = g$ on G_{r+1}. Then $0 \leq h_{r+1} \leq g$ on co $(F_{r+1} \cup G_{r+1})$, and so there exists an extension $f_{r+1} \in (F_{r+1})_{\perp}^{+}$ of h_{r+1} such that $f_{r+1} \leq g + \varepsilon/n$. We clearly have

$$\sum_{j=1}^{r+1} f_j \leq \sum_{j=1}^{r} f_j + g + \varepsilon/n = f + \frac{(r+1)\varepsilon}{n} \quad \text{on } K,$$

and

$$\sum_{j=1}^{r+1} f_j = f + \frac{r\varepsilon}{n} \geq f \quad \text{on } G_{r+1}.$$

The proof is now completed by induction.

1.4. THEOREM. *A function $f : \partial K \to \mathbb{R}$ is facially continuous if and only if $f = g|\partial K$ for some function g in the centre of $A(K)$.*

Proof. First let f be facially continuous, and let $u \in A(K)$. If we prove that for each $\delta > 0$ there exists a $v \in A(K)$ such that $|v(x) - f(x)u(x)| < \delta$ for all $x \in \partial K$, then it will follow that f has the required form.

We may assume that $u \geq 0$ and that f is not a constant function. Then if $\varepsilon > 0$ there exist open intervals $U_1, \ldots, U_n$, each of length less than ε, such that $f(\partial K) \subset \bigcup_{j=1}^{n} U_j$ while no collection of $(n-1)$ of the U_j covers $f(\partial K)$. Choose compact sets $C_j \subset U_j$, such that $\bigcup_{j=1}^{n} C_j \supset f(\partial K)$. There exist closed split faces F_j, G_j, $j = 1, \ldots, n$, such that $f^{-1}(\mathbb{R} \backslash U_j) = \partial F_j$ and $f^{-1}(C_j) = \partial G_j$, and hence $F_j \cap G_j = \varnothing$, $K = \text{co}\,(\bigcup_{j=1}^{n} G_j)$. Lemma 1.3 now gives $u_j \in (F_j)_{\perp}^{+}$ such that

$$u \leq \sum_{j=1}^{n} u_j \leq u + \varepsilon.$$

For each j choose $\alpha_j \in U_j$ and put $v = \sum_{j=1}^{n} \alpha_j u_j$. If $x \in \partial K$ then $|f(x) - \alpha_j| < \varepsilon$ when $x \in f^{-1}(U_j)$ while $u_j(x) = 0$ when $x \notin f^{-1}(U_j)$, and we can estimate

$$\begin{aligned}
|v(x) - f(x)u(x)| &\leq \left|\sum_{j=1}^{n} f(x)u_j(x) - f(x)u(x)\right| + \left|\sum_{j=1}^{n} (\alpha_j - f(x))u_j(x)\right| \\
&\leq |f(x)| \left|\sum_{j=1}^{n} u_j(x) - u(x)\right| + \varepsilon \sum_{j=1}^{n} u_j(x) \\
&\leq \varepsilon(|f(x)| + u(x) + \varepsilon).
\end{aligned}$$

Consequently the necessity of the condition in the statement is proved.

Conversely, let g be in the centre of $A(K)$. Since the centre is a subspace containing constants, we can assume $0 \le g \le 1$ and show $E = \{x \in \partial K : g(x) \ge \alpha\}$ $(0 \le \alpha \le 1)$ is facially closed.

Let L denote the space of restrictions of central functions to $X = \overline{\partial K}$. Then L is isometrically isomorphic to a closed subalgebra of $C_{\mathbb{R}}(X)$, containing constants. In particular, L is a (pointwise on X) lattice.

Let $h_n \in L$ be given by $h_n(x) = 1 \wedge (g(x) + (1-\alpha))^n$ on X. Then if $\tilde{E} = \{x \in X : g(x) \ge \alpha\}$,

$$h_n(x) = 1 \quad \text{for } x \in \tilde{E} \qquad \text{and} \qquad h_n(x) < 1 \quad \text{for } x \in X \backslash \tilde{E}.$$

Let $F = \{x \in K : h_1(x) = 1\}$. Then F is a closed face of K with

$$\partial F = F \cap \partial K = E.$$

We use 2.10.9 to show F is split. Let μ be a boundary measure in $A(K)^{\perp}$ and let $a \in A(K)$. Let $a_n \in A(K)$ with

$$a_n(x) = a(x) h_n(x) \quad \text{on } X.$$

Then $a_n \downarrow a_0$, where a_0 is u.s.c. affine and $a_0 = a$ on F, $a_0(x) = 0$ for $x \in X \backslash F$.

Thus, since $\operatorname{supp} \mu \subset X$,

$$\mu|_F(a) = \int_{F \cap X} a_0 \, \mathrm{d}\mu = \int_X a_0 \, \mathrm{d}\mu = \lim \int_X a_n \, \mathrm{d}\mu = 0.$$

As we noted in the above proof, the centre of $A(K)$ is always a sublattice containing the constants. It turns out that for simplexes the centre has a purely lattice-theoretic characterization.

1.5. Theorem. *If K is a simplex then the centre of $A(K)$ is the largest sublattice of $A(K)$ containing the constants.*

Proof. Let L be a sublattice of $A(K)$. If $a, b \in L^+$ and

$$p = \inf\{a, b\} \quad \text{on } A(K)^*_+$$

then the definition of $\check{p}$ in Chapter 2 Section 4 shows $\check{p} = a \wedge b$ and 2.4.4 shows $\check{p} = \bar{p}$. The definition of $\bar{p}$ yields the fact that

$$\bar{p}(x) = a(x) \wedge b(x) \quad \text{for } x \in \partial K.$$

Hence L is a (pointwise) sublattice of functions on ∂K.

If $f \in L$, and L contains the constants, let

$$E = \{x \in \partial K : f(x) \le \alpha\}.$$

If $g = f \vee \alpha$ then $g \le \alpha$ on K and

$$F = \{x \in K : g(x) = \alpha\}$$

is a closed face (hence split) of K with $E = \partial F$. This shows f is l.s.c. in the facial topology. Similarly, $-f$ is l.s.c. and hence continuous for the facial topology. Theorem 1.4 now shows that F belongs to the centre of $A(K)$. To obtain a further characterization of the centre of $A(K)$ when K is a simplex we need the following lemma.

1.6. LEMMA. *If F is a closed convex subset of a simplex K with $\partial F \subseteq \partial K$ then F is a face of K.*

Proof. We need to show that if μ is a maximal probability measure on K with resultant $y \in F$ then supp μ is contained in F. Suppose that there is a point $z \in (\text{supp}\,\mu)\backslash F$. Then we can find an $f \in A(K)$ such that $f(x) < 0 < f(z)$ for all $x \in F$. Therefore $\widehat{0 \vee f}(x) = (0 \vee f)(x) = 0$ for all $x \in \partial F$, and since $g = \widehat{0 \vee f}$ is affine and upper semi-continuous we have $g(x) = 0$ for all $x \in F$. This leads to $0 = g(y) = \mu(g) = \mu(0 \vee f) > 0$, which gives the required contradiction.

1.7. THEOREM. *If K is a simplex then $f \in A(K)$ belongs to the centre of $A(K)$ if and only if there exists a g in $A(K)$ such that $g(x) = f(x^2)$ for all x in ∂K.*

Proof. Let $F = \overline{\text{co}}\,\{u \in \overline{\partial K}: f(u) \geq 0\}$, and let $x \in \partial F$. By Milman's Theorem we have $x \in \overline{\partial K}$. Suppose that $x = \frac{1}{2}y + \frac{1}{2}z$, where $y, z \in K$. Then if μ_y and μ_z denote the maximal probability measures on K with resultants y and z respectively we have

$$\begin{aligned} 0 &= (g - 2f(x)f + f(x)^2)(x) \\ &= \tfrac{1}{2}\int (g - 2f(x)f + f(x)^2)\,\mathrm{d}\mu_y + \tfrac{1}{2}\int (g - 2f(x)f + f(x)^2)\,\mathrm{d}\mu_z \\ &= \tfrac{1}{2}\int (f - f(x))^2\,\mathrm{d}\mu_y + \tfrac{1}{2}\int (f - f(x))^2\,\mathrm{d}\mu_z. \end{aligned}$$

Therefore we have $f(v) = f(x)$ for all v in supp $\mu_y \cup$ supp μ_z. But then supp $\mu_y \cup$ supp μ_z is contained in $\{u \in \overline{\partial K}: f(u) \geq 0\}$, and consequently y, z belong to F. Therefore $x \in \partial K$, and Lemma 1.6 shows that F is a face of K. It follows that the set $\{u \in \partial K: f(u) \geq 0\} = \partial F$ is facially closed and, by adding constants to f, we see that $\{u \in \partial K: f(u) \geq \alpha\}$ is facially closed for all α in $\mathbb{R}$. Hence f is upper semi-continuous for the facial topology. Similarly $-f$ is upper semi-continuous for the facial topology, and the result follows from Theorem 1.4.

Notes

The notion of facial topology for a compact convex set K and the associated concept of centre for $A(K)$ were first introduced by Alfsen and Andersen

[**6, 7**]. The concepts have been extensively generalized by Alfsen and Effros [**9**] and others. We will look at some of these generalizations in Section 8. The motivation for the notion of centre comes from C^*-algebra theory and this matter will be taken up in Chapter 5. In Chapter 4 the notion of centre in relation to complex function spaces and function algebras will be discussed.

Theorem 1.5 is due independently to Fakhoury [**111**] and to Nagel [**163**]. If K is not a simplex then the conclusion of 1.5 may fail, even if a largest sublattice of $A(K)$ containing constants does exist (see Ellis [**106**]).

Lemma 1.6 is due to Mokobodzki (see Rogalski [**180**]), and Theorem 1.7 to Sternfeld [**201**].

Many other facial properties of simplexes are discussed by Goullet de Rugy [**121**].

2. FACIAL CHARACTERIZATIONS OF SIMPLEXES

We have seen (2.7.2) that a compact convex set K is a simplex if and only if every closed face of K is split. Moreover every closed face of a simplex K is a simplex and $\overline{\text{co}}\, E$ is a face whenever E is a compact subset of ∂K (see Lemma 1.6).

If K is a finite-dimensional compact convex set then it is easy to check that K is a simplex if and only if each $x \in \partial K$ is a split face, or if and only if $\overline{\text{co}}\, E$ is a face of K whenever $E \subset \partial K$ is compact. We now give some examples to show that neither of these properties characterizes simplexes in general. The first example arises naturally in the field of function algebras.

2.1. EXAMPLE. *Let $D = \{z \in \mathbb{C} : \frac{1}{2} \le |z| \le 1\}$, $A = \{f \in C_{\mathbb{C}}(D) : f$ is analytic on $D^0\}$, and let $S = \{\phi \in A^* : \phi(1) = 1 = \|\phi\|\}$ endowed with the w^*-topology. Then every proper closed face of S is a Bauer simplex and is strongly Archimedean, and every $x \in \partial S$ is a split face of S. The set S is metrizable with ∂S closed, but is not a simplex.*

Proof. The space $A(S)$ is the uniform closure of $(\text{re } A)|_S$. We identify the boundary circles γ, Γ of radii $\frac{1}{2}$, 1 with their natural embedding $X = \gamma \cup \Gamma$ in S. Denote

$$h = (f, g) \in C(X); \qquad f = h|_\gamma, \qquad g = h|_\Gamma.$$

Let λ_Γ, λ_γ denote the normalized linear Lebesgue measures on Γ, γ. Since $\lambda_\Gamma(z^n) = 0 = \lambda_\gamma(z^n)$ for any non-zero integer n and $\lambda_\Gamma(1) = 1 = \lambda_\gamma(1)$ we have $\lambda_\Gamma = \lambda_\gamma$ on A and hence, on $A(S)$. We observe next that

$$C_{\mathbb{R}}(X) = \langle A(S)|_X, |z| \rangle.$$

To see this let $h=(f,g)\in C_{\mathbb{R}}(X)$ and let

$$F(z)=\sum_{k=0}^{n} a_k z^k, \qquad G(z)=\sum_{k=0}^{n} b_k z^k$$

be polynomials with a_0, b_0 real and

$$\|\text{re}\, F-f\|_\gamma<\varepsilon \quad \text{and} \quad \|\text{re}\, G-g\|_\Gamma<\varepsilon.$$

Let

$$H(z)=c|z|+\sum_{k=-n}^{n} c_k z^k$$

where

$$a_0=(\tfrac{1}{2})c+c_0 \qquad a_k=c_k+2^{2k}\bar{c}_{-k}$$

$$b_0=c+c_0 \qquad b_k=c_k+\bar{c}_{-k}.$$

Then re $H=$ re F on γ, re $H=$ re G on Γ.

Since $c=0$ if and only if $\lambda_\Gamma(H)=\lambda_\gamma(H)$ we conclude that

$$A(S)|_X=\{(f,g)\in C_{\mathbb{R}}(X)\colon \lambda_\gamma(f)=\lambda_\Gamma(g)\}.$$

Let E_λ, E_Γ be *proper* compact subsets of γ, Γ with $E=E_\gamma\cup E_\Gamma\neq\varnothing$. Since $\lambda_\gamma(E_\gamma)$, $\lambda_\Gamma(E_\Gamma)<1$ we have

$$A(S)|_E=C_{\mathbb{R}}(E)$$

and that $F=\overline{\text{co}}\, E$ is a Bauer simplex and a strongly Archimedean face of S. In particular we see that $\partial S=X$. If $F_\gamma=\overline{\text{co}}\,\gamma$ and $F_\Gamma=\overline{\text{co}}\,\Gamma$ and λ denotes the element of S given by

$$\lambda(f,g)=\lambda_\gamma(f)=\lambda_\Gamma(g)$$

then $\lambda\in F_\gamma\cap F_\Gamma$ and the representing measures λ_γ, λ_Γ are maximal and distinct. Hence S is not a simplex. Moreover, if $\lambda\in F$, a *proper* closed face, then there is an E, compact, in either γ or Γ (say γ) such that $E\cap F=\varnothing$ and $0<\lambda_\gamma(E)$. Since $E\cap F=\varnothing$, $\overline{\text{co}}\, E\cap F=\varnothing$. Since $\lambda_\gamma=\lambda_\gamma|_E+\lambda_\gamma|_{\gamma\setminus E}$, λ_γ is a convex combination of measures, one of which represents a point in $\overline{\text{co}}\, E$. But F is a face, so that this is a contradiction. Hence the smallest closed face containing λ is S. Thus *any* proper closed face $F=\overline{\text{co}}\,(E_\gamma\cup E_\Gamma)$ where E_γ, E_Γ are proper subsets of γ,Γ ($E_\gamma=F\cap\gamma$, $E_\Gamma=F\cap\Gamma$). Finally, we invoke 2.10.9 to identify the split faces of S. Since $A(S)^\perp\cap M(X)=\langle\lambda_\gamma-\lambda_\Gamma\rangle$ we have $F=\overline{\text{co}}\,(E)$ ($E=E_\gamma\cup E_\Gamma\subset X$) is split if and only if $(\lambda_\gamma-\lambda_\Gamma)|_E\in C_{\mathbb{R}}(E)^\perp$ if and only if

$$\lambda_\gamma(E_\gamma)=0=\lambda_\Gamma(E_\Gamma).$$

In particular each $x\in\partial S$ is a split face of S.

The facts in the following example are straightforward to verify using the same kind of techniques as for 2.1.

2.2. EXAMPLE. *Let* $L=\{f\in C_{\mathbb{R}}[-1,1]: \int_{-1}^{0} f(t)\,dt=\int_0^1 f(t)\,dt,\quad f(0)=\frac{1}{2}f(1)+\frac{1}{2}f(-1)\}$ *and let* $K=\{\phi\in L^*: \phi(1)=\|\phi\|=1\}$ *with the* w^**-topology. Then* $\partial K=[-1,1]\backslash\{0\}$, *every* $x\in\partial K$ *is a split face of* K, *and every proper closed face of* K *is strongly Archimedean and also a simplex. Moreover, if* $E\subset\partial K$ *is compact then* $\overline{\text{co}}\,E$ *is a face of* K. *However* K *is not a simplex.*

It follows from 2.8.1 and 2.10.2 that for a strongly Archimedean face F in K

$$\inf\{s\geq 0: a\in A(F) \text{ implies } \exists b\in A(K),\ b|_F=a \quad\text{and}\quad \|b\|_K\leq s\|a\|_F\}<\infty.$$

This number is called the *characteristic* of F in K.

It is noticeable in the preceding example that the faces $\overline{\text{co}}\,E$ are in general not split, and indeed their characteristics are not bounded. In this connection we will prove Theorem 2.3. By a *standard* compact convex set K we will mean that ∂K is universally measurable for all regular Borel measures μ on K, and that $\mu(K\backslash\partial K)=0$ whenever μ is maximal. Of course every metrizable K is standard, and also any K for which ∂K is closed.

2.3. THEOREM. *Let K be a standard compact convex set. Then K is a simplex if either of the following conditions hold.*

(*i*) *For every compact* $E\subset\partial K$, $\overline{\text{co}}\,E$ *is a simplex and is a strongly Archimedean face of K. Moreover, the characteristics of the faces* $\overline{\text{co}}\,E$ *form a bounded set.*

(ii) *For every compact* $E\subset\partial K$, $\overline{\text{co}}\,E$ *is a split face of K.*

Proof. (i) If $E\subseteq\partial K$ is compact then, by hypothesis, $A(K)|E=C_{\mathbb{R}}(E)$ and there is a constant $\alpha>0$ such that each $f\in C_{\mathbb{R}}(E)$ has an extension $g\in A(K)$ with $\|g\|\leq\alpha\|f\|$. Let μ, ν be maximal probability measures on K with $\mu-\nu\in A(K)^{\perp}$. If $\mu\neq\nu$ then, since K is standard, there is a compact set $E'\subseteq\partial K$ and an $\varepsilon>0$ such that $|\mu(E')-\nu(E')|>\varepsilon$; say $\mu(E')-\nu(E')>\varepsilon$. Put $\beta=\varepsilon/4\alpha$ and choose a compact set E with $E'\subseteq E\subseteq\partial K$, $\mu(E)>1-\beta$ and $\nu(E)>1-\beta$; let $f\in C_{\mathbb{R}}(E)$ such that $\|f\|\leq 1$ and $\int_E f\,d\mu-\int_E f\,d\nu>\frac{1}{2}\varepsilon$. Then if $g\in A(K)$ is an extension of f satisfying $\|g\|\leq\alpha\|f\|$ we obtain $\mu(g)=\nu(g)$ and $\int_E f\,d\mu-\int_E f\,d\nu\leq 2\beta\alpha\|f\|\leq\frac{1}{2}\varepsilon$. This contradiction gives $\mu=\nu$, and hence K is a simplex.

(ii) Let μ be a boundary measure in $A(K)^{\perp}$ and let E be a compact subset of ∂K. Then, if $F=\overline{\text{co}}\,E$ we have $\mu_F\in A(K)^{\perp}$, and so $\mu(E)=0$. Since K is standard and μ is regular it follows that $\mu=0$, that is, K is a simplex.

The next result shows that K being standard is necessary in 2.3.

2.4. THEOREM. *There exists a compact convex set K which is not a simplex but which has the following properties*:

(i) ∂K *is a Borel subset of* K;
(ii) $\overline{\text{co}}\, E$ *is a split face of* K *for all compact subsets* E *of* ∂K.

Proof. Let $Y = \bigcup\{Y_\alpha : \alpha \in [0, 1]\}$, where the disjoint sets Y_α each consist of three points $\{r_\alpha, s_\alpha, t_\alpha\}$. Topologize Y so that each r_α and t_α is isolated and such that each s_α has a neighbourhood base consisting of the sets $\{s_\alpha\} \cup \bigcup\{Y_\beta : 0 < |\alpha - \beta| < \varepsilon\}$, $\varepsilon > 0$. Then Y is a compact Hausdorff space, each Y_α is discrete in the relative topology and $\alpha \to s_\alpha$ defines a homeomorphism between $[0, 1]$ and $\{s_\alpha : 0 \leq \alpha \leq 1\}$. We will identify $[0, 1]$ and $\{s_\alpha : 0 \leq \alpha \leq 1\}$ in the remainder of the proof.

Let λ denote Lebesgue measure on $[0, 1]$ and let λ_1, λ_2 be the restrictions of λ to $[0, \frac{1}{2}]$ and $[\frac{1}{2}, 1]$ respectively. Define L to be the closed linear subspace of $C_{\mathbb{R}}(Y)$ given by $L = \{f \in C_{\mathbb{R}}(Y) : f(s_\alpha) = \frac{1}{2}f(r_\alpha) + \frac{1}{2}f(t_\alpha)$ for all $\alpha \in [0, 1]$, $\int f(s_\alpha)\, d\lambda_1(\alpha) = \int f(s_\alpha)\, d\lambda_2(\alpha)\}$. Certainly L contains the constant functions. To show that L separates the points of Y we first note that the functions f_β belong to L, where $f_\beta(s_\alpha) = 0$ for all α, $f_\beta(r_\alpha) = 0 = f_\beta(t_\alpha)$ for $\alpha \neq \beta$ and $f_\beta(r_\beta) = 1 = -f_\beta(t_\beta)$. Therefore each r_α and each t_α is a peak point for L, and so we need only show that L separates the points of $\{s_\alpha : 0 \leq \alpha \leq 1\}$. But if $\alpha \neq \beta$ there exists an f in $C_{\mathbb{R}}[0, 1]$ with $f(\alpha) \neq f(\beta)$ while $\int f\, d\lambda_1 = \int f\, d\lambda_2$, and we can extend f to a function in L by putting $f(r_\lambda) = f(s_\gamma) = f(t_\gamma)$ for all γ in $[0, 1]$.

If K denotes the state space $\{\phi \in L^* : \phi(1) = \|\phi\| = 1\}$ of L then we may identify L with $A(K)$ in the natural way (see 1.4).

The Choquet boundary for L is identified with ∂K, and this clearly equals $\bigcup\{r_\alpha, t_\alpha : 0 \leq \alpha \leq 1\}$. Therefore the closure $\overline{\partial K}$ of ∂K identifies with the Šilov boundary Y for L. In particular we observe that ∂K, being the complement in Y of the compact set $\{s_\alpha : 0 \leq \alpha \leq 1\}$, is a Borel subset of Y.

Define two measures ν_1, ν_2 on Y by the equations $\nu_j(f) = \int f(s_\alpha)\, d\lambda_j(\alpha)$, $j = 1, 2, f \in C_{\mathbb{R}}(Y)$. Let B denote the family of functions of the form $g + h$, where $g, h \in C_{\mathbb{R}}(Y)$, $g = 0$ except at finitely many of the points r_α, t_α and $h(r_\alpha) = h(s_\alpha) = h(t_\alpha)$ for all α in $[0, 1]$. Then B forms a subalgebra of $C_{\mathbb{R}}(Y)$ containing constants and separating points of Y, so that B is dense in $C_{\mathbb{R}}(Y)$ by the Stone–Weierstrass Theorem. The mapping $\alpha \to \frac{1}{2}(g + h)(r_\alpha) + \frac{1}{2}(g + h)(t_\alpha) = \frac{1}{2}g(r_\alpha) + \frac{1}{2}g(t_\alpha) + h(s_\alpha)$ is measurable since g takes only finitely many values. Consequently the mapping $\alpha \to \frac{1}{2}f(r_\alpha) + \frac{1}{2}f(t_\alpha)$ is also measurable for $f \in C_{\mathbb{R}}(Y)$, using the density result which has just been established. We therefore obtain, for $j = 1, 2$,

$$\int \tfrac{1}{2}((g + h)(r_\alpha) + (g + h)(t_\alpha))\, d\lambda_j(\alpha) = \int h(s_\alpha)\, d\lambda_j(\alpha) = \int (g + h)\, d\nu_j,$$

and hence for all f in $C_{\mathbb{R}}(Y)$

$$\int \tfrac{1}{2}(f(r_\alpha) + f(t_\alpha))\, d\lambda_j(\alpha) = \int f\, d\nu_j. \tag{1}$$

We now wish to find all measures in $A(K)^{\perp} \cap M(Y)$. Clearly the measure $\mu = \nu_1 - \nu_2$ belongs to $A(K)^{\perp} \cap M(Y)$, and the definition of L implies that $A(K)^{\perp} \cap M(Y)$ is the linear span of μ and N, where N is the w^*-closed linear span of the measures $\{\varepsilon_{s_\alpha} - \frac{1}{2}\varepsilon_{r_\alpha} - \frac{1}{2}\varepsilon_{t_\alpha} : 0 \leq \alpha \leq 1\}$ in $M(Y)$. Suppose that $\{\eta^{(a)}\}$ is a net w^*-convergent to η in N, where

$$\eta^{(a)} = \sum_{j=1}^{N(a)} b_j^{(a)} (\varepsilon s_j^{(a)} - \tfrac{1}{2}\varepsilon r_j^{(a)} - \tfrac{1}{2}\varepsilon t_j^{(a)}).$$

Then for each $\alpha \in [0, 1]$, we can find $h_n^{(\alpha)}$ in $A(K)$, $\|h_n^{(\alpha)}\| \leq 1$, such that $h_n^{(\alpha)}$ converges pointwise to $\chi\{r_\alpha\} \cup \{s_\alpha\} \cup \{t_\alpha\}$, we see that $\eta(\{r_\alpha\}) + \eta(\{s_\alpha\}) + \eta(\{t_\alpha\}) = 0$. Using the fact that there is a $g_\alpha \in A(K)$ with $g_\alpha = 0$ except for $g_\alpha(r_\alpha) = 1 = -g_\alpha(t_\alpha)$ we see that $\eta(\{r_\alpha\}) = \eta(\{t_\alpha\}) = \frac{1}{2}\eta(\{s_\alpha\})$, and hence the discrete part η_d of η has the form

$$\sum_{j=1}^{\infty} \eta(s_{\alpha_j})(\varepsilon_{s_{\alpha_j}} - \tfrac{1}{2}\varepsilon_{r_{\alpha_j}} - \tfrac{1}{2}\varepsilon_{t_{\alpha_j}}).$$

Consequently η_d, and also $\eta - \eta_d$, belong to $A(K)^{\perp}$. Since $\eta - \eta_d$ has no atoms and since the sets $\{r_\alpha : 0 \leq \alpha \leq 1\}$ and $\{t_\alpha : 0 \leq \alpha \leq 1\}$ are discrete it follows from regularity that $\eta - \eta_d$ is concentrated on $\{s_\alpha : 0 \leq \alpha \leq 1\}$. But then $(\eta - \eta_d)(h) = 0$ for all $h \in C_{\mathbb{R}}[0, 1]$, and so $\eta = \eta_d$.

This description of N shows that the only possible measures in $\partial M(K) \cap A(K)^{\perp}$ are the multiples of μ, because no maximal measure can have an atom lying outside ∂K. Now if E is a compact subset of ∂K then E must be finite, and it is easy to construct a function in $A(K)$ which peaks on a face F with $\partial F = E$. Since $\mu(E) = 0$ for a finite set E we have $\mu'|E \in A(K)^{\perp}$ whenever $\mu' \in A(K)^{\perp} \cap \partial M(K)$, and hence $F = \overline{\text{co}}\, E$ is a split face of K.

To complete the proof we need to show that K is not a simplex, and this will be the case if we show that μ is a boundary measure. We first note that $\frac{1}{2}(\varepsilon_{r_\alpha} + \varepsilon_{t_\alpha})$ is the only maximal probability measure on K with resultant s_α. Indeed, if μ' is another such measure then $\frac{1}{2}(\varepsilon_{r_\alpha} + \varepsilon_{t_\alpha}) - \mu' \in \partial M(K) \cap A(K)^{\perp}$ and so $\mu' = \frac{1}{2}(\varepsilon_{r_\alpha} + \varepsilon_{t_\alpha}) + b\mu$ for some real number b. Since μ' is a probability measure and μ has no atoms we can deduce that $b = 0$.

The measures ν_j will be maximal measures if and only if $\nu_j(f) = \nu_j(\hat{f})$ for each continuous convex function f. Using (i) we have

$$\nu_j(f) = \int \tfrac{1}{2}(f(r_\alpha) - f(t_\alpha))\, d\lambda_j(\alpha)$$

$$= \int \hat{f}(s_\alpha)\, d\lambda_j(\alpha) = \nu_j(\hat{f}),$$

where the second equality follows from 1.6.2 and the fact that $\frac{1}{2}(\varepsilon_{r_\alpha} + \varepsilon_{t_\alpha})$ is the only maximal measure representing s_α. Therefore $\mu = \nu_1 - \nu_2$ is a boundary measure.

If F is a face of K then the complementary set F' consists of the union of all faces of K which are disjoint from F. Each x in K may be decomposed $x=\lambda y+(1-\lambda)z$, with $0\leq\lambda\leq 1$, $y\in F$, $z\in F'$ (see [**2**, Proposition 2.6.5]). If F' is itself a face of K and if the constant λ is uniquely determined by x then F is called a *parallel* face of K (see 2.10). A closed face F of K is parallel if $\mu(F)=0$ whenever $\mu\in A(K)^\perp\cap\partial M(K)$, or equivalently if $\hat{\chi}_F$ is an affine function, where for any bounded real-valued function f on K we write $\hat{f}(x)=\inf\{g(x): g\in A(K), g\geq f\}$, $x\in K$.

A closed subset D of K is said to be *dilated* if whenever μ is a maximal probability measure on K with resultant $x\in D$ then μ is supported by D. Evidently every closed face of K is dilated, and every compact subset of ∂K is dilated. If D is a dilated subset of K then a continuous real-valued function f on D is called *affine* if $\mu(f)=f(x)$ whenever μ is a maximal probability measure on K with resultant $x\in D$. If D is a compact subset of ∂K then it is clear that every continuous function on D is affine. It is easy to verify that if D is a closed face of K then a continuous function on D is affine in the preceding sense precisely if $f\in A(D)$.

In the case when K is standard, Theorem 2.3(i) shows that K is a simplex if there is a constant C such that whenever E is a compact subset of ∂K then $\overline{\mathrm{co}}\, E$ is a simplex and a face of K and each $f\in A(\overline{\mathrm{co}}\, E)^+$ has an extension $g\in A(K)^+$ with $\|g\|\leq C\|f\|$. This result can be rephrased to say that K is a simplex if and only if *every* continuous positive function f on E has an extension $g\in A(K)^+$ with $\|g\|\leq C\|f\|$, where C is a constant independent of E. Theorem 2.4 shows that the hypothesis that K is standard may not be dropped in this result of Rogalski.

We present below an analogue of Theorem 2.3(i) which is valid for all sets K. In the case when K is standard the compact subsets of ∂K play a role similar to that played by the dilated sets when K is not standard; this is because when K is standard the supports of the maximal measures on K can be approximated by compact subsets of ∂K. $A^b(K)$ stands for the space of bounded affine functions on K.

2.5. THEOREM. *The following statements are equivalent for compact convex sets K.*

(i) *K is a simplex.*

(ii) *If D is a dilated subset of K and if $f\geq 0$ is continuous and affine on D then f has an extension $g\in A(K)^+$; moreover for each closed face F of K, $\inf\{a(x): a\in A^b(K),\ 0\leq a\leq 1,\ a|F=1\}=0$ for all x in F'.*

(iii) *The convex hull of any finite number of closed faces of K is a face of K, and for each closed face F of K, $\inf\{a(x): a\in A^b(K), 0\leq a\leq 1, a|F=1\}=0$ for all x in F'.*

(iv) *Every closed face of K is parallel.*

Proof. The implication (i) implies (iv) holds because every split face is parallel.

(i) implies (ii). The second statement in (ii) holds trivially for split faces, so it is sufficient to prove the first statement of (ii).

If D is a dilated set then $\overline{\text{co}}\,D$ is a face of K. Indeed if $x \in \partial(\overline{\text{co}}\,D)$ then $x \in D$, so that the maximal representing measure μ for x is supported by D. But then $\mu = \varepsilon_x$ by the extremity of x, so that $x \in \partial K$. Lemma 1.6 now shows that $\overline{\text{co}}\,D$ is a face of K. To complete the proof it will be sufficient to show that every function f which is continuous and affine on D has an extension in $A(\overline{\text{co}}\,D)$. This will follow if $\mu_1(f) = \mu_2(f)$ whenever μ_1 and μ_2 are probability measures on D with the same resultant $y \in \overline{\text{co}}\,D$.

For each $x \in \overline{\text{co}}\,D$ let λ_x denote the unique maximal probability measure on $\overline{\text{co}}\,D$ representing x. Given $\varepsilon > 0$ there exists a continuous concave function $u \geq f$ such that $\lambda_y(u) < \lambda_y(\hat{f}) + \varepsilon = \lambda_y(f) + \varepsilon$. Since f is affine on D we have, for x in D, $f(x) = \lambda_x(f) \leq \lambda_x(u) = \check{u}(x)$ (see 1.6.2), and because $\check{u}$ is convex we obtain

$$\mu_1(f) \leq \mu_1(\check{u}) \leq \lambda_y(\check{u}) = \lambda_y(u) < \lambda_y(f) + \varepsilon.$$

It follows that $\mu_1(f) \leq \lambda_y(f)$, and applying the same argument to $-f$ we obtain $\mu_1(f) = \lambda_y(f) = \mu_2(f)$ as required.

(ii) implies (iii). It is sufficient to prove the first part of (iii) for two closed faces F and G of K. If x belongs to $\partial K \backslash (F \cup G)$ then it is evident that $F \cup G \cup \{x\}$ is a dilated subset of K and that the function f_x is affine and continuous on $F \cup G \cup \{x\}$, where $f_x(x) = 1$ and $f_x = 0$ on $F \cup G$. Therefore there exists on extension $g_x \in A(K)^+$ of f_x. The set $H = \bigcap\{g_x^{-1}(0)\colon x \in \partial K \backslash (F \cup G)\}$ is a closed face of K containing $F \cup G$, whose extreme points belong to $F \cup G$. Therefore $H = \text{co}\,(F \cup G)$ and (iii) is proved.

(iii) implies (i). To prove that K is a simplex we will show that $A(K)^*$ is a vector lattice. Using the Bishop–Phelps Theorem it will be sufficient to prove that ρ^+ exists whenever $\rho \in A(K)^*$ attains its norm at some $f \in A(K)$ with $\|f\| = 1$. (Indeed if ρ_n^+ exists in $A(K)^*$ for $n \geq 1$, and if $\rho_n \to \rho$ in norm, then $\rho_n^+ \to \rho^+$ in norm (see [**108**]).) If μ is a representing measure for ρ on $A(K)$ with $\|\mu\| = \|\rho\|$ then $\|\mu\| = \|\mu^+\| + \|\mu^-\| = \rho(f) = \int f\, d\mu^+ - \int f\, d\mu^-$, and so it follows that $\text{supp}\,\mu^+ \subseteq F$, $\text{supp}\,\mu^- \subseteq G$ where F and G are the closed faces of K given by $F = \{x \in K : f(x) = 1\}$, $G = \{x \in K : f(x) = -1\}$. Without loss of generality we can assume that $\mu^+ \neq 0$, $\mu^- \neq 0$. Define $\sigma(g) = \int g\, d\mu^+$ for $g \in A(K)$, so that clearly $\sigma \geq \rho, 0$. To show that $\sigma = \rho^+$ we need to prove that $\phi \geq \sigma$ whenever $\phi \geq \rho, 0$ in $A(K)^*$. By the Bishop–Phelps Theorem, for each natural number n there exists $\psi_n \in A(K)^*$ which attains its norm and is such that $\|\psi_n - (\phi - \sigma)\| < 1/n$. If λ_n is a representing measure for ψ_n, with $\|\lambda_n\| = \|\psi_n\|$, then there are disjoint closed faces H_n^+, H_n^- of K such that $\text{supp}\,\lambda_n^+ \subseteq H_n^+$, $\text{supp}\,\lambda_n^- \subseteq H_n^-$. If $\eta_n \in A(K)^*$ is defined by $\eta_n(g) = \int g\, d\lambda_n^-$

for $g \in A(K)$, then we can decompose $\eta_n = \|\eta_n\|(\alpha_n x_n + (1-\alpha_n)y_n)$ where $x_n \in G \cap H_n^-$, $y_n \in (G \cap H_n^-)'$. We will show that $\eta_n \to 0$ as $n \to \infty$.

If $(1-\alpha_n) \neq 0$ then $y_n \in H_n^-$, and hence y_n belongs to the complementary set for the face co $(H_n^+ \cup G)$. Therefore, by (iii), we can find an $f_n \in A^b(K)$ with $0 \leq f_n \leq 1$, $f_n = 0$ on co $(H_n^+ \cup G)$ and $f_n(y_n) > 1 - 1/n$. Identifying $A(K)^{**}$ with $A^b(K)$ as usual, we have

$$0 \leq (\phi - \rho)(f_n) = (\sigma - \rho)(f_n) + (\phi - \sigma)(f_n) < (\psi_n + \sigma - \rho)(f_n) + \frac{1}{n}$$

$$= -\eta_n(f_n) + \frac{1}{n} = -\|\eta_n\|(1-\alpha_n) f_n(y_n) + \frac{1}{n}$$

$$\leq -\|\eta_n\|(1-\alpha_n)\left(1 - \frac{1}{n}\right) + \frac{1}{n}.$$

It follows that in general we have $\|\eta_n\|(1-\alpha_n) \to 0$. Similarly we can find $g_n \in A^b(K)$ with $0 \leq g_n \leq 1$, $g_n = 0$ on co $(F \cup H_n^+)$, $g_n(x_n) > 1 - 1/n$, and hence

$$0 \leq \phi(g_n) < (\psi_n + \sigma)(g_n) + \frac{1}{n} = \frac{1}{n} - \eta_n(g_n)$$

$$\leq \frac{1}{n} - \|\eta_n\| \alpha_n g_n(x_n) \leq \frac{1}{n} - \|\eta_n\| \alpha_n \left(1 - \frac{1}{n}\right).$$

We obtain $\|\eta_n\|\alpha_n \to 0$, and consequently $\eta_n \to 0$. It now follows that if $h \in A(K)^+$ then

$$(\phi - \sigma)(h) = \lim \psi_n(h) = \lim \left\{ \int h \, \mathrm{d}\lambda_n^+ - \eta_n(h) \right\} \geq 0,$$

and hence $\phi \geq \sigma$ as required.

(iv) implies (iii). Let F and G be closed, and hence parallel, faces of K and let $H = \text{co}\,(F \cup G)$. To show that H is a face of K we need to show that if λ is a maximal probability measure on K with resultant $x \in H$ then supp $\lambda \subseteq H$. We have $x = ty + (1-t)z$, where $0 \leq t \leq 1$, $y \in F$ and $z \in G$. Let μ, ν be maximal measures on K with resultant y, z respectively. We have $\eta = \lambda - t\mu - (1-t)\nu \in A(K)^\perp \cap \partial M(K)$, and since F, G and $F \cap G$ are parallel faces $\eta(F) = \eta(G) = \eta(F \cap G) = 0$. Therefore

$$\lambda(F \cup G) = \lambda(F) + \lambda(G) - \lambda(F \cap G) = t\mu(F \cup G) + (1-t)\nu(F \cup G) = 1,$$

because μ and ν are supported by the faces F and G respectively, so that supp λ is contained in H.

The second condition of (iii) is satisfied since, for any closed parallel face F, the function $\hat{\chi}_F$ belongs to $A^b(K)$ with $\hat{\chi}_F = 1$ on F and $\hat{\chi}_F = 0$ on F'.

The following result gives two ways in which facial structure distinguishes Bauer simplexes amongst simplexes. We will say that a compact convex set

K satisfies *Størmer's axiom* if $\overline{\mathrm{co}}\,(\bigcup_\alpha F_\alpha)$ is a closed split face of K whenever $\{F_\alpha\}$ is any family of closed split faces of K.

2.6. THEOREM. *Let K be a compact simplex. Then K is a Bauer simplex if and only if either of the two following conditions hold.*
(i) *The closure of each face of K is a face of K;*
(ii) *K satisfies Størmer's axiom.*

Proof. The necessity of the two conditions is trivial. (i) implies (ii): Let $\{F_\alpha\}$ be closed faces of K, and hence split faces of K. Since the convex hull of finitely many of the F_α is a face of K it is easy to check that $\mathrm{co}\,(\bigcup_\alpha F_\alpha)$ is a face of K. Consequently $\overline{\mathrm{co}}\,(\bigcup_\alpha F_\alpha)$ is a face of K; and so K satisfies Størmer's axiom.

Now assume that K satisfies Størmer's axiom, and let $E \subseteq \partial K$ be relatively closed. Then there exists a compact set $G \subseteq K$ such that $E = G \cap \partial K$, and Milman's Theorem shows that $\partial(\overline{\mathrm{co}}\, E) \subseteq G$. Now each $x \in E$ is a (split) face of K, and hence $\overline{\mathrm{co}}\, E$ is a face of K because of Størmer's axiom. But then $\partial(\overline{\mathrm{co}}\, E) \subseteq G \cap \partial K$, so that $E = \partial(\overline{\mathrm{co}}\, E)$. Therefore E is facially closed and it follows from the proof of 1.2 that K is a Bauer simplex.

Although faces of simplexes are simplexes it is far from true that cross-sections of simplexes are simplexes, as the following result amply illustrates.

2.7. THEOREM. *Let S be the $w^*(l^1, c)$-compact simplex $\{\mathbf{y} \in l^1 : \|\mathbf{y}\| = 1, y_n \geq 0$, for all $n\}$, and let K be any metrizable compact convex subset of a locally convex Hausdorff space. Then there exists a linear subspace M of l^1 such that $M \cap S$ is affinely homeomorphic to K.*

Proof. Let $\{f_n\}$ be a sequence which is dense in $A(K)^+$, $f_n \neq 0$ for all n, and let $g_n = f_n(2^n\|f_n\|)^{-1}$, $n \geq 1$, and $g_0 = 1 - \sum_{n=1}^\infty g_n$. Define a linear operator $T: c \to A(K)$ by $T\mathbf{x} = \sum_{n=0}^\infty x_n g_n$ where $\mathbf{x} = \{x_n\}$, $n \geq 0$, belongs to c. Let M be the range of T^* in l^1.

Since $T(c)$ is dense in $A(K)$, T^* is one-to-one; also T^* is positive since T is positive. If $\phi \in K$ and if $\mathbf{e} = \{1\} \in c$, then we have

$$(T^*\phi)(\mathbf{e}) = (T\mathbf{e})(\phi) = \left(\sum_{n=0}^{\infty} g_n\right)(\phi) = 1.$$

Therefore T^* maps K into $M \cap S$. Suppose that $\Psi \in A(K)^*$ and that $T^*\Psi \in M \cap S$. Then we have

$$\Psi(1) = \Psi(T\mathbf{e}) = (T^*\Psi)(\mathbf{e}) = 1,$$

and

$$\Psi(g_n) = \Psi(T\mathbf{e}_n) = (T^*\Psi)(\mathbf{e}_n) \geq 0, \quad n \geq 1,$$

where $\mathbf{e}_n$ is the nth coordinate vector in c. Hence we have $\Psi(f_n) \geq 0$ for all n, and the density of $\{f_n\}$ in $A(K)^+$ ensures that $\Psi \geq 0$ and consequently that $\Psi \in K$. Therefore T^* is an affine homeomorphism of K onto $M \cap S$.

Notes

Examples with similar properties to those of 2.1 and 2.2 have been given by Rogalski [**181**], to whom Theorem 2.3 is due. It appears to be unknown whether, in the standard case, the condition that $\overline{\text{co}}\, E$ is a simplex can be removed from 2.3(i).

Theorem 2.4 is due to Ellis and Roy [**110**] and it involves the Bishop–de Leeuw porcupine topology construction [**40**]; it is related to examples studied by Alfsen [**5**, 1.4, 2.3].

The fact that condition (i) of 2.5 implies the first statement of (ii) is due to Effros [**94**, 3.3; **95**, 2.3]. The equivalence of (i) and (iv) of 2.5 is due to Briem [**54**]. Theorem 2.5, in the form given here, is due to Ellis and Roy [**110**]. The first statement of 2.5(ii) certainly does not characterize simplexes since it does not imply uniform boundedness of the affine extensions involved. Similarly the first statement of 2.5(iii) does not characterize simplexes; an example of this phenomenon is given by McDonald [**159**, 1.9].

In the duality between $A(K)^*$ and $A^b(K)$ the set $F^\# = \{x \in K : \inf\{a(x) : a \in A^b(K), \chi_F \leq a \leq 1\} = 0\}$ is called the *quasi-complement* of the face F (see Alfsen and Shultz [**10**, p. 15] and also Section 5.6 (below). The second statement of 2.5(ii) and 2.5(iii) can be re-written $F^\# = F'$. This condition does not imply that F' is convex, but it does imply that for each $x \in K$ the constant $\lambda \in [0, 1]$ in the decomposition $x = \lambda y + (1-\lambda)z$, $y \in F$, $z \in F'$ is unique.

Theorem 2.6 is due to Størmer [**206**]. Of course every finite-dimensional compact convex set satisfies condition 2.6(i) since every face is closed. It is not know precisely which infinite-dimensional compact convex sets satisfy this condition. Some results of a local nature on this topic have been obtained by Roy [**184**].

Theorem 2.7 is due to Lazar [**146**]. In the same paper he shows that every infinite-dimensional Banach space contains a compact set which is contained in no bounded simplex lying in the space.

3. IDEALS IN SIMPLEX SPACES

Recall that a subspace I of an ordered space (E, P) is an order ideal ($a \leq b \leq c$ for $a, c \in I$ implies $b \in I$) if and only if $I^+ \doteq I \cap P$ is a face of P. It is convenient in the context of $A(K)$ spaces to also require that I be positively generated. Thus we will say I is an *ideal* in $A(K)$ if I^+ is a face of $A(K)^+$ and $I = I^+ - I^+$. The first result is a direct reformulation of 2.8.15.

3.1. THEOREM. *Let K be a compact simplex and let I be a closed linear subspace of $A(K)$. Then I is an ideal in $A(K)$ if and only if $I = F_\perp = \{f \in A(K): f(x) = 0, \forall x \in F\}$ for some closed face F of K.*

It is worth noting that if I is a (not necessarily closed) ideal that $I^\perp = I^0 \cap K$ is still a closed face F and hence $F_\perp = I^{00} = \bar{I}$ is a closed ideal.

If $\{I_\alpha\}$ is a family of closed ideals in $A(K)$ then, if $F_\alpha = I_\alpha^\perp$, we have $\bigcap_\alpha I_\alpha = (\overline{\text{co}} \bigcup_\alpha F_\alpha)^\perp$. Hence if K is a Bauer simplex, so that Størmer's axiom holds, then $\bigcap_\alpha I_\alpha$ is a closed ideal in $A(K)$. This fails in general, as the following example shows.

3.2. EXAMPLE. *Let $L = \{f \in C_{\mathbb{R}}[-1, 1]: f(0) = \frac{1}{2}f(-1) + \frac{1}{2}f(1)\}$, $K = \{\phi \in L^*: \phi(1) = \|\phi\| = 1\}$ with the w^*-topology, and let $I_n = \{f \in A(K): f(1/n) = 0\}$. Then $A(K)$ is a simplex space and I_n is closed ideal for each n, while $\bigcap_{n=1}^\infty I_n$ is not positively generated.*

An interesting problem is whether Bauer simplexes are characterized among simplexes K by the property that the intersection of an arbitrary family of closed ideals in $A(K)$ is again an ideal. In the metrizable case, at least, the answer is affirmative, and we now discuss this result. We first consider a theorem which leads to the construction of many interesting examples of simplex spaces. $\partial M_q^+(K)$ will denote the maximal probability measures on K with resultant q.

3.3. THEOREM. *Let K be a compact metrizable simplex and let $q \in \overline{\partial K} \backslash \partial K$. Suppose that Y is a closed subset of $\overline{\partial K}$ such that $Y \cap (\overline{\partial K} \backslash \partial K) = \{q\}$ and $\pi(\partial K \backslash Y) \neq 0$, where $\{\pi\} = \partial M_q^+(K)$. Then $A = \{f \in A(K): f(y) = \pi(f), \forall y \in Y\}$ is a simplex space $A(H)$ where ∂H can be identified with $\partial K \backslash Y$.*

Proof. If $\alpha = \pi(Y)$ then $0 < \pi(\partial K \backslash Y) \leq 1 - \alpha$. Let

$$L = \{f \in C_{\mathbb{R}}(\overline{\partial K}): f(y) = \pi(f), \forall f \in Y\}$$

and put

$$X = \{\varepsilon_y - \pi: y \in Y\}$$

so that X is homeomorphic to Y. To each $\mu \in M(X)$ there corresponds naturally a measure $\lambda \in M(Y)$ so that

$$\int_X x(f)\, d\mu = (\lambda - \lambda(Y)\pi)(f)$$

for each $f \in C_{\mathbb{R}}(\overline{\partial K})$. Therefore, if

$$N = \{\lambda - \lambda(Y)\pi: \lambda \in M(Y)\},$$

by approximating λ by point measures we have $N = \text{lin}(\overline{\text{co}}\, X)$. If $\mu = \lambda - \lambda(Y)\pi \in N$ then (for each Borel set $B \subseteq \overline{\partial K}$) we have $\mu(B) = \lambda(B \cap Y) - \lambda(Y)\pi(B)$, so that for each Borel set $A \subseteq Y$ we obtain

$$\lambda(A) = \mu(A) + \lambda(Y)\pi(A) = \mu(A) + \pi(A)\mu(Y)(1-\alpha)^{-1}.$$

Hence λ is uniquely determined by μ, and also

$$\|\lambda\| \le |\mu|(Y) + \pi(Y)|\mu|(Y)(1-\alpha)^{-1} \le \|\mu\|\left(1 + \frac{\alpha}{1-\alpha}\right).$$

It follows easily now that N is norm-closed, and hence w^*-closed (1.3.5 since N is w^*-compactly generated) so that $N = L^\perp$.

We can write

$$A(K)^\perp = \Big\{\mu \in M(\overline{\partial K}) \colon \mu(f) = \int_{\overline{\partial K} \backslash \partial K} (\varepsilon_x - \mu_x)(f)\,\mathrm{d}\nu \quad \text{some } \nu \in M(\overline{\partial K}) \qquad \forall f \in C_{\mathbb{R}}(\overline{\partial K})\Big\},$$

where $\{\mu_x\} = M_x^+(\partial K)$ for $x \in K$; in fact $(\varepsilon_x - \mu_x)(f) = 0$ when $f \in A(K)|\overline{\partial K}$, while if $\mu \in A(K)^\perp$ and $f \in C_{\mathbb{R}}(\overline{\partial K})$ the function $h(x) = \mu_x(f)$ satisfies $\int h\, \mathrm{d}\nu = 0$ (cf. Alfsen [**2**] Proposition 2.3.14). Since $A = L \cap A(K)$, $A^\perp$ equals the w^*-closure of $A(K)^\perp + L^\perp$, and

$$A(K)^\perp + L^\perp = \Big\{\mu \in M(\overline{\partial K}) \colon \mu(f) = \int_{\overline{\partial K} \backslash \partial K} (\varepsilon_x - \mu_x)(f)\,\mathrm{d}\nu + (\lambda - \lambda(Y)\pi(f),\ \nu \in M(\overline{\partial K}),\ \lambda \in M(Y),\ \forall f \in C_{\mathbb{R}}(\overline{\partial K})\Big\}; \qquad (*)$$

this set is w^*-closed if and only if it is norm-closed (1.3.5). In the representation of $\mu \in A(K)^\perp + L^\perp$ we may, by transferring an atom if necessary, assume that $\lambda(\{q_0\}) = 0$; hence for any Borel subset C of $Y \backslash \{q_0\}$ we have

$$\mu(C) = \nu(C \cap (\overline{\partial K} \backslash \partial K)) - \int_{\overline{\partial K} \backslash \partial K} \mu_x(C)\,\mathrm{d}\nu + \lambda(C) - \lambda(Y)\pi(C)$$

so that $\mu|(\overline{\partial K} \backslash \partial K) = \nu|(\overline{\partial K} \backslash \partial K)$ and consequently

$$\lambda(C) = \mu(C) + \lambda(Y)\pi(C) + \int_{\overline{\partial K} \backslash \partial K} \mu_x(C)\,\mathrm{d}\mu,$$

and

$$(1-\alpha)\lambda(Y) = \mu(Y) + \int_{\overline{\partial K} \backslash \partial K} \mu_x(Y)\,\mathrm{d}\nu.$$

We obtain the inequality

$$\|\lambda\| \le \|\mu\|(1 + 2\alpha/(1-\alpha)),$$

and as before, it follows that $A(K)^\perp + L^\perp$ is norm-closed.

Let τ denote the quotient map of $M(\overline{\partial K})$ into $M(\overline{\partial K})/A^{\perp}$, the dual space of A. Define, for $\nu \in M(\overline{\partial K})$, $\lambda(\nu) \in M(Y)$ by

$$\lambda(\nu)(\{q_0\}) = 0, \qquad \lambda(\nu)(A) = \int_{\overline{\partial K}\backslash \partial K} \mu_x(A)\, \mathrm{d}\nu + \nu(A) + \pi(A)\lambda(\nu)(Y)$$

for each Borel subset A of $Y\backslash\{q_0\}$, and let $\mu(\nu)$ be as in $(*)$ so that $\tau(\mu(\nu)) = 0$. It is easily verified that $\mu(\nu)(B) = \nu(B)$ for each Borel subset of $(\overline{\partial K}\backslash \partial K) \cup Y$. Therefore

$$|\mu(\nu) - \nu|((\overline{\partial K}\backslash \partial K) \cup Y) = 0, \quad \text{and} \quad \tau(\nu - \mu(\nu)) = \tau(\nu).$$

If

$$\nu_1 \in M(\overline{\partial K}), \qquad \tau(\nu_1) = \tau(\nu), \quad \text{and} \quad |\nu_1|((\overline{\partial K}\backslash \partial K) \cup Y) = 0$$

then

$$\nu_2 = \nu - \mu(\nu) - \nu_1 \in A^{\perp} \cap S, \quad \text{where } S = \{\gamma \in M(\overline{\partial K}) : |\gamma((\overline{\partial K}\backslash \partial K) \cup Y) = 0\}$$

and so there is a $\lambda \in M(Y)$ with

$$\nu_2(B) = \nu_2(B \cap (\overline{\partial K}\backslash \partial K)) - \int_{\overline{\partial K}\backslash \partial K} \mu_x(B)\, \mathrm{d}\nu_2 + \lambda(B \cap Y) - \lambda(Y)\pi(B)$$

for each Borel subset B of $\overline{\partial K}$ (reasoning as above). Since $\nu_2(\overline{\partial K}\backslash \partial K) = 0$ we obtain

$$\nu_2(B) = \lambda(B \cap Y) - \lambda(Y)\pi(B)$$

and putting $B = Y$ we have $\lambda(Y) = 0$; this together with the fact that $\nu_2|Y = 0$ gives $\lambda = 0$ and consequently $\nu_2 = 0$. This argument shows that the map $\phi: A^* \to S$, such that $\phi(\tau(\nu)) = \nu - \mu(\nu)$, is well defined.

Clearly ϕ is linear and onto, and if $\phi(\tau(\nu_1)) = \phi(\tau(\nu_2))$ then $\nu_1 - \mu(\nu_1) = \nu_2 - \mu(\nu_2)$ so that $\tau(\nu_1) = \tau(\nu_2)$.

Let $\nu \geq 0$ belong to $M(\overline{\partial K})$ and let B be a Borel subset of $\overline{\partial K}$. If $B \subseteq (\overline{\partial K}\backslash \partial K) \cup Y$ then $\tau(\nu)(B) = 0$, and if $B \subseteq \partial K\backslash Y$ we have

$$\mu(\nu)(B) = -\int_{\overline{\partial K}\backslash \partial K} \mu_x(B)\, \mathrm{d}\nu - \lambda(\nu)(Y)\pi(B) \leq 0,$$

so that $\phi(\tau(\nu)) \geq 0$.

We have, for $\tau(\nu) \in A^*$,

$$\|\phi(\tau(\nu))\| = \|\nu - \mu(\nu)\| \geq \|\tau(\nu)\| = \inf\{\|\nu^1\| : \tau(\nu^1) = \tau(\nu)\}.$$

For $\nu \in M(\overline{\partial K})$ and B a Borel subset of $\partial K\backslash Y$ we see from the definition of $\lambda(\nu)(Y)$ that

$$\mu(\nu)(B) = -\int_{\overline{\partial K}\backslash \partial K} (\mu_x(B) + \mu_x(Y)\pi(B)(1-\alpha)^{-1})\, \mathrm{d}\nu$$
$$-\nu(Y\backslash q_0)\pi(B)(1-\alpha)^{-1}.$$

Therefore

$$\begin{aligned}\|\phi(\tau(\nu))\| &= |\phi(\tau(\nu))|(\partial K\setminus Y) \leq |\nu|(\partial K\setminus Y) + |\mu(\nu)|(\partial K\setminus Y)\\ &\leq |\nu|(\partial K\setminus Y)\int_{\overline{\partial K}\setminus\partial K}(\mu_x(\partial K\setminus Y)+\mu_x(Y))d|\nu| + |\nu|(Y\setminus q_0)\\ &\leq |\nu|(\partial K\setminus Y)+|\nu|(\overline{\partial K}\setminus\partial K)+|\nu|(Y\setminus q_0) = \|\nu\|,\end{aligned}$$

so that $\|\phi(\tau(\nu))\| \leq \|\nu\|$, and ϕ is an isometry.

We have proved that A^* is isometrically order isomorphic to S. It is evident that S is an L-space and that the extreme points of the state space of S can be identified with $\overline{\partial K}\setminus(\overline{\partial K}\setminus\partial K)\cup Y) = \partial K\setminus Y$.

3.4. COROLLARY. *Let K be a compact metrizable simplex such that whenever $\{I_n\}$ is a sequence of closed ideals in $A(K)$ their intersection is also an ideal. Then K is a Bauer simplex.*

Proof. If K is not a Bauer simplex then there is a $q \in \overline{\partial K}\setminus\partial K$ such that supp (π) has at least two distinct points p_1 and p_2, where $\{\pi\} = M_q^+(\partial K)$. Let $\{x_n\}$ be a sequence in $\partial K\setminus\{p_1, p_2\}$ which converges to q, and let $Y = \{x_n\}\cup\{q\}$. If A is defined as in 3.3 then A is a simplex space $A(H)$ with $\partial H = \partial K\setminus Y$. Since co (p_1, p_2) is a closed face of H we can find $g_1, g_2 \in A(H)$ with $g_i(p_j) = \delta_{ij}$, $i, j = 1, 2$ and $g_i \geq 0$. We must have $g_i(q) > 0$ since $p_1, p_2 \in$ supp (π) and so we can define $f = g_1 - (g_1(q)/g_2(q))g_2 \in A$. Since $f(q) = 0$ and $f \in A$ we have $f(x_n) = 0$ for all n. Put $I_n = \{x_n\}^\perp$ in $A(K)$ so that I_n is a closed ideal and $f \in \bigcap_{n=1}^\infty I_n = I$. Because I is an ideal there exists a $g \in I$ with $g \geq f, 0$, and so $g(q) = \lim_{n\to\infty} g(x_n) = 0$. Therefore p_1 belongs to the closed face $g^{-1}(0)$ of K which contains q, so that $0 = g(p_1) \geq f(p_1) = 1$. This contradiction shows that K is a Bauer simplex.

If K is a compact simplex with $0 \in \partial K$ then $A_0(K) = \{f \in A(K): f(0) = 0\}$ is a partially ordered Banach space with normal, generating cone and such that $A_0(K)^*$ is a vector lattice. The space $A_0(K)$ is also called a simplex space (without a unit) and Theorem 3.1 has a complete analogue for $A_0(K)$.

Notes

Theorem 3.1 is due to Effros [**93**]. Examples of simplex spaces for which the family of closed ideals is not closed under arbitrary intersections have been given by Bunce [**58**] and Perdrizet [**167**]. The example presented here in 3.2 is due to A. Gleit.

Gleit [**116, 117**] proved Theorem 3.3, and also proved Corollary 3.4 for simplex space $A_0(K)$, with K metrizable, showing that $A_0(K)$ is an M-space if and only if the intersection of an arbitrary family of closed ideals in $A_0(K)$ is a closed ideal.

The maximal ideals in $A_0(K)$ are in one-to-one correspondence with the points of $\partial K\backslash\{0\}$; in fact an ideal I is maximal if and only if $I=\{x\}_\perp$ for some $x\in\partial K\backslash\{0\}$. Effros [**93**] developed a hull-kernel topology for the maximal ideal space of $A_0(K)$ and also of $A(K)$ which is essentially the facial topology for $\partial K\backslash\{0\}$.

4. STONE–WEIERSTRASS THEOREMS FOR SIMPLEX SPACES

4.1. THEOREM. *Let L be a simplex space and a closed linear subspace of $A(K)$ containing the constants, with Choquet boundary ∂K. Then for each $x\in\partial K$ the sets*

$$Q(x)=\{y\in K: f(x)=f(y),\ \forall f\in L\}$$

are closed faces of K, and

$$L=\{f\in A(K): f \text{ is constant on } Q(x),\ \forall x\in\partial K\}.$$

Consequently, if L separates the points of ∂K then $L=A(K)$.

Proof. Let $x\in\partial K$ and suppose that $\lambda y+(1-\lambda)z\in Q(x)$ for some y, $z\in K$ and $0<\lambda<1$. If y is not in $Q(x)$ then there is an $f\in L$ such that $f(x)=0$ and $f(y)=1$. Since $\{g\in L: g(x)=0\}$ is a strongly Archimedean order ideal in L; by the hypothesis that $\partial_L K=\partial K$, there exists for each $\varepsilon>0$ a $g\in L$ with $g\geq f$, 0 and $g(x)<\varepsilon$. Hence we have

$$\varepsilon>g(x)=g(\lambda y+(1-\lambda)z)\geq\lambda f(y)=\lambda,$$

and since $\varepsilon>0$ was arbitrary we obtain a contradiction; so $Q(x)$ is a closed face of K.

Let $f\in A(K)$ be constant on $Q(x)$ for all x in ∂K, and define $\hat{f}(y)=\inf\{g(y): g\in L, g\geq f\}$ for each $y\in K$. We evidently have $\hat{f}\geq f$ and also for $x\in K$

$$\hat{f}(x)=\sup\{\mu(f): \mu\in P_x\},$$

(by 1.6.2) where P_x is the set of probability measures on K with L-resultant x.

Therefore if $x\in\partial K$ and $\mu\in P_x$ we have supp μ contained in the face $Q(x)$, so that $\hat{f}(x)=f(x)$.

Let F denote the family

$$\{g\in L: g>f\},$$

let $g_1, g_2\in F$ and write u for the pointwise minimum of g_1 and g_2. Since L is a simplex space the family

$$F_1=\{v\in L: v<u\}$$

is directed upwards with supremum $\check{u}$ such that $\check{u}(x) = u(x)$ for all $x \in \partial K$, by arguing as above. Therefore $\check{u} - f$ is lower semi-continuous and affine on K and so attains its infimum on ∂K, and hence $u \geq \check{u} > f$ on K. Consequently for each $x \in K$ there is a $v_x \in F_1$ with $v_x(x) > f(x)$. A simple compactness argument together with the fact that L has the Riesz separation property now shows that there is a $v \in L$ with $u > v > f$ on K, and hence F is directed downwards.

It follows that $\hat{f}$ is affine and upper semi-continuous and hence that $\hat{f} = f$ on K. Therefore f is a uniform limit of a sequence of functions in F, so that $f \in L$.

Finally, if L separates the points of ∂K then, since

$$\partial Q(x) = Q(x) \cap \partial K = \{x\},$$

for each $x \in \partial K$, we obtain $L = A(K)$.

The classical Stone–Weierstrass Theorem corresponds to the special case of Theorem 4.1 where K is a Bauer simplex and L is a closed sublattice of $A(K)$ containing the constants; in the case $L \cap x^{-1}(0)$ is a lattice ideal in L for each $x \in \partial K$, and hence the Choquet boundary $\partial_L K$ of L equals ∂K. Simple finite-dimensional examples show that the conditions that L be a simplex space and that $\partial_L K = \partial K$ cannot be dropped from the hypotheses of Theorem 4.1.

The next result is a generalization of the separable case of the Stone–Weierstrass Theorem, in the same spirit as Theorem 4.1.

4.2. THEOREM. *Let $1 \in L \subseteq M$, where L and M are closed linear subspaces of $C_{\mathbb{R}}(X)$ which separate points of the compact metric space X, and such that $\partial_L X = \partial_M X$. Then $L = M$ if and only if whenever $l_r < f < l_s$, $r = 1, \ldots, n$, $s = 1, \ldots, m$, where $l_j \in L$ and $f \in M$, there exists an $l \in L$ with $l_r < l < l_s$.*

Proof. Note that L can be identified with a closed linear subspace of $M = A(K)$, separating the points of ∂K and containing the constants, and that $\partial K = \partial_M X = \partial_L K$. Therefore, if $f \in A(K)$ and if $\hat{f}$ and $\check{f}$ are defined as in the proof of 4.1, we have $\hat{f}(x) = \check{f}(x) = f(x)$ for each $x \in \partial K$. Since X is metrizable there exists a dense sequence $\{l_m\}$ in L, and hence we can choose increasing sequences $\{g_n\}$ and $\{-h_n\}$ of functions which are pointwise suprema of finitely many of the l_m, and such that $g_n < f < h_n$ and $g_n(x) \to \check{f}(x)$, $h_n(x) \to \hat{f}(x)$ for all $x \in X$. The hypothesis implies that there exist functions $k_n \in L$ with $g_n < k_n < h_n$, and hence $k_n(x) \to f(x)$ boundedly for each $x \in \partial K$. Choquet's Theorem and the Dominated Convergence Theorem now show that f belongs to the closed subspace L.

Notes

Theorem 4.1 is due to Edwards and Vincent-Smith [**91**]. Generalizations to the case of simplex spaces without constants have been given by Edwards [**89**]. Applications of Theorem 4.1 to uniform approximation of harmonic functions have been given by Vincent-Smith [**212**]. For example, let D be an open subset of $\mathbb{R}^n$ with closure $\bar{D}$ and boundary D^*, and let $M = \{f \in C_{\mathbb{R}}(\bar{D}) : f$ is harmonic in $D\}$, $L = \{f \in C_{\mathbb{R}}(\bar{D}) : f$ extends to a function harmonic in a neighbourhood U_f of $\bar{D}\}$. Then M can be identified with a space $A(K)$, K a simplex, and ∂K identifies with the set of regular points of D^*. The Choquet boundary for L coincides with the set of stable points of D^*. In this set up Theorem 4.1 gives the result of Deny [**77**] that L is uniformly dense in M if and only if every regular point of D^* is stable. Further applications in potential theory have been given by Vincent-Smith [**212**] and de la Pradelle [**174, 175**].

Theorem 4.2 was proved by Alfsen [**4**].

5. PRIME SIMPLEXES AND ANTILATTICES

A compact convex set K is said to be *prime* if whenever F and G are (closed) semi-exposed faces of K such that $\mathrm{co}\,(F \cup G) = K$, then either $F = K$ or $G = K$. $A(K)$ is called an *antilattice* if whenever f and g belong to $A(K)$ and $f \wedge g$ exists in $A(K)$ then either $f \leq g$ or $g \leq f$.

5.1. THEOREM. *A compact convex set K is prime if and only if $A(K)$ is an antilattice.*

Proof. Suppose first that $A(K)$ is an antilattice and that $K = \mathrm{co}\,(F \cup G)$, where F and G are proper semi-exposed faces of K. Then there exist $f, g \geq 0$ in $A(K)$, $f, g \neq 0$, such that $f(x) = 0 = g(y)$ for all $x \in F$ and $y \in G$. It is evident that $0 = f \wedge g$ in $A(K)$, so that either $f \leq g$ or $g \leq f$. But then either $f = 0$ or $g = 0$, which gives a contradiction. Hence K is prime.

Conversely, suppose that K is prime and that $f, g \in A(K)$ such that $f \wedge g$ exists in $A(K)$. Without loss of generality we can take $g = 0$, and we note that if $h(x) = \min\{f(x), 0\}$ for $x \in K$, then $(f \wedge 0)(x) = h(x)$ for each $x \in \partial K$ (cf. 1.6.1(ix)). Therefore if $F = \{x \in K : (f \wedge 0)(x) = f(x)\}$ and $G = \{x \in K : (f \wedge 0) = 0\}$ then F and G are exposed faces of K such that $\partial K \subseteq F \cup G$. Hence $K = \mathrm{co}\,(F \cup G)$ so that either $F = K$ or $G = K$, that is either $f \wedge 0 = f$ or $f \wedge 0 = 0$. Thus either $f \leq 0$ or $f \geq 0$ so that $A(K)$ is an antilattice.

5.2. COROLLARY. *If K is a compact convex set such that $\overline{\partial K} = K$ then K is prime.*

Proof. Let $f, g \in A(K)$ such that $f \wedge g$ exists in $A(K)$. Then $(f \wedge g)(x) = f(x) \wedge g(x)$ for each $x \in \partial K$, and hence for each $x \in \overline{\partial K} = K$ by continuity. Since $f \wedge g$ is affine this implies that either $f \leq g$ or $g \leq f$, so that $A(K)$ is an antilattice and K is prime.

An important example of a metrizable simplex K with $K = \overline{\partial K}$ will be given in Section 7. We now give an example of a naturally occurring prime simplex S with $\overline{\partial S} \neq S$. By the *state space* S of A we mean $S = \{\phi \in A^*: \phi(1) = \|\phi\| = 1\}$ with the relative w^*-topology.

5.3. EXAMPLE. *The state space S of the function algebra $A = \{f \in C(\Delta): f(0) = f(1),\ f$ analytic on $\Delta^0\}$, where $\Delta = \{z \in \mathbb{C}: |z| \leq 1\}$, is a prime simplex.*

Proof. The extreme points of S coincide with the peak points for A (cf. [**114**]), and these are the points of $\Gamma \backslash \{1\}$, where $\Gamma = \{z \in \mathbb{C}: |z| = 1\}$. The space $A(S)$ can be realized, when restricted to Γ as

$$\left\{f \in C_{\mathbb{R}}(\Gamma): f(1) = \frac{1}{2\pi} \int_0^{2\pi} f(e^{i\theta})\, d\theta\right\}.$$

Evidently S is a simplex. Suppose that $f \in A(S)$ and that $f \wedge 0$ exists in $A(S)$; then $(f \wedge 0)(z) = f(z) \wedge 0$ for all $z \in \Gamma$. Now $A(S)$ can be identified with the uniform closure of re A (see Chapter 4) and hence $g = f \wedge 0$ has a continuous extension to Δ which is harmonic in Δ^0. In particular we have

$$\begin{aligned} f(1) \wedge 0 &= g(1) = g(0) \\ &= \frac{1}{2\pi} \int_0^{2\pi} g(e^{i\theta})\, d\theta = \frac{1}{2\pi} \int_0^{2\pi} (f(e^{i\theta}) \wedge 0)\, d\theta \\ &= f(0) \wedge 0 = \left(\frac{1}{2\pi} \int_0^{2\pi} f(e^{i\theta})\, d\theta\right) \wedge 0. \end{aligned}$$

If $f(0) \leq 0$ we have $\int_0^{2\pi} (f(e^{i\theta}) - f(e^{i\theta}) \wedge 0)\, d\theta = 0$, which implies that $f = g$, that is $f \leq 0$. If $f(0) \geq 0$ then $\int_0^{2\pi} (f(e^{i\theta}) \wedge 0)\, d\theta = 0$, which implies that $f \geq 0$. Therefore $A(S)$ is an antilattice and S is a prime simplex.

Notes

The notion of prime simplex K was introduced by Effros and Kazdan [**98**], and they showed that K is a prime simplex if and only if $A(K)$ is an antilattice. Amongst many other interesting results they prove that if $A(K)$ represents the solutions of Laplace's equation for a bounded region in $\mathbb{R}^n$ then K is either a Bauer simplex or a prime simplex.

The definition of a prime set for non-simplexes, and Theorem 5.1, is due to Chu [**65**]. The application of these results to C^*-algebras will be discussed in Chapter 5. If, in the definition of a prime set K, we replace the semi-exposed faces F and G by closed split faces we obtain the notion of a weakly prime set; for example a plane rectangle is weakly prime but not prime. The relevance of this notion to function algebras and C^*-algebras is studied in Ellis [**107**].

6. TOPOLOGICAL CHARACTERIZATION OF THE EXTREME BOUNDARY OF A COMPACT SIMPLEX

Bauer simplexes are, by 2.7.5, in one-to-one correspondence with compact Hausdorff spaces. The extreme points ∂K of a compact simplex K, endowed with the relative w^*-topology, is a much more general topological space. The precise classification of topological spaces which have the form ∂K with K a compact metrizable simplex is given below. We recall that a separable complete metrizable space is called a *Polish space.*

6.1. THEOREM. *A topological space X is homeomorphic to ∂K, where K is a compact metrizable simplex, if and only if X is a Polish space.*

Proof. If X is homeomorphic to ∂K, where K is a compact metrizable convex set, then since ∂K is a G_δ-subset of the Polish space K (1.6.6) X must be Polish (see [**48**, 9, 6.1]).

Conversely, let X be a Polish space and let (T, d) be a compact metric space in which X can be embedded topologically as a dense G_δ-subset (cf. [**48**, 6.1]). Hence we can write $X = \bigcap_{n=0}^{\infty} G_n$, where the sets G_n are open and decreasing, with $G_0 = T$.

Let U be an open subset of T and for each positive integer n put

$$W_n = \{x \in T\colon d(T\backslash U, x) \geq 1/n\}.$$

Choose open subsets $U_n^1, \ldots, U_n^{k_n}$ of $\{x\colon d(T\backslash U, x) > 1/(n+1)\}$ covering W_n, and let $\{V_k\}$ be the sequence of all sets U_n^j, for all n. Then $\{V_k\}$ covers U and so there exists a partition of unity $\{g_k\} \in C_{\mathbb{R}}(U)$, subordinate to $\{V_k\}$ (cf. [**48**, 4.3]). Since each V_k has a positive distance from U there exists an $h_k \in C_{\mathbb{R}}(T)$ with $h_k = 0$ on $T\backslash U$ and $h_k = g_k$ on U. Therefore we have $\sum_{k=1}^{\infty} h_k = \chi_U$.

Applying this result to each G_n in place of U we obtain functions $h_k^n \in C_{\mathbb{R}}^+(T)$ and constant $\beta_k^n \in (0, 1)$, with

$$\beta_k^n \to 0 \quad \text{as } k \to \infty, \qquad \sum_{k=0}^{\infty} h_k^n = \chi_{G_n}$$

and

$$\operatorname{diam}(\operatorname{supp} h_k^n) \leq 2^{-n}\beta_k^n$$

for each n. Putting

$$p_k^n = h_k^n \chi_{T\setminus G_{n+1}}$$

we obtain

$$\sum_{n=0}^{\infty}\left\{\sum_{k=0}^{\infty} p_k^n\right\} = \chi_{T\setminus X}. \qquad (*)$$

If h_k^n is not identically 0 on $T\setminus X$ then it is different from 0 on an open subset of $T\setminus X$ and since X is dense in T, the set $X \cap \operatorname{supp} h_k^n$ has infinitely many points. We can therefore choose for each n and k with $p_k^n \neq 0$ distinct points x_k^n and y_k^n on $X \cap \operatorname{supp} h_k^n$. Now define $\gamma: M(T) \to M(T)$ by

$$\gamma(\mu) = \mu - \tfrac{1}{2}\sum_{n=0}^{\infty}\sum_{k=0}^{\infty}\left(\left(\int p_k^n \, d\mu\right)(\varepsilon_{x_k^n} + \varepsilon_{y_k^n})\right),$$

and let

$$M = \{\gamma(\mu): \mu \in M(T\setminus X)\}.$$

The next step is to prove that M is w^*-closed in $M(T)$ or equivalently that the unit ball M_1 is w^*-compact. If $\mu \in M(T\setminus X)$ then we have

$$\|\gamma(\mu)\| \geq \|\mu\|$$

because $x_k^n, y_k^n \in X$, and so if $\{\sigma_i\}$ is a sequence in M_1 we can find

$$\{\mu_i\} \quad \text{in} \quad M(T\setminus X)_1$$

such that $\sigma_i = \gamma(\mu_i)$. For each n we can, by w^*-compactness of $M(T\setminus G_{n+1})_1$, find a w^*-convergent subsequence of $\{\mu_i|_{T\setminus G_{n+1}}\}$, and hence a diagonal argument gives a subsequence $\{\nu_i\}$ of $\{\mu_i\}$ such that $\{\nu_i|_{T\setminus G_{n+1}}\}$ is, for each n, w^*-convergent to some $\nu^n \in M(T\setminus G_{n+1})$. Defining

$$\nu = \sum_{n=0}^{\infty} \nu^n|_{(G_n\setminus G_{n+1})} = \sum_{n=0}^{\infty}\sum_{k=0}^{\infty} p_k^n \nu^n$$

we see that $\|\nu\| \leq 1$; in fact if $f \in C_{\mathbb{R}}(T)$, $\|f\| \leq 1$, then

$$\left|\sum_{n=0}^{\infty}\sum_{k=0}^{\infty}\int p_k^n f \, d\nu_i\right| \leq \int \sum_{n=0}^{\infty}\sum_{k=0}^{\infty} p_k^n \, d|\nu_i| = \|\nu_i\| \leq 1,$$

and since

$$p_k^n \nu_i = h_k^n(\nu_i|_{T\setminus G_{n+1}}) \to h_k^n \nu^n = p_k^n \nu^n$$

in the w^*-topology it follows that $|\nu(f)| \leq 1$. If we can show that $\gamma(\nu_i) \to \gamma(\nu)$ for the w^*-topology then M_1 will be w^*-sequentially compact and hence

w^*-compact. For $f \in C_{\mathbb{R}}(T)$, $\mu \in M(T\backslash X)$, we have

$$\int f \,\mathrm{d}\gamma(\mu) = \int f \,\mathrm{d}\mu - \tfrac{1}{2} \sum_{n=0}^{\infty} \sum_{k=0}^{\infty} \left(\int p_k^n \,\mathrm{d}\mu \right) (f(x_k^n) + f(y_k^n)) = \sum_{n=0}^{\infty} \sum_{k=0}^{\infty} \alpha_k^n(\mu),$$

where

$$\alpha_k^n(\mu) = \int p_k^n (f - \tfrac{1}{2}[f(x_k^n) + f(y_k^n)]) \,\mathrm{d}\mu = \int [f - \tfrac{1}{2}(f(x_k^n) + f(y_k^n)] \,\mathrm{d}p_k^n \mu).$$

Since $p_k^n \nu_i \to p_k^n \nu$ for the w^*-topology we have $\alpha_k^n(\nu_i) \to \alpha_k^n(\nu)$ for all n, k. Given f and $\varepsilon > 0$ there exists a $\delta > 0$ such that $|f(s) - f(t)| < \varepsilon$ whenever $d(s, t) < \delta$. Choose N satisfying $2^{-N} < \delta$, and for each $n \leq N$ choose $K(n)$ such that $\beta_k^n < \delta$ for $k > K(n)$. If $n > N$ then diam (supp $h_k^n) < 2^{-N}$ so that $d(x_k^n, t)$, $d(y_k^n, t) < \delta$ whenever $p_k^n(t) \neq 0$; if $k > K(n)$ and $n \leq N$ then the same inequalities hold. For such values of n and k we have $|\alpha_k^n(\mu)| \leq \varepsilon \int p_k^n \,\mathrm{d}|\mu|$, and therefore

$$\left| \int f \,\mathrm{d}\gamma(\mu) - \sum_{n=0}^{N} \sum_{k=0}^{K(n)} \alpha_k^n(\mu) \right| \leq \varepsilon \int \left(\sum_{n=0}^{\infty} \sum_{k=0}^{\infty} p_k^n \right) \mathrm{d}|\mu| = \varepsilon \|\mu\|.$$

Therefore the series $\sum_{n=0}^{\infty} \sum_{k=0}^{\infty} \alpha_k^n(\mu)$ converges uniformly on $M(T\backslash X)_1$ to $\int f \,\mathrm{d}\gamma(\mu)$, and since $\alpha_k^n(\nu_i) \to \alpha_k^n(\nu)$ for all n and k we obtain $\int f \,\mathrm{d}\gamma(\nu_i) \to \int f \,\mathrm{d}\gamma(\nu)$. This completes the proof that M is w^*-closed in $M(T)$.

Let $\phi : M_1^+(T) \to M(T)/M$ be the quotient map and let $K = \phi(M_1^+(T))$. If $\mu \in M_1^+(T)$ satisfies $\phi(\mu) = \phi(\varepsilon_x)$, for some $x \in X$, then $\mu = \varepsilon_x \in M$, and so if $\mu \neq \varepsilon_x$ we can find $\lambda \in M_1^+(T)$ with $\lambda - \varepsilon_x \in M$ and $\lambda(\{x\}) = 0$. However, if $\mu^1 \in M(T\backslash X)$ with

$$\lambda - \varepsilon_x = \gamma(\mu^1) = \mu^1 - \tfrac{1}{2} \sum_{n=0}^{\infty} \sum_{k=0}^{\infty} \left(\left(\int p_k^n \,\mathrm{d}\mu \right) (\varepsilon_{x_k^n} + \varepsilon_{y_k^n}) \right)$$

then $\lambda - \varepsilon_x$ can have at most $\frac{1}{4}$ of the mass at any single point of X. This contradiction shows that the map $\phi \circ \Psi : X \to K$, where $\Psi(x) = \varepsilon_x$ for $x \in X$, is a continuous injection. Moreover, if E is closed in X then $E = S \cap X$ for some compact set $S \subseteq T$ so that $\phi \circ \Psi(E) = \phi \circ \Psi(S) \cap \phi \circ \Psi(X)$, that is $\phi \circ \Psi(E)$ is closed in $\phi \circ \Psi(X)$. Therefore X is homeomorphic to $\phi \circ \Psi(X)$.

We finally show that $\partial K = \phi \circ \Psi(X)$ and that K is a simplex. If $\eta \in \partial K$ then $\phi^{-1}(\eta)$ is a closed face of $M_1^+(T)$ and hence intersects ext $M_1^+(T)$, so that ∂K is contained in $\phi \circ \Psi(T)$. However if $t \in T\backslash X$ then $t \in G_n \backslash G_{n+1}$ for some n so that

$$\gamma(\varepsilon_t) = \varepsilon_t - \tfrac{1}{2} \sum_{n=0}^{\infty} \sum_{k=0}^{\infty} \left(\left(\int p_k^n \,\mathrm{d}\varepsilon_t \right) (\varepsilon_{x_k^n} + \varepsilon_{y_k^n}) \right) = \varepsilon_t - \lambda \in M,$$

where

$$\lambda = \tfrac{1}{2} \sum_{k=0}^{\infty} p_k^n(t) (\varepsilon_{x_k^n} + \varepsilon_{y_k^n})$$

so that $\lambda \in M_1^+(T)$. Since

$$\phi \circ \Psi(t) = \phi(\lambda) = \tfrac{1}{2} \sum_{k=0}^{\infty} p_k^n \phi(\varepsilon_{x_k^n}) + \tfrac{1}{2} \sum_{k=0}^{\infty} p_k^n \phi(\varepsilon_{y_k^n}),$$

$\phi(\lambda)$ can only belong to ∂K if

$$\sum_{k=0}^{\infty} p_k^n(t)(\varepsilon_{x_k^n} - \varepsilon_{y_k^n}) \in M,$$

which is impossible since this measure has no mass in $T \backslash X$. Therefore it follows that $\partial K \subseteq \phi \circ \Psi(X)$.

Let

$$\mu \in A(X)^{\perp} \cap M(\phi \circ \Psi(X)),$$

and put

$$\nu = \mu \circ \phi \circ \Psi \in M(X).$$

Since ν is concentrated on X it will follow that $\nu = 0$, and hence that K is a simplex, if we show that $\nu \in M = (M_0)^0$. If $g \in M_0$ then the function $h : K \to \mathbb{R}$, given by $h(k) = \int_T g \, \mathrm{d}(\phi^{-1}(k))$, is well defined and belongs to $A(K)$. Therefore we have

$$0 = \int_K h \, \mathrm{d}\mu = \int_{\phi \circ \Psi(X)} h(\phi \circ \Psi(x)) \, \mathrm{d}\mu(\phi \circ \Psi(x)) = \int_X g(x) \, \mathrm{d}\nu(x)$$

and so $\nu \in M$. Finally, if $\phi \circ \Psi(x) = w$ does not belong to ∂K, for some $x \in X$, then there is a $\mu^1 \in \partial M_1^+(K)$ with resultant w so that

$$\varepsilon_w - \mu^1 \in A(X)^{\perp} \cap M(\phi \circ \Psi(X)).$$

But then $\varepsilon_w = \mu^1$ by the preceding argument, which gives a contradiction. Therefore we obtain $\partial K = \phi \circ \Psi(X)$.

Notes

The topological classification of ∂K, where K is a compact metrizable simplex is given by Choquet in [**62**, p. 183]. The proof given here is due to Haydon [**124**]. Some of the techniques used in the proof of Theorem 6.1 have been applied elsewhere. For example, Lazar [**148**] has shown that for any uncountable Polish space X there exists a bounded closed convex body K in l^2 with ∂K homeomorphic to X.

Andenaes [**15**] has extended the result of 6.1 by showing that every complete, metrizable locally separable space is homeomorphic to ∂K for some compact simplex K. A characterization of ∂K, where K is a standard simplex, has been given by Stacey [**199**].

7. POULSEN'S SIMPLEX—ITS UNIQUENESS AND UNIVERSALITY

The following important example was given by Poulsen [**173**], and is called *Poulsen's simplex.*

7.1. THEOREM. *There exists a compact simplex K in l^2 with $\overline{\partial K}=K$.*

Proof. We construct simplexes S_n in Euclidean n-space $\mathbb{R}^n$, embedded in l^2 in the natural way.

Let

$$S_1=[0,\tfrac{1}{2}]$$

and let

$$S_2=\mathrm{co}\,(S_1\cup\{(0,\tfrac{1}{2})\})\subseteq\mathbb{R}^2.$$

Next choose points $y_3,\ldots,y_k\in S_2$ forming a 2^{-2}-net for S_2, and let

$$z_l=y_l+2^{-l}e_l\in\mathbb{R}^l,\qquad 3\leq l\leq k,$$

where e_l denotes the lth coordinate vector in $\mathbb{R}^l$; define

$$S_l=\mathrm{co}\,(S_2\cup\{z_3,\ldots,z_l\})\subseteq\mathbb{R}^l\quad\text{for } 3\leq l\leq k.$$

Now choose a 2^{-k}-net for S_k and define $S_{k+1},\ldots$ by the same procedure. Evidently each S_n is an n-dimensional simplex and $P_nS_m=S_n$ for $m\geq n$, where P_n is the projection of l^2 onto $\mathbb{R}^n$. If we define

$$K=\mathrm{cl}\left(\bigcup_{n=1}^{\infty}S_n\right)$$

then the construction gives $\overline{\partial K}=K$, and K is compact in l^2.

For each $x\in K$, $P_nx\to x$ and hence $f\circ P_n\to f$ weakly for each $f\in A(K)$. Consequently we have

$$A(K)=\overline{\bigcup_{n=1}^{\infty}A_n}$$

where

$$A_n=\{f\circ P_n: f\in A(K)\}.$$

Therefore to show that $A(K)^*$ is a Banach lattice, and hence that K is a simplex (see 2.5.4, 5.5) it is sufficient to show that $\bigcup_{n=1}^{\infty}A_n$ has the Riesz Separation Property, and that is clearly the case because each S_n is a simplex.

Recently some remarkable results concerning the uniqueness and universality of Poulsen's simplex have been proved. To present some of these results we need first to consider representing matrices for simplex spaces $A(K)$.

7.2. THEOREM. *If K is a compact metrizable simplex there exists an increasing sequence $\{E_n\}$ of finite-dimensional subspaces of $A(K)$ such that, for each n, E_n is isometrically isomorphic to some $l^{\infty}_{m_n}$ space, and with $A(K)=\text{cl}\bigcup_{n=1}^{\infty} E_n$.*

Proof. It will be sufficient to show that, given a $g\in A(K)^+$ and a linear subspace L of $A(K)$ which contains 1 and is isometrically isomorphic to a space l^{∞}_n, there exists for any $\varepsilon>0$ a linear subspace E of $A(K)$ containing L with $d(g, E)<\varepsilon$ and such that E is isometrically isomorphic to a space l^{∞}_m. Indeed if such a process can be carried out we can inductively take g to be a member of a countable dense set in the positive part of the unit ball of $A(K)$ and a sequence of ε's converging to 0, obtaining an increasing sequence $\{E_n\}$ of subspaces of $A(K)$ with $E_1=\text{lin}\{1\}$ and $d(g_n, E_n)<\varepsilon_n$ and such that E_n is isometrically isomorphic to a space $l^{\infty}_{m_n}$. Clearly we will then have $\text{cl}\bigcup_{n=1}^{\infty} E_n=A(K)$.

Suppose then that L, g and $\varepsilon>0$ are given, with $\|g\|=1$. Since L is isometrically isomorphic to l^{∞}_n and 1 is an extreme point of the unit ball S of L with

$$S+1=\{f\in A(K)^+: \|f\|\le 2\},$$

it is easy to see that there exists a basis $f_1,\ldots,f_n$ for L with $f_j\ge 0$ for each j and $f_1+\cdots+f_n=1$. Define

$$T:K\to\mathbb{R}^{n+1} \quad \text{by } Tx=(g(x), f_1(x),\ldots,f_n(x)),$$

and let W be the range of T. If q denotes the projection of $\mathbb{R}^{n+1}$ onto the last n coordinates, then

$$qW=\left\{\mathbf{t}\in\mathbb{R}^n: t_j\ge 0, \sum_{j=1}^{n} t_j=1\right\}.$$

Each $\mathbf{t}\in\partial(qW)$ is of the form $\mathbf{t}=q\mathbf{s}$ for some $\mathbf{s}\in\partial W$ and we can choose a finite collection $\mathbf{s}_1,\ldots,\mathbf{s}_m\in\partial W$ such that $qW=qW'$, where $W'=\text{co}\{\mathbf{s}_1,\ldots,\mathbf{s}_m\}$, and such that for each $(\alpha,\mathbf{y})\in W$ there is a point $(\alpha',\mathbf{y})\in W'$ with $|\alpha-\alpha'|<\varepsilon$. (In fact we can decompose qW into finitely many non-overlapping polyhedra $P_1,\ldots,P_r$ such that h_1 and h_2 have oscillations on each P_j less than $\varepsilon/2$, where $h_1, h_2: qW\to\mathbb{R}$ are defined by

$$h_1(\mathbf{y})=\sup\{\alpha: (\alpha,\mathbf{y})\in W\}, \qquad h_2(\mathbf{y})=\inf\{\alpha: (\alpha,\mathbf{y})\in W\}.$$

Let E be the finite set of points in W of the form $(h_i(\mathbf{y}),\mathbf{y})$, $i=1,2$, $y\in\partial P_j$, $j=1,\ldots,r$. If E' is a finite subset of ∂W such that $\text{co}\,E'\supseteq\text{co}\,E$ then we can define $W'=\text{co}\,E'$.)

Let $\lambda_1,\ldots,\lambda_m: W'\to[0,1]$ such that

$$\mathbf{w}=\sum_{j=1}^{m}\lambda_j(\mathbf{w})\mathbf{s}_j$$

and

$$\sum_{j=1}^{m} \lambda_j(\mathbf{w}) = 1$$

for each $w \in W'$, while $\lambda_j(\mathbf{s}_j) = 1$, $1 \le j \le m$. For each $w \in W \backslash W'$ define $\lambda_j(\mathbf{w}) = \lambda_j(\mathbf{w}')$ for some $\mathbf{w}' \in W'$ such that $q\mathbf{w}' = q\mathbf{w}$ and $|p(\mathbf{w}') - p(\mathbf{w})| < \varepsilon$, where p denotes the projection of $\mathbb{R}^{n+1}$ onto the first coordinate. For each $x \in K$ we will have

$$\sum_{j=1}^{m} \lambda_j(Tx) = 1, \qquad f_i(x) = \sum_{j=1}^{m} \lambda_j(Tx) s_j^{i+1}, \qquad 1 \le i \le n$$

and

$$\left| g(x) - \sum_{j=1}^{m} \lambda_j(Tx) s_j^1 \right| < \varepsilon.$$

We will show, in the lemma below, that these facts imply the existence of functions

$$l_1, \ldots, l_m \in A(K)^+$$

such that

$$\sum_{j=1}^{m} l_j = 1, \qquad f_i = \sum_{j=1}^{m} l_j s_j^{i+1}, \qquad 1 \le i \le n, \qquad \|l_j\| = 1, \qquad 1 \le j \le m$$

and

$$\left\| g - \sum_{j=1}^{m} l_j s_j^1 \right\| < \varepsilon.$$

If L is the linear span of $\{l_j\}$ then it is easy to see that L will satisfy the requirements. The proof of 7.2 will therefore be completed by applying the following lemma, with $u_j = \lambda_j \circ T$ and $x_j \in \partial K$ such that $Tx_j = \mathbf{s}_j$.

7.3. LEMMA. *Let $f_i, g_i \in A(K)$, $\alpha_{ij} \in \mathbb{R}$ and $x_r \in \partial K$ for $1 \le i \le k$, $1 \le j \le m$, $1 \le r \le s$ and let $u_j : K \to \mathbb{R}$ be functions such that $g_i \le \sum_{j=1}^{m} \alpha_{ij} u_j \le f_i$ for each i. Then there exist functions $l_j \in A(K)$ such that*

$$g_i \le \sum_{j=1}^{m} \alpha_{ij} l_j \le f_i \quad \text{and} \quad l_j(x_r) = u_j(x_r) \quad \text{for each } j \text{ and } r.$$

Proof. If $m = 1$ we may as well assume that each $\alpha_{ij} = 1$. Since $F = \mathrm{co}\,(\{x_r\colon 1 \le r \le s\})$ is a closed face of K we can find an $h \in A(F)$ with $h(x_r) = u_1(x_r)$, $1 \le r \le s$, and hence the result follows in this case from the Edwards Separation Theorem (2.7.6).

Now suppose that $m>1$ and that the lemma is true with $m-1$ replacing m. We can clearly assume that each α_{im} is either 0 or 1. If $\alpha_{im}=1$ we have

$$g_i-\sum_{j=1}^{m-1}\alpha_{ij}u_j\le u_m\le f_i-\sum_{j=1}^{m-1}\alpha_{ij}u_j,$$

and hence, if $\alpha_{i'm}=1$ also, we have

$$g_i-f_{i'}\le\sum_{j=1}^{m-1}(\alpha_{ij}-\alpha_{i'j})u_j\le f_i-g_{i'};$$

if $\alpha_{i_0m}=0$ we have

$$g_{i_0}\le\sum_{j=1}^{m-1}\alpha_{i_0j}u_j\le f_{i_0}.$$

The induction hypothesis now gives the existence of $l_1,\ldots,l_{m-1}\in A(K)$ satisfying

$$g_i-f_{i'}\le\sum_{j=1}^{m-1}(\alpha_{ij}-\alpha_{i'j})l_j\le f_i-g_{i'},$$

and

$$g_{i_0}\le\sum_{j=1}^{m-1}\alpha_{i_0j}l_j\le f_{i_0},\qquad l_j(x_r)=u_j(x_r),\qquad 1\le j\le m-1,\qquad 1\le r\le s.$$

We can apply Edwards Separation Theorem again to obtain an $l_m\in A(K)$ such that, for each i and i' with $\alpha_{im}=\alpha_{i'm}=1$ we have

$$g_i-\sum_{j=1}^{m-1}\alpha_{ij}l_j\le l_m\le f_{i'}-\sum_{j=1}^{m}\alpha_{i'j}l_j.$$

while $l_m(x_r)=u_m(x_r)$ for $1\le r\le s$. The functions $l_1,\ldots,l_m$ now satisfy the requirements of the lemma.

In the course of the proof of the last theorem we constructed positive isometries $T:l_n^\infty\to l_m^\infty$ with $T(1,1,\ldots,1)=(1,1,\ldots,1)$. It is easy to see that we can re-order the coordinate basis of l_m^∞ to obtain a basis $\boldsymbol{\nu}_1,\ldots,\boldsymbol{\nu}_m$ such that, if $\mathbf{u}_1,\ldots,\mathbf{u}_n$ is the coordinate basis for l_n^∞, we have

$$T\mathbf{u}_j=\boldsymbol{\nu}_j+\sum_{i=n+1}^{m}a_{ji}\boldsymbol{\nu}_i,\qquad 1\le j\le n,$$

where $a_{ji}\ge0$ and $\sum_{j=1}^{n}a_{ji}=1$ for each $i=n+1,\ldots,m$. If $m>n+1$ then we can define, for $1\le j\le n$,

$$\mathbf{w}_j=\boldsymbol{\nu}_j+\sum_{i=n+2}^{m}a_{ji}\boldsymbol{\nu}_i,$$

and

$$\mathbf{w}_{n+1}=\boldsymbol{\nu}_{n+1}.$$

It is easy to check that lin $\{\mathbf{w}_1, \ldots, \mathbf{w}_{n+1}\}$ is isometrically isomorphic to l^∞_{n+1} and contains l^∞_n. Therefore in Theorem 7.2 we can actually obtain spaces E_n isometrically isomorphic to l^∞_n. Moreover the isometric embedding T of l^∞_n in l^∞_{n+1} takes the form

$$T\mathbf{u}_j = \boldsymbol{\nu}_j + a_j \boldsymbol{\nu}_{n+1}, \qquad \text{where } a_j \geq 0, \quad 1 \leq j \leq n, \quad \text{and} \quad \sum_{j=1}^{n} a_j = 1.$$

We call a finite set $\{f_1, \ldots, f_n\}$ in $A(K)^+$ a *partition of unity* for $A(K)$ if $\sum_{i=1}^{n} f_i(x) = 1$, for all $x \in K$; if, in addition, $\|f_i\| = 1$ for each i then we refer to a *peaked partition of unity*.

7.4. LEMMA. *Let K be a compact metrizable simplex, F a proper closed face of K and $\{e_1, \ldots, e_n\}$ a peaked partition of unity for $A(K)$.*

(i) *If $\|e_i|_F\| = 1$, $i = 1, \ldots, n$, there exists an affine continuous projection Q from K onto F such that $e_j = e_j \circ Q$ for $j = 1, \ldots, n$.*

(ii) *Let $\{f_1, \ldots, f_{n+1}\}$ be a partition of unity for $A(F)$ and let $a_1, \ldots, a_n \in \mathbb{R}^+$ such that $\sum_{i=1}^{n} a_i = 1$ and $e_i|_F = f_i + a_i f_{n+1}$, $1 \leq i \leq n$. Then if $\overline{\partial K} = K$, and if $\varepsilon > 0$, there exists a peaked partition of unity $\{g_1, \ldots, g_{n+1}\}$ for $A(K)$ such that $g_i|_F = f_i$ for $i = 1, \ldots, n+1$, and $\|e_i - (g_i + a_i g_{n+1})\| < \varepsilon$ for $i = 1, \ldots, n$.*

Proof. (i) Let $s_i \in \partial F$ be peak points for e_i, $1 \leq i \leq n$, and define

$$P : K \to F \quad \text{by} \quad Px = \sum_{i=1}^{n} e_i(x) s_i, \quad x \in K;$$

so P is an affine continuous projection from K into F which satisfies $e_i = e_i \circ P$ for $i = 1, \ldots, n$. Since $A(F)$ is separable there exists a dense sequence $\{u_n\}$ in $A(F)_1$ (the closed unit ball of $A(F)$), and hence

$$\rho(\phi, \phi') = \sum_{n=1}^{\infty} 2^{-n} |\phi(u_n) - \phi'(u_n)|$$

defines a metric in $A(F)^*$ which induces the w^*-topology on $A(F)^*_1$. Let E be the ρ-completion of $A(F)^*$ and define $\Psi : K \to \mathscr{C}(E)$ (the non-empty closed convex subsets of E) by $\Psi(x) = P^{-1}(Px) \cap F$. It is evident that Ψ is affine, and also that $P : K \to PK$ is an open map. Consequently, if $U \subseteq E$ is open then

$$\{x \in K : \Psi(x) \cap U \neq \varnothing\} = P^{-1}(P(U \cap F))$$

is also open, so that Ψ is lower semi-continuous. By Lazar's Selection Theorem [**145**, Corollary 3.4] there exists a continuous affine map $Q : K \to E$ such that $Qx \in \Psi(x)$ for each x in K, while $Qy = y$ for each y in F. Therefore Q is a projection from K onto F, and $(P \circ Q)(x) = Px$ for each $x \in K$. Finally we have $e_j \circ Q = (e_j \circ P) \circ Q = e_j \circ P = e_j$ for $1 \leq j \leq n$.

(ii) Choose $s_1, \ldots, s_n \in \partial K$ such that $e_i(s_i) = 1$. Since F is a proper face of K, $\partial K \backslash F$ is dense in K and so there exists a point $t_{n+1} \in \partial K \backslash F$ such that

$$\left| e_i(t_{n+1}) - e_i\left(\sum_{j=1}^{n} a_j s_j\right) \right| < \varepsilon$$

for $1 \leq i \leq n$. If $1 \leq j \leq n$ and if $s_j \notin F$ we put $t_j = s_j$, while if $s_j \in F$ we choose $t_j \in \partial K \backslash F$ such that $|e_i(t_j) - e_i(s_j)| < \varepsilon$ for $1 \leq i \leq n$. Let $F' = \operatorname{co}(F, \{t_1, \ldots, t_{n+1}\})$, so that F' is a closed face of K containing $s_1, \ldots, s_n$. By part (i) there exists a projection $Q: K \to F'$ such that $e_i \circ Q = e_i$ for $1 \leq i \leq n$. We define $g_1, \ldots, g_{n+1} \in A(K)$ by putting $g_i = f_i$ on F, $g_i(t_j) = \delta_{ij}$, and $g_i(x) = g_i(Qx)$ for each x in K. Then for each i we have $\|g_i\| = 1$, $\sum_{i=1}^{n+1} g_i = 1$, $g_i|_F = f_i$, and finally

$$\begin{aligned} \|e_i - (g_i + a_i g_{n+1})\| &= \sup_{x \in K} \{|e_i(x) - (g_i(x) + a_i g_{n+1}(x))|\} \\ &= \sup_{x \in K} |e_i(Qx) - (g_i(Qx) + a_i g_{n+1}(Qx))| \\ &= \max_{1 \leq j \leq n+1} |e_i(t_j) - (g_i + a_i g_{n+1})(t_j)| < \varepsilon. \end{aligned}$$

7.5. THEOREM. *Let K_1, K_2 be compact metrizable simplexes with $\overline{\partial K_j} = K_j$, $j = 1, 2$, and let F_1, F_2 be closed faces of K_1, K_2 respectively and $\phi: F_2 \to F_1$ a surjective affine homeomorphism. Then there exists a surjective affine homeomorphism $\Psi: K_2 \to K_1$ such that $\Psi|_{F_2} = \phi$.*

Proof. Let $u_n \in A(K_1)$, $v_n \in A(K_2)$ such that $\|u_n\| = \|v_n\| = 1$ for each $n = 1, 2, \ldots$, and such that $\operatorname{lin}(\{u_n\})$ and $\operatorname{lin}(\{v_n\})$ are dense in $A(K_1)$ and $A(K_2)$ respectively.

We will construct inductively peaked partitions of unity $\{e^j_{i,m}\}_{1 \leq i \leq m}$ for $A(K_1)$ and $\{f^j_{i,m}\}_{1 \leq i \leq m}$ for $A(K_2)$, $j \geq m$, $m = 1, 2, \ldots$, and non-negative numbers $\{a_{i,m}\}_{1 \leq i \leq m}$ such that $\sum_{i=1}^{m} a_{i,m} = 1$ for $m = 1, 2, \ldots$, with the following properties:

(i) $e_{i,m} = e^j_{i,m+1} + a_{i,m} e^j_{m+1,m+1}$, (i)′ $f^j_{i,m} = f^j_{i,m+1} + a_{i,m} f^j_{m+1,m+1}$, for $1 \leq i \leq m$, $j \geq m+1$, $m = 1, 2, \ldots$;

(ii) $e^j_{i,m}(\phi(x)) = f^j_{i,m}(x)$, $x \in F_2$, $1 \leq i \leq m$, $j \geq m$, $m = 1, 2, \ldots$;

(iii) $\|e^j_{i,m} - e^{j+1}_{i,m}\| < 2^{-j}$, (iii)′ $\|f^j_{i,m} - f^{j+1}_{i,m}\| < 2^{-j}$, for $1 \leq i \leq m$, $j \geq m$, $m = 1, 2, \ldots$;

(iv) if $E_m = \operatorname{lin}(e^m_{1,m,\ldots,} e^m_{m,m})$ then, for each $n > 0$, there exists an $m > 0$ such that $d(U_k, E_m) < 2^{-n}$ for $1 \leq k \leq n$, (iv)′ if $F_m = \operatorname{lin}(F^m_{1,m,\ldots,} F^m_{m,m})$ then, for each $n > 0$, there exists an $m > 0$ such that $d(v_k, F_m) < 2^{-n}$ for $1 \leq k \leq n$.

Firstly we define $e^1_{1,1} = 1 = f^1_{1,1}$. Now assume that $\{e^j_{i,k}\}_{1 \leq i \leq k}$ and $\{f^j_{i,k}\}_{1 \leq i \leq k}$ have been constructed for $k = 1, \ldots, m$, $k \leq j \leq m$ so that (i), (i)′, (ii), (iii), and (iii)′ hold, and also so that (iv) and (iv)′ hold for some integer n. By Theorem 7.2 and the remarks following it, there exists a subspace E_{m+r} of $A(K_1)$ isometrically isomorphic to l^∞_{m+r} such that $E_m \subset E_{m+r}$, $d(u_k, E_{m+r}) < 2^{-(n+1)}$ for $1 \le k \le n+1$. Moreover there exist non-negative numbers $\{a_{i,k}\}_{1\le i\le k}$, $k = m, \ldots, m+r-1$ with $\sum_{i=1}^k a_{i,k} = 1$, and peaked partitions of unity $\{e_{i,k}\}_{1\le i\le k}$ for $A(K_1)$, $k = m, \ldots, m+r-1$, such that

$$e^m_{i,m} = e_{i,m+1} + a_{i,m}e_{m+1,m+1}, \qquad 1 \le i \le m$$

$$e_{i,k} = e_{i,k+1} + a_{i,k}e_{k+1,k+1}, \qquad 1 \le i \le k, \qquad m+1 \le k \le m+r-1.$$

Put $e^j_{i.k} = e^m_{i,k}$ for $k \le m$, and $e^j_{i,k} = e_{i,k}$ for $k = m+1, \ldots, m+r-1, 1 \le i \le k$, $k \le j \le m+r$. Then (i) and (ii) hold up to $m+r-1$ and (iv) holds for E_{m+r} and $n+1$.

Now $\{e^{m+1}_{i,m+1} \circ \phi\}_{1\le i\le m+1}$ is a partition of unity for $A(F_2)$ and for $1 \le i \le m$, we have $f^m_{i,m}|F_2 = e^{m+1}_{i,m+1} \circ \phi + a_{i,m}e^{m+1}_{m+1,m+1} \circ \phi$. By 7.4(ii), for each $\varepsilon > 0$ there exists a peaked partition of unity $\{f_i\}_{1\le i\le m+1}$ for $A(K_2)$ such that

$$f_i(x) = e^{m+1}_{i,m+1}(\phi(x)), \qquad x \in F_2, \qquad 1 \le i \le m+1,$$

and

$$\|f^m_{i,m} - (f_i + a_{i,m}f_{m+1})\| < \varepsilon \quad \text{for } 1 \le i \le m.$$

Hence if $\varepsilon < 2^{-2(m+1)}$ and if we define, for $1 \le i \le k$ and $1 \le k \le m+1$, $f^{m+1}_{i,k}$ by $f^{m+1}_{i,m+1} = f_i$, $f^{m+1}_{i,m} = f^{m+1}_{i,m+1} + a_{i,m}f^{m+1}_{m+1,m+1}, \ldots, f^{m+1}_{1,1} = f^{m+1}_{1,2} + a_{1,1}f^{m+1}_{2,2}$, it follows that $\|f^{m+1}_{i,k} - f^m_{i,k}\| < 2^{-m}$ for $1 \le i \le k$ and $1 \le k \le m$.

In precisely the same way we construct $f^j_{i,k}$ for $1 \le i \le k$, $k \le j \le m+r$ and $k = m+2, \ldots, m+r-1$. Again, by Theorem 7.2 there exists a subspace F_{m+r+s} of $A(K_2)$, containing F_{m+r} which is isometrically isomorphic to l^∞_{m+r+s} and satisfies $d(v_k, F_{m+r+s}) < 2^{-(n+1)}$ for $1 \le k \le n+1$. Now construct $\{f_{i,k}\}_{1\le i\le k}$ for $k \le j \le m+r+s$ and $m+r+1 \le k \le m+r+s-1$ by the above procedure, interchanging the roles of $A(K_1)$ and $A(K_2)$, and then define $\{e^j_{i,k}\}_{1\le i\le k}$ using $\{f^j_{i,k}\}_{1\le i\le k}$, $k \le j \le m+r+s$, $m+r+1 \le k \le m+r+s-1$. In this way we ensure that (i), (i)′, (ii), (iii), (iii)′ hold up to $m+r+s-1$, while (iv) and (iv)′ hold for $n+1$. This completes the inductive construction.

Let $e_{i,m} = \lim_{j\to\infty} e^j_{i,m}$ and $f^j_{i,m} = \lim_{j\to\infty} f^j_{i,m}$ for $1 \le i \le m$ and $m \ge 1$. Because $\|e^m_{i,m} - e_{i,m}\| < 2^{-m+1}$, it follows easily from (iv) that the linear span of $\{e_{i,m}\}$ is dense in $A(K_1)$; similarly, the linear span of $\{f_{i,m}\}$ is dense in $A(K_2)$. Moreover, using (i), (i)′ and (ii), we obtain for $1 \le i \le m$ and $m = 1, 2, \ldots$, $e_{i,m} = e_{i,m+1} + a_{i,m}e_{m+1,m+1}$, $f_{i,m} = f_{i,m+1} + a_{i,m}F_{m+1,m+1}$, and $e_{i,m}(\phi(x)) = f_{i,m}(x)$ for all $x \in F_2$.

Since, for each $j \ge m$, $\{e^j_{i,m}\}_{1\le i\le m}$ is a peaked partition of unity for $A(K_1)$ so is $\{e_{i,m}\}_{1\le i\le m}$, and in particular $e_{1,m}, \ldots, e_{m,m}$ are linearly independent. For each m and $1 \le i \le m$ we define $Te_{i,m} = f_{i,m}$. The equations obtained above show that T is well-defined on a dense subspace of $A(K_1)$. Since for each fixed m lin $(e_{1,m}, \ldots, e_{m,m})$ is isometric to l_m^∞ it is easy to see that T is an isometry on this subspace. Consequently T extends to an isometry from $A(K_1)$ onto $A(K_2)$, and satisfies $Tg(x) = g(\phi(x))$ for each $x \in F_2$ and $g \in A(K_1)$. Since $T1 = 1$ we can now define $\Psi: K_2 \to K_1$ by $\Psi = T^*|_{K_2}$. It is clear that Ψ is a surjective affine homeomorphism with $\Psi|_{F_2} = \phi$.

An immediate consequence of the above theorem, taking F_1 and F_2 to be singletons, is that K_1 and K_2 are affinely homeomorphic to each other, and also to the Poulsen simplex. We will denote the Poulsen simplex by S_p.

7.6. THEOREM. *If K is any compact metrizable simplex then K is affinely homeomorphic to a closed face of S_p.*

Proof. If we can construct a compact metrizable simplex S with $\overline{\partial S} = S$ and such that K is affinely homeomorphic to a closed face of S, then the result will follow from Theorem 7.5.

We may assume that K is a compact convex subset of a Banach space E. (We can take E to be the completion of $A(K)^*$ for the metric ρ defined in the proof of Lemma 7.4.) We identify K with the subset $\{(x, \mathbf{0}): x \in K\}$ of $E \times l^2$, and write $e_n^1 = (0, \mathbf{e}_n) \in E \times l^2$, where $\mathbf{e}_n$ denotes the nth coordinate vector in l^2. Using the method of proof of Theorem 7.1 we can easily construct a sequence of points $\{z_n\}$ and simplexes $\{S_n\}$ in $E \times l^2$ with the properties:

(i) $S_0 = K$;

(ii) $S_{n+1} = \text{co}\,(z_{n+1} \cup S_n)$ for $n = 0, 1, 2, \ldots$;

(iii) $P_n S_m = S_n$ for $m \geq n$, where P_n denotes the projection of $E \times l^2$ onto $E \times \mathbb{R}^n$;

(iv) $\partial S_n \subset \partial S_m$ for $m \geq n$;

(v) for each $\varepsilon > 0$ there exists an n such that ∂S_n is an ε-net for S_n.

If we now define $S = \overline{\text{co}} \bigcup_{n=1}^\infty S_n$ then, as in the proof of Theorem 7.1, it follows that S is a compact simplex with $\overline{\partial S} = S$, and clearly K is a closed face of S.

A consequence of Theorem 7.5 is that ∂S_p is homogeneous. Moreover, every Polish space is, by Theorems 7.1 and 7.6, homeomorphic to a closed subset of ∂S_p.

In particular $[0, 1]$ is homeomorphic to an arc in ∂S_p. Using Theorem 7.5 it now follows easily that ∂S_p is arcwise-connected.

Notes

Theorem 7.2 was proved by Lazar and Lindenstrauss [**149**]. As we noted above, the spaces E_n appearing in 7.2 can be chosen to be isometrically isomorphic to l_n^∞, and the isometric embedding of l_n^∞ into l_{n+1}^∞ takes the form

$$T_n\mathbf{u}_{j,n} = \mathbf{v}_{j,n+1} + a_{j,n}\mathbf{v}_{n+1,n+1} \quad \text{where } a_{j,n} \geq 0, \sum_{j=1}^{n} a_{j,n} = 1$$

and

$$\{\mathbf{u}_{j,n}\}, \{\mathbf{v}_{k,n+1}\}, 1 \leq j \leq n, 1 \leq k \leq n+1$$

are suitable bases of l_n^∞ and l_{n+1}^∞ respectively. The triangular matrix $\{a_{j,n}\}$, $1 \leq j \leq n$, $n = 1, 2, \ldots$, is called a *representing matrix* for $A(K)$. The representing matrices for $A(K)$ are not unique, but in some cases they do determine certain properties of K. For example, Lazar and Lindenstrauss [**149**] show that K is a Bauer simplex with ∂K totally disconnected if and only if $A(K)$ can be represented by a matrix $\{a_{j,n}\}$ where, for each n, $a_{j_n,n} = 1$, $a_{j,n} = 0$ for $j \neq j_n$, for some j_n with $1 \leq j_n \leq n$.

Sternfeld [**201**] gives characterizations of Bauer simplexes, and of the Poulsen simplex, in terms of representing matrices.

The Uniqueness Theorem 7.5 and the Universality Theorem 7.6, for the Poulsen simplexes, are due to Lindenstrauss, Olsen and Sternfeld [**154**]. They also prove that ∂S_p is homeomorphic to l^2. The proof of 7.5 uses ideas similar to those of Lusky [**155**] in proving that the Gurarij space is unique. (A *Gurarij space* is a separable predual of an L_1-space such that for any $\varepsilon > 0$ and for any finite-dimensional Banach spaces $E \subseteq F$ and linear isometry $T: E \to G$, there exists a linear extension $\tilde{T}: F \to G$ of T with $(1-\varepsilon)\|x\| \leq \|\tilde{T}x\| \leq (1+\varepsilon)\|x\|$ for all x in F.) Some analogies between $A(S_p)$ and G have been given by Lusky [**156**].

8. NON-COMPACT SIMPLEXES AND CONVEX SETS

Let K be a convex set in a real vector space E such that K is linearly compact, that is every line in E meets K in a (possibly empty) compact subset of the line. Without loss of generality we can suppose that K is contained in a hyperplane of the form $e^{-1}(1)$, and also that $\text{lin}\, K = E$. Let C be the cone generated in E by K, with vertex 0, and partially order E so that

$$C = \{x \in E : x \geq 0\}.$$

In these circumstances we say that K is a *simplex* if E is a vector lattice. A result of Choquet and Kendall (see Kendall [**139**]) shows that a linearly compact set K is a simplex if and only if K has the property that, whenever $\lambda \geq 0$ and $x \in E$, there exists $\mu \geq 0$ and $y \in E$ such that $K \cap (\lambda K + x) = \mu K + y$.

8.1. THEOREM. *Let K be a linearly compact simplex embedded in the partially ordered vector space (E, C).*

(i) *The set* $\sum = \mathrm{co}\,(K \cup -K)$ *is the closed unit ball for a lattice norm in* E, *such that the norm is additive on* C.

(ii) *If* $-f$, g *are bounded convex functions on* K *with* $f \geq g$, *then there exists a bounded affine function* h *on* K *with* $f \geq h \geq g$.

(iii) *If* E *is complete for the norm in* (i) *then every closed face of* K *is split.*

Proof. (i) If p is the gauge for $\sum$ then p is evidently a semi-norm, and since $C \cap \sum = \{x \in C : e(x) \leq 1\}$ it follows that p and e coincide on C; in particular p is additive on C. If $x \in E$ then we have

$$p(x) = p(x^+ - x^-) \leq p(x^+) + p(x^-) = p(x^+ + x^-) = p(|x|).$$

If $p(x) < 1$ then $x \in \sum$, so that $x = \lambda y - (1-\lambda)z$ with $0 \leq \lambda \leq 1$ and $y, z \in K$. Therefore $|x| \leq \lambda y + (1-\lambda)z$ and so $p(|x|) \leq p(\lambda y + (1 = \lambda)z) = 1$. We have therefore shown that $p(x) = p(|x|) = e(|x|) > 0$ if $x \neq 0$, and that $p(x^+) + p(x^-) = p(x)$. The statement (i) now follows

If we define $f(\alpha x) = \alpha f(x)$, $g(\alpha x) = \alpha g(x)$ for $\alpha \geq 0$, $x \in K$, then $-f$, g are subadditive positive homogeneous functionals on C. The methods of Theorem 2.4.2. can now be used to obtain the required h. (iii) If E is complete for the norm in (i) then E is an L-space (2.6.2) and hence is an order-complete vector lattice. If F is a closed face of K then the continuity of the lattice operations in E implies that $\mathrm{lin}\, F$ is an order-complete lattice ideal in E, and consequently there exists an ideal L in E such that E is the order-direct sum of $\mathrm{lin}\, F$ and L (see Peressini [**168**, p. 39]). If $F' = L \cap K$ it is now easy to check that F and F' are complementary split faces of K.

Statement (i) of Theorem 8.1 does not hold in general for non-simplexes, as the example (2.1.6) shows. Statement (ii) is a simple analogue of Edwards Separation Theorem for compact simplexes (Theorem 2.7.6), but this property does not characterize linearly compact simplexes. In fact if E consists of the real polynomials on $[0, 1]$ and if $K = \{f \in E^+ : \int_0^1 f(t)\, \mathrm{d}t = 1\}$, then E possesses the Riesz Decomposition Property and so statement (ii) holds; however since E is not a vector lattice K is not a simplex. We now give an example to show that statement (iii) may fail if E is not a Banach space.

8.2. EXAMPLE. *Let* $E = C_{\mathbb{R}}[0, 1]$, $K = \{f \in E^+ : \int_0^1 f(t)\, \mathrm{d}t = 1\}$ *and let*

$$F = \left\{ f \in K : \int_0^{1/2} f(t)\, \mathrm{d}t = 0 \right\}.$$

Then K *is a linearly compact simplex and* F *is a closed face of* K *for the induced norm in* E. *However,* F *is not a split face of* K.

Proof. All the statements, except the final one, are evident. Suppose that F is a split face of K with complementary face F'. Then, since $f(\frac{1}{2}) = 0$ for all

$f \in F$ there must exist a $u \in F'$ with $u(\frac{1}{2}) > 0$. But then we can easily decompose $u = \lambda g + (1-\lambda)h$ with $0 < \lambda < 1$ and $g \in F$, $h \in K$, and since F' is a face we conclude that $g \in F \cap F'$. This contradiction shows that F is not a split face of K.

Let K be a linearly compact convex set, not necessarily a simplex, but embedded in E such that $\sum = \text{co}(K \cup -K)$ is linearly bounded, that is every line in E intersects $\sum$ in a bounded (possibly empty) subset of that line. Then the gauge of $\sum$ is a norm for E which is additive on C, and the dual space E^* is order and isometrically isomorphic to the Banach space $A^b(K)$ of all bounded affine real-valued functions on K, with the supremum norm. If $\sum$ is the closed unit ball for E then the proof of Theorem 2.7.2 shows that K is a simplex if every closed face of K is split.

We now illustrate, in a very special case, some results concerning the structure of Banach spaces.

Let K be a linearly compact convex set such that $\sum$ is the closed unit ball in E for a complete norm in E. A linear projection $P: E \to E$ such that $0 \leq P \leq I$, for the pointwise ordering, will be called an *L-projection*, and the range of an L-projection will be called an *L-ideal.*

8.3. THEOREM.

(i) *I is an L-ideal in E if and only if $I = \text{lin}\, F$ for some split face F of K.*

(ii) *The L-projections on E form a commutative subset of $B(E)$ the bounded linear operators on E, and they are precisely the extreme points of the set $\{T \in B(E): 0 \leq T \leq I\}$.*

(iii) *If $\{I_\alpha\}$ is a family of L-ideals in E then $\bigcap_\alpha I_\alpha$ and $\overline{\bigcup_\alpha I_\alpha}$ are also L-ideals in E.*

(iv) *If $\{F_\alpha\}$ is a family of split faces of K then $\bigcap_\alpha F_\alpha$ and $\overline{\text{co}} \bigcup_\alpha F_\alpha$ are also split faces of K.*

Proof. (i) Let F be a split face of K with complementary face F', so that E is the direct sum of lin F and lin F'. If P is the projection of E onto lin F then it is clear that $0 \leq P \leq I$, so that lin F is an L-ideal.

Conversely, let P be an L-projection and let $F = K \cap P(E)$ and $F' = K \cap (I-P)(E)$. Then since $0 \leq P \leq I$ and since $\text{lin}\, F \cap \text{lin}\, F' = \{0\}$; it is easy to check that F and F' are complementary split faces of K.

(ii) Let P_1 and P_2 be L-projections on E, and let $F_j = K \cap P_j(E)$ for $j = 1, 2$. Now $F_1 \cap F_2$ is a split face of K (see the proof of 1.1) and so there is an associated L-projection Q. If $x \in K$ then x can be uniquely decomposed as

$$x = \sum_{j=1}^{4} \lambda_j y_j, \quad \text{where } \lambda_j \geq 0, \quad \sum_{j=1}^{4} \lambda_j = 1$$

and

$$y_1 \in F_1 \cap F_2, \qquad y_2 \in F_1 \cap F_2', \qquad y_3 \in F_2 \cap F_1', \qquad y_4 \in F_1' \cap F_2'.$$

But then

$$Qx = \lambda_1 y_1 = P_1 P_2 x = P_2 P_1 x,$$

so that the linear operators Q, P_1P_2 and P_2P_1 agree on K, and hence are equal. Therefore, since clearly $\|P\| = 1$ for any L-projection $P \neq 0$, the L-projections form a commutative subset of $B(E)$.

Let $\Lambda = \{T \in B(E): 0 \leq T \leq I\}$ and suppose that P is an extreme point of Λ. Then $P^2 \in \Lambda$ and also

$$2P - P^2 = P + P(I-P) \in \Lambda.$$

The equation $P = \frac{1}{2}P^2 + \frac{1}{2}(2P - P^2)$ now gives $P^2 = P$, that is P is an L-projection. Conversely, let P be an L-projection and suppose that $P = \frac{1}{2}T_1 + \frac{1}{2}T_2$ for some T_1, $T_2 \in \Lambda$. Then we have

$$P = \tfrac{1}{2}PT_1 + \tfrac{1}{2}PT_2 \quad \text{and} \quad 0 = \tfrac{1}{2}(I-P)T_1 + \tfrac{1}{2}(I-P)T_2,$$

and because $PT_j \leq P$ and $(I-P)T_j \geq 0$, for $j = 1, 2$, it follows that $T_j = PT_j = P$ for $j = 1, 2$. Consequently $P \in \partial\Lambda$.

(iii) Let $\{I_\alpha\}$ be a family of L-ideals. To prove that $\bigcap_\alpha I_\alpha$ is an L-ideal we can assume that the family is decreasing, because the intersection of finitely many I_α is an L-ideal (using (i)). If $\{P_\alpha\}$ is the corresponding family of L-projections we have $\{P_\alpha x\}$ decreasing in C for each $x \in C$; since E is complete and the norm is additive on C, $\{P_\alpha x\}$ is norm-convergent to a limit $Px \in C$. Therefore the net $\{P_\alpha\}$ converges strongly to an L-projection P such that $P(E) = \bigcap_\alpha I_\alpha$. Moreover the complementary projections $\{I - P_\alpha\}$ converge strongly to $I - P$ and $(I-P)(E) = \operatorname{cl} \bigcup_\alpha (I - P_\alpha)(E)$. The final part of (iii) now follows.

(iv) The required facts follow by direct verification using the results in (i) and (iii).

By the above result the L-projections on E generate a commutative Banach subalgebra of $B(E)$, with identity I. This Banach algebra is denoted by $\mathscr{C}(E)$ and is called the *Cunningham algebra* for E.

An operator $T \in B(E)$ is said to be *order-bounded* if $-\lambda I \leq T \leq \lambda I$ for some constant $\lambda > 0$, and we denote infimum of such λ by $\|T\|_1$. It is easy to see that $\|\cdot\|_1$ is a norm and that

$$\|T\| = \sup\{\|Tx\|: x \in K\} \leq \|T\|_1.$$

We can identify $A^b(K)$ with $A(K_1)$ where K_1 is the state space of the complete order unit normed space $A^b(K)$. Suppose that $T \in B(E)$ satisfies $\alpha I \leq T \leq (1-\alpha)I$, where $0 < \alpha < 1/\alpha$. Then $\alpha I \leq T^* \leq (1-\alpha)I$, and so if

$x \in \partial K_1$ we have $\alpha \leq (T^*1)(x) \leq (1-\alpha)$. Putting $\lambda = (T^*1)(x)$ we obtain

$$x = \lambda y + (1-\lambda)z \quad \text{where } y, z \in K_1 \text{ satisfy}$$

$$\lambda f(y) = (T^*f)(x), \qquad (1-\lambda)f(z) = ((I - T^*)f)(x) \quad \text{for all } f \in A(K_1),$$

and hence $x = y = z$, that is $(T^*f)(x) = (T^*1)(x)f(x)$. Consequently for any order-bounded $T \in B(E)$ we see that

$$(T^*f)(x) = (T^*1)(x)f(x) \quad \text{for all } x \in \partial K_1, \qquad f \in A(K_1).$$

Therefore

$$\begin{aligned}\|T\| = \|T^*\| &= \sup \{|(T^*1)(x)|;\ x \in \partial K_1\} \\ &= \inf \{\beta > 0\colon -\beta I \leq T^* \leq \beta I\} \\ &= \inf \{\beta > 0\colon -\beta I \leq T \leq \beta I\} = \|T\|_1.\end{aligned}$$

It is now clear that every $T \in \mathscr{C}(E)$ is order-bounded. The converse inclusion and a representation of $\mathscr{C}(E)$ is given in the next theorem.

8.4. THEOREM. *$\mathscr{C}(E)$ coincides with the order-bounded operators on E, and is order and isometrically isomorphic to $C_{\mathbb{R}}(\Omega)$ for some hyperstonean space Ω.*

Proof. Let $\mathscr{O}(E)$ denote the order bounded operators on E. Then the above reasoning shows $\mathscr{O}(E)$ is a complete Banach algebra with identity I, and that I is an order unit defining the norm in $\mathscr{O}(E)$. Moreover, if $S, T \in \mathscr{O}(E)$ then S^*T^*, T^*S^* are order bounded on $A(K_1)$, so that for $f \in A(K_1)$, $x \in \partial K_1$

$$\begin{aligned}(S^*T^*f)(x) &= (S^*1)(x) \cdot (T^*f)(x) \\ &= (S^*1)(x) \cdot (T^*1)(x) \cdot f(x) = (T^*S^*f)(x).\end{aligned}$$

Therefore $S^*T^* = T^*S^*$, so that $ST = TS$ and $\mathscr{O}(E)$ is commutative. It follows that $\mathscr{O}(E)$ is order and isometrically isomorphic to $C_{\mathbb{R}}(\Omega)$ for some compact Hausdorff space Ω (cf. Kadison [**136**]).

Arguing as in the proof of 8.3(iii) we see that $\mathscr{O}(E)$ is boundedly order complete, and hence Ω is extremally disconnected. The L-projections in $\mathscr{O}(E)$ correspond to functions χ_G where G is open and closed in Ω, and linear combinations of these functions will separate points of Ω, and will belong to $\mathscr{C}(E)$. Therefore the Stone–Weierstrass Theorem gives $\mathscr{C}(E) = \mathscr{O}(E)$.

For each $x \in K$ we can define a probability measure μ_x on Ω such that, identifying $\mathscr{C}(E)$ with $C_{\mathbb{R}}(\Omega)$,

$$\mu_x(T) = 1(Tx), \qquad T \in \mathscr{C}(E).$$

If $\{T_\alpha\}$ is an increasing net in $\mathscr{C}(E)$ with supremum T then $T_\alpha x$ converges strongly to Tx for each $x \in K$, and so

$$\mu_x(T) - \mu_x(T_\alpha) = 1(Tx - T_\alpha x) = \|Tx - T_\alpha x\| \to 0.$$

Thus each μ_x is a normal measure on Ω and, since these measures separate the functions in $C_{\mathbb{R}}(\Omega)$, the space Ω is hyperstonean.

The algebra $\mathcal{O}(E^*)$ of order bounded operators on

$$E^* = A^b(K) = A(K_1)$$

is a commutative Banach algebra with identity I and, for

$$T \in \mathcal{O}(E^*),$$

$$\|T\| = \|T\|_1 = \inf\{\lambda > 0: -\lambda I \leq T \leq \lambda I\}.$$

We can also deduce that the map $\phi : \mathcal{O}(E^*) \to A(K_1)$ defined by $\phi(T)(x)$ for $x \in \partial K_1$, is an algebraic, order and isometric isomorphism of $\mathcal{O}(E^*)$ on to the centre of $A(K_1)$. For this reason $\mathcal{O}(E^*) = \mathcal{O}(A^b(K))$ is called the *centralizer* of $A^b(K)$.

8.5. THEOREM. *The mapping $T \rightsquigarrow T^*$ is an algebraic, order and isometric isomorphism of the Cunningham algebra $\mathscr{C}(E)$ onto the centralizer $\mathcal{O}(A^b(K))$. The centre of $A(K_1) = A^b(K)$ is w^*-closed.*

Proof. If $T \in \mathscr{C}(E)$ then $-\lambda I \leq T \leq \lambda I$ for some $\lambda > 0$, and hence $-\lambda I \leq T^* \leq \lambda I$ so that $T^* \in \mathcal{O}(E^*)$. If we can show that every $S \in \mathcal{O}(E^*)$ is of the form $S = T^*$ for some $T \in \mathscr{C}(E)$ then the first statement of the Theorem will follow.

It is obvious that $A^b(K)$ is boundedly order-complete, and consequently $\mathcal{O}(A^b(K))$ is boundedly order-complete. Therefore $\mathcal{O}(A^b(K))$ is the closed linear span of the projections $P \in \mathcal{O}(A^b(K))$ such that $0 \leq P \leq I$, and so it will be sufficient for the first statement to show each such projection P is w^*-continuous on $A^b(K)$. Hence it will be sufficient to prove that $P(A^b(K))$ and $(I-P)(A^b(K))$ are w^*-closed.

Let $Q \in \mathcal{O}(A^b(K))$ be a projection such that $0 \leq Q \leq I$ and let $L = Q(A^b(K)) \cap \Delta$, where Δ is the closed unit ball of $A^b(K)$. If $\{f_\alpha\}$ is a net in L which is w^*-convergent to $f \in \Delta$ then $-1 \leq f_\alpha \leq 1$, so that $-Q(1) \leq Q(f_\alpha) = f_\alpha \leq Q(1)$ and consequently $-Q(1) \leq f \leq Q(1)$ since $(A^b(K))^+$ is w^*-closed. Therefore we have

$$0 = (I-Q)(-Q(1)) \leq f - Qf \leq (I-Q)(Q(1)) = 0,$$

so that $Qf = f$ and L is w^*-closed. The Krein-Šmulian Theorem now implies that $\mathcal{O}(A^b(K))$ is w^*-closed. Putting $Q = P$ and $Q = (I-P)$, in turn, the first statement is proved.

Theorem 8.4 and its proof show now that $\mathcal{O}(A^b(K))$ can be identified with $C_{\mathbb{R}}(\Omega)$ for some hyperstonean space Ω, and that the measure μ_x on Ω, given by

$$\mu_x(T^*) = (T^*1)(x), \qquad T^* \in \mathcal{O}(A^b(K)),$$

is a normal measure for each $x \in K$; thus μ_x is a w^*-continuous linear functional on $C_{\mathbb{R}}(\Omega)$ for every $x \in K$, and hence for every $x \in E$. If $\pi(x) = \mu_x$ for $x \in E$, then

$$(\pi^*(T^*))(x) = \mu_x(T^*) = (T^*1)(x) \quad \text{for } T^* \in \mathcal{O}(A^b(K)),$$

so that

$$\pi^*(T^*) = T^*1.$$

Since

$$\|T^*\| = \|T^*1\| \quad \text{for all } T^* \in \mathcal{O}(A^b(K))$$

we obtain

$$\{T^*1 \colon T^* \in \mathcal{O}(A^b(K)), \|T^*1\| \le 1\} = \pi^*(\{T^* \in \mathcal{O}(A^b(K)) \colon \|T^*\| \le 1\}.$$

Finally, because π^* is w^*-continuous, the Krein–Šmulian Theorem implies that the centre of $A^b(K)$, that is $\{T^*1 \colon T^* \in \mathcal{O}(A^b(K))\}$, is w^*-closed.

We conclude this chapter with a brief mention of some important recent work concerning non-compact versions of Choquet's Theorem.

A Banach space B is said to have the *Radon–Nikodym property* (RNP) if whenever $(X, \mathcal{F}, \mu)$ is a probability space and $m: \mathcal{F} \to B$ is a countably additive measure (convergence in norm) which is absolutely continuous with respect to μ and has bounded variation, then there is a Borel-measurable function $F: X \to B$ such that $m(E) = \int_E F \, d\mu$ for each $E \in \mathcal{F}$, where the integral is a Bochner integral. For example, every weakly compactly generated Banach dual space has the RNP.

Every Banach space B with the RNP has the *Krein–Milman property*, that is every bounded closed convex subset of B is the closed convex hull of its extreme points. The converse is true for Banach dual spaces but is unknown in general.

If K is a bounded closed convex subset of a separable Banach space then ∂K is universally measurable, but need not be a Borel set.

8.6. THEOREM. *Let K be a bounded closed separable convex set in a Banach space B having the* RNP. *The following statements hold.*

(i) *For each $a \in K$ there is a probability measure μ on K with $\mu(\partial K) = 1$ and $\int_K x \, d\mu(x) = a$ (as a Bochner integral).*

(ii) *K is a simplex if and only if for each $a \in K$ there is a unique probability measure μ on K with $\mu(\partial K) = 1$ and $\int_K x \, d\mu(x) = a$.*

Notes

The first part of Theorem 8.1 was proved by D. A. Edwards [**87**]. Parts (ii) and (iii) appear in Asimow and Ellis [**32**].

Theorems 8.3 and 8.4 are due to Alfsen and Effros [**8**] and to Wils [**216**]. In their fundamental paper [**8**] Alfsen and Effros develop a structure theory for real Banach spaces. There is a very extensive literature extending the results of Alfsen and Effros, in particular to complex Banach spaces and to Lindenstrauss spaces. We mention only the paper of Lima [**152**] which develops some connections between L-ideals, M-ideals and intersection properties of balls in Banach spaces.

Some applications of the structure theory to function algebras and to C^*-algebras will be discussed in Chapters 4 and 5. Some applications to other Banach algebras have recently been given by Smith and Ward [**196**, **197**].

Theorem 8.5 is due to Alfsen and Effros [**8**] and to D. A. Edwards [**90**]; see also Cunningham *et al.* [**69**].

Interpretation and application of the above material in quantum mechanics has been given by C. M. Edwards [**85**, **86**]. Much more information concerning the RNP in Banach spaces can be found in the books of Diestel [**78**] and Diestel and Uhl [**79**]. An example of a bounded closed convex subset K of a separable Banach space such that ∂K is not Borel has been given by Jayne and Rogers [**135**]. Theorem 8.6 is due to Bourgin and Edgar [**49**, **83**, **84**]. The same authors have also obtained more technical results in the non-separable case. In this context Mankiewicz [**158**] has shown that if K is a closed, bounded convex subset of a Banach space X with the RNP, then each $y \in K$ is the resultant of a separably-supported complete Borel probability measure μ on K with the property that $\mu(\partial(K \cap Y)) = 1$ for every closed separable subspace Y of X containing supp μ.

CHAPTER 4

Complex Function Spaces

0. THE COMPLEX STATE SPACE

Let A be a complex Banach space and let S be a w^*-compact convex subset of A_1^*. Let

$$Z = \text{co}\,(S \cup -iS)$$

$$B = \text{co}\,(Z \cup -Z)$$

$$N = \bigcup_{n=1}^{\infty} nB \quad \text{(the real linear span of } Z\text{)}$$

and consider $A_0(B)$, the space of real affine homogeneous w^*-continuous functions on B, with the supremum norm

$$\|f\| = \sup\{|f(x)|: x \in B\} = \sup\{|f(x)|: x \in Z\}.$$

Let

$$\theta: A \to A_0(B); \qquad (\theta a)(x) = \text{re}\,(a, x).$$

Then θ is a real linear map with dense range and θ^* identifies the real dual $A_0(B)^*$ with (N, B) (cf. Chapter 2 Section 8, for example). Now, Theorems 1.3.5 and 2.8.1 yield the following result.

0.1. THEOREM. *The following are equivalent*:

(i) *θ is a real isomorphism with $\|\theta\| \leq \|\theta^{-1}\| \leq c\|\theta\|$,*

(ii) *$N_1 \subset cB$ ($N_1 = \{x \in N: \|x\| \leq 1\}$).*

If S is contained in $a^{-1}(1)$ for some $a \in A$ then $A_0(B)$ is isometrically isomorphic to $A(Z)$ since $S \subset a^{-1}(1)$ implies $Z \subset [\theta(a + ia)]^{-1}(1)$.

Example 1. Let $1 \in A$ be a closed separating subspace of $C(X)$, X compact Hausdorff, and let S be the state space of A. Then, by a simple computation,

$$\|\theta a\| \leq \|a\| \leq \sqrt{2}\|\theta a\|$$

so that Theorem 0.1 shows A is real isomorphic to $A(Z)$, $Z = \text{co}\,(S \cup -iS)$, the *complex state space.*

Example 2. Let A be a complex unital Banach algebra and again, take S to be the state space arising from the unit u of A. A theorem of Bohnenblust and Karlin [**43**] shows that (i) holds with $c = \sqrt{2}e$. The fact that (i) implies (ii) in this case was proved by Moore [**162**].

To illustrate the utility of the complex state space we demonstrate the algebraic significance of S being a split face in Z. Thus, as in the above examples, let A be real isomorphic to $A(Z)$. We denote

$$H(A) = \{a \in A\colon (a, s) \text{ real for all } s \in S\}.$$

These are the real functions in A in Example 1, and, in Example 2, are called the *Hermitian* elements of A.

0.2. THEOREM. *The following are equivalent*:
(i) *S is a split face of Z*;
(ii) $A = H(A) \oplus iH(A)$.

Proof. Since $Z = \text{co}\,(S \cup -iS)$ and both S and $-iS$ are closed we see that S is split in Z if and only if S and $-iS$ are complementary split faces. Thus, we can apply Theorem 2.10.5(i) which states that S and $-iS$ are complementary split faces if and only if

$$A(Z) = A_1 \oplus A_2,$$
$$A_1 = \{\theta a\colon \theta a|_S \equiv 0\}$$
$$A_2 = \{\theta a\colon \theta a|_{-iS} \equiv 0\}.$$

But $\theta a \in A_2$ if and only if $\text{im}\,(a, s) = 0$ if and only if $a \in H(A)$, and similarly $\theta a \in A_1$ if and only if $ia \in H(A)$. Thus the equivalence of (i) and (ii) follows.

In the case of Example 1 we say A is *self-adjoint* if Theorem 0.2(ii) holds, in which case $H(A) = \text{re}\, A$.

In Example 2 this provides a geometric version of the Vidav–Palmer Theorem [**46**] which states that the Banach algebra A is a B^*-algebra if and only if $A = H(A) \oplus iH(A)$.

One is sometimes only interested in the real linear span of the state space $S \subset A^*$. This is denoted $H(A^*)$ and is identified as usual with $A(S)^*$. This induces the norm on $H(A^*)$,

$$\|x\|_S = \inf\{\alpha + \beta\colon x = \alpha s_1 - \beta s_2;\ \alpha, \beta \geq 0,\ s_i \in S\}$$

(as in Theorem 2.2.3). If $H(A^*)$ is a real dual L-space the representation of $x \in \text{co}\,(S \cup -S)$ (for $\|x\| = 1$) is unique. Grothendieck [**123**] showed that this is also the case if A is a C^*-algebra. But Sinclair [**195**] noted an instance where the uniqueness failed to hold with A a unital Banach algebra. This uniqueness property does not characterize B^*-algebras however. In the commutative case any Dirichlet algebra A (one in which re A is dense in $C_{\mathbb{R}}(X)$) gives rise to a simplicial state space and hence has $H(A^*)$ a dual L-space. Of course, A need not be self-adjoint, considering, for example, the disc algebra.

Notes

The complex state space was first introduced by Asimow [**24**] in the context of Example 1 in order to discuss geometric peak-point conditions related to the Bishop Theorem for function algebras. We discuss this in greater detail in Section 4. The formulation of Theorem 0.1 and its application to unital Banach algebras appears in Asimow and Ellis [**32**].

The theorem of Bohnenblust and Karlin as well as the developments leading to what has come to be called the Vidav–Palmer Theorem are discussed in Bonsall and Duncan [**46**].

1. COMPLEX REPRESENTING MEASURES

Let X be a compact Hausdorff space and let $C(X)$ denote the space of continuous complex-valued functions on X. We let $M(X)$ denote the dual space of regular (complex) Borel measures on X. As we have seen, the real subspace of $M(X)$ of real measures is an L-space. The order structure also applies to a complex measure μ by means of the total variation measure, $|\mu|$, given by

$$|\mu|(f) = \sup\{|\mu(g)| : |g| \leq f\} \quad (\text{for } f \geq 0).$$

If E is a Borel set then

$$|\mu|(E) = \sup\left\{\sum_{k=1}^{n} |\mu(E_k)| : (E_k)_{k=1}^{n} \text{ a Borel partition of } E\right\}$$

(see Rudin [**185**]).

It follows that $\|\mu\| = \||\mu|\| = |\mu|(X)$. Also

$$d\mu = h\, d|\mu|$$

where h is a Borel function on X such that

$$|h| = 1 \quad (\text{a.e. } |\mu|).$$

Finally, if ρ is a Borel map from X to Y then for each $\mu \in M(X)$ there is a $\nu \in M(Y)$ given by $\nu = \mu \circ \rho^{-1}$ and

$$\int_Y g\, d\nu = \int_X (g \circ \rho)\, d\mu$$

for all bounded Borel functions g on Y.

Let A be a closed separating subspace of $C(X)$ containing the constants. We identify X with its embedding in the state space S_A and say $\mu \in M(X)$ is a *boundary measure* if either $\mu = 0$ or $|\mu|/\|\mu\|$ is a maximal probability measure on S_A. We denote boundary measures by $\partial_A M(X)$, or just $\partial M(X)$ if A is understood.

We let S denote the state space of $C(X)$, in other words, the probability measures on X, so that $\partial_A S$ (not to be confused with ext S) denotes the maximal measures on $X \subset S_A$. The Choquet–Bishop–de Leeuw Theorem asserts that for each $L \in S_A$ there is a (maximal) $\mu \in \partial_A S$ such that

$$L(f) = \int_X f \, d\mu$$

for all $f \in A$. This measure is unique if and only if S_A is a simplex. In this section we derive the complex version of this result: for each $L \in A_1^*$ there is a (complex) $\mu \in \partial M(X)$ such that

$$\|\mu\| = \|L\|$$

and

$$L(f) = \int_X f \, d\mu$$

for all $f \in A$. We also discuss the uniqueness question. This is accomplished by identifying complex measures on $X \subset S_A$ with real measures on $TX \subset A_1^*$ where

$$T = \{t \in \mathbb{C}: |t| = 1\} \quad \text{and} \quad TX = \{tx: t \in T, x \in X \subset S_A\}.$$

To do this we identify $C(X)$ with the subspace M in $C(X \times T)$ given by

$$M = \{f \in C(X \times T): f(x, t) = f(x, 1) = f(x), \text{ for all } t \in T\}.$$

Let $\phi \in C(X \times T)$ be given by $\phi(x, t) = t$ and define

$$\Psi: C(X \times T) \to C(X \times T) \quad \text{by } \Psi f = \phi f.$$

Since $|\phi| \equiv 1$, Ψ is an isometric-isomorphism and so is

$$\Psi^*: M(X \times T) \to M(X \times T); \qquad \Psi^* \tilde{\mu} = \bar{\mu} \quad \text{with } d\bar{\mu} = \phi \, d\tilde{\mu}.$$

Clearly $|\Psi^* \tilde{\mu}| = |\tilde{\mu}|$.

We note that the restriction of $\bar{\mu} \in M(X \times T)$ to the subspace M defines an element $\mu \in M(X)$. In fact it is easy to see that

$$\mu = \bar{\mu} \circ \pi^{-1}; \qquad \pi: X \times T \to X \quad \text{with } \pi(x, t) = x.$$

We now define, for $\tilde{\mu} \in M(X \times T)$,

$$\mu \doteq H\tilde{\mu} = \Psi^* \tilde{\mu}|_M.$$

Thus

$$d\mu = \phi \, d(\tilde{\mu} \circ \pi^{-1}).$$

It follows that for f a bounded Borel function on X,

$$\int_X f(x) \, d\mu(x) = \int_{X \times T} tf(x) \, d\tilde{\mu}(x, t).$$

1.1. PROPOSITION. *Let $\tilde{\mu}$ be a probability measure on $X \times T$ and let $\mu = H\tilde{\mu}$. If μ represents $L \in A^*$ with $\|L\| = 1$ then*

$$|\mu| = \tilde{\mu} \circ \pi^{-1}.$$

Proof. It is straightforward that $|\mu| \leq |\tilde{\mu}| \circ \pi^{-1} = \tilde{\mu} \circ \pi^{-1}$. Now

$$1 = \|L\| = \sup \{|L(f)| : f \in A_1^*\} = \sup \{|\mu(f)| : f \in A_1^*\} \leq \|\mu\| \leq \|\tilde{\mu}\| = 1.$$

Hence $|\mu|(X) = \tilde{\mu} \circ \pi^{-1}(X)$ so that equality must hold for any measurable subset E of X.

If $\mu \in M(X)$ with $d\mu = h \, d|\mu|$ $(|h| = 1)$ we can invert H as follows: let

$$\sigma_h : X \to X \times T; \qquad \sigma_h(x) = (x, h(x))$$

so that the graph G of h, given by

$$G = \{(x, t) : t = h(x)\} = \{(x, t) : \bar{t}h(x) = 1\}$$

is a Borel set. We consider $|\mu| \circ \sigma_h^{-1} \in M(X \times T)$.

1.2. PROPOSITION. *Let $L \in A^*$ with $\|L\| = 1$ and let $\mu \in M(X)_1$ represent L. Then $\mu = H\tilde{\mu}$ for $\tilde{\mu}$ a probability measure on $X \times T$ if and only if*

$$\tilde{\mu} = |\mu| \circ \sigma_h^{-1}.$$

Proof. Given $\mu \in M(X)_1$ representing L and $\tilde{\mu} = |\mu| \circ \sigma_h^{-1}$ we have, for $f \in M$,

$$(H\tilde{\mu})(f) = \int_{X \times T} tf(x) \, d\tilde{\mu}(x, t) = \int_{X \times T} (\Psi f) \, d(|\mu| \circ \sigma_h^{-1}) = \int_X (\Psi f) \circ \sigma_h \, d|\mu|$$

$$= \int_X h(x) f(x) \, d|\mu|(x) = \mu(f).$$

Thus $H\tilde{\mu} = \mu$. If $H\tilde{\mu} = \mu$ representing L, then Proposition 1.1 shows

$$|\mu| = \tilde{\mu} \circ \pi^{-1}.$$

Hence

$$1 = |\mu|(X) = \int_X \overline{h(x)} \, d\mu(x) = \int_{X \times T} \overline{th(x)} \, d\tilde{\mu}(x, t)$$

so that $\overline{th(x)} = 1$ a.e. $(\tilde{\mu})$. In other words $\tilde{\mu}(G) = 1$. But then if g is a bounded Borel function on $X \times T$

$$\int_{X\times T} g \,\mathrm{d}(|\mu| \circ \sigma_h^{-1}) = \int_X g(x, h(x)) \,\mathrm{d}|\mu|(x)$$

$$= \int_{X\times T} g(x, t) \,\mathrm{d}\tilde{\mu}(x, t) = \int_{X\times T} g \,\mathrm{d}\tilde{\mu}.$$

Let U denote A_1^* and observe that

$$\operatorname{ext} U = T \cdot (\operatorname{ext} S_A),$$

a subset of the compact set TX which is homeomorphic to $X \times T$. An element $f \in C(U)$ is called *invariant* if

$$f(tx) = f(x) \quad \text{for all } t \in T.$$

Define $\operatorname{inv} f(x) = \int_T f(tx) \,\mathrm{d}t$, where $\mathrm{d}t$ denotes normalized linear Lebesgue measure on T. Recall (Theorem 1.6.5) that a probability measure μ on U is maximal if and only if $\mu(B_f) = 1$ for all $f \in Q(U)$ (continuous convex functions on U), where

$$B_f = B(f) = \{x \in U : \hat{f}(x) = f(x)\}.$$

Let μ_t denote the measure on U given by

$$\mu_t(f) = \int_U f(tx) \,\mathrm{d}\mu(x).$$

1.3. PROPOSITION. *Let $f \in C(U)$ and $\mu \in M_1^+(U)$.*

(i) $\operatorname{inv} f$ *is invariant and if f is invariant,* $\operatorname{inv} f = f$.
(ii) *If f is convex then so is* $\operatorname{inv} f$.
(iii) *If f is invariant then so is $\hat{f}$.*
(iv) *If μ represents x then μ_t represents tx.*
(v) *If f is convex then $B(\operatorname{inv} f) \subset B(f)$.*

Proof. (i)–(iv) are straightforward. For (v) we recall (Theorem 1.6.1(vii))

$$\hat{f}(x) = \sup \{\mu(f) : \mu \in M_1^+(U) \text{ represents } x\}$$

and, since f is convex, we have

$$f(x) \le \mu(f) \le \hat{f}(x) \quad (\mu \sim x).$$

Let $\hat{f}(x) - f(x) > 0$ and choose μ representing x such that

$$\mu(f) - f(x) > 0.$$

If $g : T \to \mathbb{R}$ is given by

$$g(t) = \mu_t(f) - f(tx)$$

then g is continuous with $g \geq 0$ and $g(1) > 0$. Hence

$$0 < \int_T g(t)\,\mathrm{d}t = \int_T \int_U f(ty)\,\mathrm{d}\mu(y)\,\mathrm{d}t - \operatorname{inv} f(x)$$
$$= \mu(\operatorname{inv} f) - \operatorname{inv} f(x) \leq (\operatorname{inv} f)\hat{}\,(x) - \operatorname{inv} f(x).$$

If $\tilde{\mu}$ is a probability measure on $X \times T$ we can regard $\tilde{\mu}$ as a measure on the homeomorphic image $TX \subset U = A_1^*$. Since $\operatorname{ext} U \subset TX$, any maximal measure on U is supported by TX.

1.4. PROPOSITION. *Let $\tilde{\mu}$ be a probability measure on TX and let $\mu = H\tilde{\mu}$. Then*

(i) $\int_X f\,\mathrm{d}\mu = \int_{TX} f\,\mathrm{d}\tilde{\mu}$ *for all* $f \in A$;
(ii) $\tilde{\mu}$ *is maximal on* U *if and only if* $\tilde{\mu} \circ \pi^{-1}$ *is maximal on* S_A.

Proof. If $f \in A$ then

$$\int_{TX} f\,\mathrm{d}\tilde{\mu} = \int_{TX} f(tx)\,\mathrm{d}\tilde{\mu}(x, t) = \int_{TX} tf(x)\,\mathrm{d}\tilde{\mu}(x, t) = \int_X f\,\mathrm{d}(H\tilde{\mu}) = \int_X f\,\mathrm{d}\mu.$$

If f is an invariant Borel function on U then

$$\int_U f\,\mathrm{d}\tilde{\mu} = \int_{TX} f(x)\,\mathrm{d}\tilde{\mu}(x, t) = \int_X f \circ \pi\,\mathrm{d}\tilde{\mu} = \int_{S_A} f\,\mathrm{d}(\tilde{\mu} \circ \pi^{-1}).$$

Thus, if f is continuous and convex on U then Proposition 1.3(ii) and (iii) gives

$$\tilde{\mu}(B(\operatorname{inv} f)) = \tilde{\mu} \circ \pi^{-1}(B(\operatorname{inv} f) \cap S_A).$$

Hence by Proposition 1.3(v) $\tilde{\mu}$ is maximal on U if and only if $\tilde{\mu} \circ \pi^{-1}$ is maximal with support in S_A on U. But $\tilde{\mu} \circ \pi^{-1}$ is supported by $X \subset S_A$, a face of U, so that $\tilde{\mu} \circ \pi^{-1}$ is maximal on U if and only if it is maximal on S_A.

1.5. THEOREM. *Let $L \in A^*$ with $\|L\| = 1$.*

(i) *There is an element $\mu \in \partial_A M(X)_1$ such that μ represents L.*
(ii) *The measure μ in* (i) *is unique in the boundary measures of norm one if and only if L is represented by a unique maximal probability measure $\tilde{\mu}$ on $U = A_1^*$.*

Proof. Given $L \in A^*$ with $\|L\| = 1$ we can choose, by the real Choquet–Bishop–de Leeuw Theorem a maximal $\tilde{\mu}$ on $TX \subset U$ representing L. Let $\mu = H\tilde{\mu}$. Then $\mu = M(X)_1$ and represents L by Proposition 1.4(i). Since $\tilde{\mu} \circ \pi^{-1}$ is maximal on S_A by Proposition 1.4(ii) it follows from Proposition 1.1 that $|\mu| = \tilde{\mu} \circ \pi^{-1}$ is maximal. Hence μ is a boundary measure. The uniqueness equivalence of statement (ii) now follows from Proposition 1.2.

We say the subspace $A \subset C(X)$ has the *uniqueness property* if each $L \in A^*$ is represented by a unique boundary measure μ with $\|\mu\| = \|L\|$. We characterize this in terms of the geometry of A_1^* in the following corollary. Further applications of uniqueness to Lindenstrauss spaces will be discussed in Section 9.

We call a (not necessarily closed) proper face F of A_1^* a *simplex* if the cone spanned by F in A^* is a lattice. This is of course consistent with the usual definition of simplex in case F is w^*-compact (Chapter 2, Section 7). Since F is proper,

$$F \subset \{x \in A_1^* : \|x\| = 1\}.$$

1.6. COROLLARY. *The subspace $A \subset C(X)$ has the uniqueness property if and only if each proper face of A_1^* is a simplex.*

Proof. Let $U = A_1^*$ and let F be a proper face of U. If A has the uniqueness property then Theorem 1.5 shows the resultant map is one-to-one from a face of the maximal measures in $M_1^+(U)$ onto F. But the maximal measures themselves form a face of the simplex $M_1^+(U)$ (see the remark preceding Theorem 2.7.3) so that F is affinely equivalent to a simplex. For the converse let $\|x\| = 1$ in U and let F be the (algebraic) face of U generated by x. Then

$$F \subset \{y \in U : \|y\| = 1\}$$

is a simplex. We show that maximal measures on U representing x must agree on the cone $Q(U)$ of continuous convex functions in $C_{\mathbb{R}}(U)$. Since the linear span of $Q(U)$ is dense in $C_{\mathbb{R}}(U)$ this will establish the uniqueness property, by Theorem 1.5(ii). Thus let $f \in Q(U)$ and let μ be a maximal measure representing x. By Theorem 1.6.1(viii) there is a $-g \in Q(U)$ with

$$f \leq g$$

and

$$\int g \, d\mu < \int \hat{f} \, d\mu + \varepsilon$$

for given $\varepsilon > 0$. Since the cone spanned by F has the Riesz decomposition property it follows from Chapter 2, Section 4 (cf. Theorem 2.7.3) that $\hat{f}|_F$, given by

$$\hat{f}(y) = \sup\left\{\sum_{i=1}^{n} \lambda_i f(y_i) : y = \sum \lambda_i y_i\right\},$$

is affine on F. Similarly, $\check{g}|_F$ is affine and

$$\hat{f} \leq \check{g} \quad \text{on } F.$$

Thus, using Proposition 1.6.2 and Theorem 1.6.5 we have

$$\hat{f}(x) \geq \int f \, d\mu = \int \hat{f} \, d\mu > \int g \, d\mu - \varepsilon \geq \check{g}(x) - \varepsilon \geq \hat{f}(x) - \varepsilon.$$

Since $\varepsilon > 0$ is arbitrary we have

$$\int f \, d\mu = \hat{f}(x).$$

Notes

The Representation Theorem (Theorem 1.5(i)) is referred to as the Hustad Theorem. Hustad [**132**] constructed the map H and used it to obtain the representing measure μ. Hirsberg [**127**] showed that in general $H\tilde{\mu}$ is a boundary measure if $\tilde{\mu}$ is maximal on A_1^*. The construction of the inverse map in Proposition 1.2 is due to Fuhr and Phelps [**112**] who proved the uniqueness statement Theorem 1.5(ii). They also proved the geometric version of the uniqueness property in Corollary 1.6. Unit balls with that property are referred to as *simplexoids.* Many of the techniques herein employed, and in particular those of Proposition 1.3, are due to Effros [**96**]. For the analogous theory for subspaces A not containing constants we refer to Phelps [**172**].

2. INTERPOLATION SETS

Let A be a closed subspace of $C_{\mathbb{C}}(X)$ and let E be a closed subset of X. If Ψ is the restriction map from $C(X)$ to $C(E)$ then the Tietze Extension Theorem shows Ψ is onto and in fact maps the unit ball U of $C(X)$ onto the unit ball V of $C(E)$. Then Ψ^* embeds $(M(E), V^0)$ onto the w^*-closed subspace $(N, U^0 \cap N) \subset (M(X), U^0)$. If $\pi_1\mu = \mu|_E$, $\pi_2\mu = \mu|_{X \setminus E}$ then π_1, π_2 are complementing projections with $N = \text{range } \pi_1$.

Let $\theta = \Psi|_A$ and denote range θ by $A|_E$. We say E is an *interpolation set* for A if $A|_E$ is closed in $C(E)$. Since $\Psi U = V$, we have a subset B of $C(E)$ is closed if and only if $\Psi^{-1}(B)$ is closed in $C(X)$. In particular $A|_E$ is closed if and only if $A + M$ is closed in $C(X)$, where

$$M = \ker \Psi = \{f \in C(X) : f|_E \equiv 0\}.$$

If q_X, q_E denote the quotient maps

$$q_X : M(X) \to M(X)/A^\perp = A^*$$

$$q_E : M(E) \to M(E)/(A|_E)^\perp = (A|_E)^*$$

then $\theta^* \circ q_E = q_X \circ \Psi^*$.

Let $\hat{U} = q_X(U^0)$, $\hat{N} = q_X N$. Then θ^* identifies $\hat{V} = q_E(V^0)$ with

$$q_X(U^0 \cap N) = (U^0 \cap N)\hat{}.$$

We can now formulate Theorem 1.3.5 in this context.

2.1. THEOREM. *Let E be a closed subset of X and let A be a closed separating subspace of $C(X)$ with constants. The following are equivalent:*

(i) *E is an interpolation set for A;*
(ii) $(A+M)^- \cap (U+M) \subset r(A \cap U+M)^-$ *for some $r \geq 1$;*
(iii) $(A+M) \cap (\operatorname{int} U+M) \subset r(A \cap \operatorname{int} U+M)$ *for some $r \geq 1$;*
(iv) *$\hat{N}$ is w^*-closed in A^*;*
(v) *$\hat{N}$ is norm-closed in A^*;*
(vi) $\hat{U} \cap \hat{N} \subset r(U^0 \cap N)\hat{}$ *for some $r \geq 1$.*

In particular, if E is an interpolation set then for each $g \in (A|_E)^-$ and $\alpha > 1$ (using (iii)) there is an $f \in A$ such that $f|_E = g$ and $\|f\|_X \leq \alpha r\|g\|_E$.

Another formulation of 2.1(vi) is useful. By taking inverse images under q_X we see that (vi) is equivalent to

$$(U^0 + A^\perp) \cap (N + A^\perp) \subset r(U^0 \cap N + A^\perp),$$

or

$$N \cap (U^0 + A^\perp) \subset rV^0 + A^\perp \cap N.$$

Since $\theta^* \circ q_E = q_X \circ \Psi^*$, the last statement is the same as

$$\|\mu + A^\perp \cap N\| = \|\mu + (A|_E)^\perp\| \leq r\|\mu + A^\perp\| \quad \text{for all } \mu \in M(E).$$

2.2. COROLLARY. *The following are equivalent:*

(i) *E is an interpolation set for A;*
(ii) $\|\mu + (A|_E)^\perp\| \leq r\|\mu + A^\perp\|$ *for some $r \geq 1$ and all $\mu \in N$;*
(iii) $\|\pi_1 m + (A|_E)^\perp\| \leq r\|\pi_2 m\|$ *for all $m \in A^\perp$.*

Proof. As noted above, (ii) is a reformulation of 2.1(vi). If (ii) holds and $m \in A^\perp$ then

$$\begin{aligned}\|\pi_1 m + (A|_E)^\perp\| &\leq r\|\pi_1 m + A^\perp\| = r\|m - \pi_2 m + A^\perp\| \\ &= r\|-\pi_2 m + A^\perp\| \leq r\|\pi_2 m\|.\end{aligned}$$

If (iii) holds and $\mu \in N$, $m \in A^\perp$, then

$$\begin{aligned}\|\mu + m\| &= \|\pi_2 m + (\pi_1 m + \mu)\| = \|\pi_2 m\| + \|\pi_1 m + \mu\| \\ &\geq (1/r)\|\pi_1 m + (A|_E)^\perp\| + (1/r)\|\pi_1 m + \mu\| \\ &\geq (1/r)\|(\mu + \pi_1 m) - (\pi_1 m + (A|_E)^\perp)\| = (1/r)\|\mu + (A|_E)^\perp\|.\end{aligned}$$

Hence

$$r\|\mu + A^\perp\| \geq \|\mu + (A|_E)^\perp\|.$$

We say E is a *full interpolation set* for A if $A|_E$ in fact equals $C(E)$. This is equivalent to E being an interpolation set for which $(A|_E)^\perp = \{0\}$ in $N = M(E)$.

2.3. COROLLARY. *The following are equivalent*:

(i) *E is a full interpolation set for A*;

(ii) $\|\mu\| \le r\|\mu + A^\perp\|$ *for some* $r \ge 1$ *and all* $\mu \in N$;

(iii) $\|\pi_1 m\| \le r\|\pi_2 m\|$ *for all* $m \in A^\perp$.

Proof. The above remark shows (i) implies (ii). The proof of (ii) implies (iii) is the same as in Corollary 2.2. If (iii) holds then, in particular, E is an interpolation set. If $\mu \in (A|_E)^\perp$ then $\mu \in A^\perp \cap N$ and

$$\|\mu\| = \|\pi_1 \mu\| \le r\|\pi_2 \mu\| = 0$$

so that $(A|_E)^\perp = \{0\}$ and E is a full interpolation set.

We will say a subset B of $M(X)$ is *A-stable* if $\hat{B} = q_X(B) = q_X(B \cap \partial_A M(X))$. We will denote $B \cap \partial_A M(X)$ by $\partial_A B$ so that B is A-stable if and only if

$$\hat{B} = (\partial B)\hat{}.$$

Hustad's Theorem shows U^0 is A-stable. If, in addition, N is A-stable then we can modify the measure theoretic conditions of Corollary 2.2 and Corollary 2.3 by restricting attention to the boundary annihilating measures, $\partial A^\perp$.

2.4. COROLLARY. *Let $N = M(E)$ be A-stable. Then the following are equivalent*:

(i) *E is an interpolation set for A*;

(ii) $\|\pi_1 m + (A|_E)^\perp\| \le r\|\pi_2 m\|$ *for some* $r \ge 1$ *and all* $m \in \partial A^\perp$.

Proof. If N is A-stable then for $\mu \in \partial M(X)$,

$$\|\mu + A^\perp\| = \|\mu + \partial A^\perp\|$$

since

$$\mu = s\nu + m; \qquad \nu \in U^0, \qquad m \in A^\perp$$

implies, by Hustad's Theorem, that

$$\mu = s\nu' + m'; \qquad \nu' \in \partial U, \qquad m' \in A^\perp.$$

Since $\mu, \nu' \in \partial M(X)$, it follows that $m' \in \partial A^\perp$. Thus, if E is an interpolation set then Corollary 2.2 shows

$$\|\mu+(A|_E)^\perp\|\le r\|\mu+\partial A^\perp\|$$

for all $\mu\in\partial N$. Thus (ii) follows for $m\in\partial A^\perp$ since $\pi_1 m\in\partial A^\perp$ also.

Conversely, if (ii) holds and $x\in\hat{N}\cap\hat{U}$, choose $\mu\in\partial N$ with $\hat{\mu}=x$ and let m be any element of $\partial A^\perp$. Then the computation of Corollary 2.2(iii) implies 2.2(ii) shows

$$r\|\mu+m\|\ge\|\mu+(A|_E)^\perp\|$$

and hence

$$\|\mu+(A|_E)^\perp\|\le r\|\mu+\partial A^\perp\|=r\|\mu+A^\perp\|.$$

Thus $x\in r(U^0\cap N)\hat{}$ so that

$$\hat{N}\cap\hat{U}\subset r(U^0\cap N)\hat{}$$

and E is an interpolation set.

2.5. Corollary. *If $N=M(E)$ is A-stable then the following are equivalent*:

(i) *E is a full interpolation set for A;*

(ii) *$\|\pi_1 m\|\le r\|\pi_2 m\|$ for some $r\ge 1$ and all $m\in\partial A^\perp$.*

Let $E^\perp=M\cap A$, where $M=\{f\in C(X): f|_E=0\}$. Then $E^{\perp\perp}$ is the w^*-closed (complex) subspace of A^* spanned by E as a subset of $S_A\subset A^*$. We define the *A-hull* $h(E)$ to be

$$E^{\perp\perp}\cap X\subset S_A.$$

Thus

$$h(E)=\{x\in X: f(x)=0 \text{ for all } f\in E^\perp\}.$$

Similarly, the *A-convex hull* $k(E)$ is given by

$$k(E)=(\overline{\text{co}}\,E)\cap X.$$

Clearly $E\subset k(E)\subset h(E)$.

2.6. Proposition. *The set $k(E)=\{x\in X: |f(x)|\le\|f\|_E$ for all $f\in A\}$.*

Proof. The Separation Theorem shows (using rotation by $t\in T$)

$$k(E)=\{x\in X: \operatorname{re} f(x)\le\sup_{y\in E}\operatorname{re} f(y) \text{ for all } f\in A\}$$

$$=\{x\in X: f(x)\in\overline{\text{co}}\,f(E) \text{ for all } f\in A\}.$$

Thus $x\in k(E)$ and $f\in A$ implies $|f(x)|\le\|f\|_E$. If $x\notin k(E)$ then $x\notin\overline{\text{co}}\,f(E)$ for some $f \in A$. Hence there is a disc $D(\alpha; r)$ about $\alpha \in \mathbb{C}$ of radius r such that

$$f(E) \subset D(\alpha; r) \quad \text{and} \quad x \notin D(\alpha; r).$$

If $t \in T$ is chosen so that $t(f(x)-\alpha) \geq 0$ and $g = t(f-\alpha)$ then $g(E) \subset D(0; r)$ with $|g(x)| = g(x) > r$. Thus

$$|g(x)| > \|g\|_E.$$

We close this section by observing some instances where $N = M(E)$ is A-stable.

2.7. PROPOSITION. *If E is a closed interpolation set for A such that either*
(i) $E \subset \partial_A X$, *or*
(ii) $E = F \cap X$, *F a closed face of S_A, then $M(E)$ is A-stable.*

Also, (ii) *implies $E = k(E)$.*

Proof. If (i) holds then each element of $M_1^+(E)$ is maximal (cf. Theorems 1.6.3 and 1.6.5) so that $M(E) \subset \partial M(X)$. If (ii) holds and μ is an element of $\partial M_1^+(X)$ representing $x \in F$ the $\mu \in M(E)$ (Proposition 2.10.6). Since $(M(E))\hat{}$ is spanned by F we have $M(E)$ is A-stable.

Notes

As indicated in Chapter 2 the closed range conditions of Theorem 2.1 have been recognized in various forms by many researchers. The seminal work in this regard is contained in the 1962 paper by Glicksberg [**118**]. In particular he gives the equivalences of Corollary 2.2(i) and (ii) as well as Corollary 2.3. The equivalence of Theorem 2.1(i), (iv)–(vi) appears in Edwards [**87**] for the real $A(K)$ case where N is spanned by a w^*-closed face of K. Alfsen [**3**] also discusses this situation and exhibits the constant r in statement (vi) as a bound on the norm of extensions. General formulations of this principle for $A(K)$ and for complex function spaces are given in Asimow [**23, 26**].

The formulation of Corollary 2.2(iii) in the general case is given by Gamelin [**113**]. The use of boundary measures of course awaited the maturation of Choquet theory and first appears in Alfsen–Hirsberg [**9**] where stability is guaranteed by the hypothesis that E is a compact subset of $\partial_A X$ (cf. Proposition 2.7(i)). They prove (ii) implies (i) in Corollary 2.4 for the case $r = 0$. This very important case will be discussed in detail in Section 4. Corollary 2.4 is proved in its present form by Briem [**50**] where the same stability condition ($E \subset \partial_A X$) is assumed. The alternate stability condition (Proposition 2.7(ii)) appears in Hirsberg [**126**].

Further abstract versions of the closed range conditions can be found in Andô [**19**], Roth [**182**] and Asimow [**28, 29**]. Roth [**182**] also discusses conditions involving boundary measures in a quite general setting.

The equivalence of the geometric definition of $k(E)$ and the statement of Proposition 2.6 has been well known in the context of function algebras. It is shown in Alfsen and Hirsberg [**9**] for the case $\overline{\text{co}}\, E$ a parallel face of S_A and in general by Briem [**50**].

Examples of closed subsets that are not interpolation sets are plentiful. We show, for example, in Section 6 that for A the *disc algebra* and E a proper closed arc of T we have $A|E$ dense in, but not equal to, $C(E)$.

3. GAUGE DOMINATED EXTENSIONS AND COMPLEX STATE SPACE

If E is a closed interpolation set in X for a subspace $A \subset C(X)$ then Theorem 2.1 shows there is an $r \geq 1$ such that each $g \in (A|_E)^-$ with $|g| < 1$ on E is the restriction of an $f \in A$ with $|f| < r$ on X.

If we let

$$\rho_r : X \to \mathbb{R}^+; \qquad \rho_r = \begin{cases} 1 \text{ on } E \\ r \text{ on } X \backslash E \end{cases}, \qquad r \geq 1$$

then ρ_r is a strictly positive bounded l.s.c. function on X such that if $|g| < \rho_r$ on E then $f|_E = g$ and $|f| < \rho_r$ on X. The smallest such r is related to a number called the extension constant, $e(A, E)$, which we will discuss in detail in Section 6.

In this section we introduce machinery for dealing with questions of extending gauge-dominated functions on E to X and take a closer look at complex state spaces in this framework. To obtain results of greater generality we can require the *dominator*, ρ, to be sensitive to argument as well as modulus in the complex plane. Thus, let

$$\rho : X \times T \to \mathbb{R}^+ \cup \{\infty\} \qquad (T = \{z \in \mathbb{C} : |z| = 1\})$$

be a strictly positive l.s.c. function and let

$$U = \{f \in C(X) : \text{re } tf(x) \leq \rho(x, t);\ (x, t) \in X \times T\}.$$

We will denote the Minkowski functional $p(U)$ by $\|\cdot\|_\rho$. Thus

$$\|f\|_\rho = \sup \{\text{re } tf(x)/\rho(x, t) : (x, t) \in X \times T\}.$$

Since ρ is strictly positive and l.s.c. ρ is bounded away from 0. Hence there is a constant k such that

$$\|f\|_\rho \leq k\|f\|$$

where $\|\cdot\|$ denotes the usual uniform norm in $C(X)$. We shall refer to a $\|\cdot\|_\rho$ generated in this way as a *gauge* on $C(X)$.

If in addition ρ is bounded above then $\|\cdot\|_\rho$ is equivalent to $\|\cdot\|$. In general U is a closed convex neighbourhood of 0 so that $\|\cdot\|_\rho$ is sub-additive and positive homogeneous.

We say the *ρ-interpolation problem* has *approximate solutions* for A, E if for each $g \in (A|_E)^-$ and $\varepsilon > 0$ there is an $f \in A$ with $f|_E = g$ and $\|f\|_\rho \leq \|g\|_\rho + \varepsilon$. Here $\|g\|_\rho$ refers to $\rho|_{E\times T}$. We say E is an *approximate ρ-interpolation set* for A if approximate solutions exist. In particular E is an interpolation set for A. We say the ρ-interpolation problem has *exact solutions* if the f can be chosen with $\|f\|_\rho = \|g\|_\rho$. Thus, if E is an interpolation set for A then Theorem 2.1 says that for a minimum $r < \infty$ E is an approximate ρ_r-interpolation set.

We take up some special cases in the remainder of this section and return to interpolation problems in general in Section 5.

First, we indicate how results for real spaces, in particular $A(K)$ spaces, can be derived in this context. If A_0 is a subspace of $C_{\mathbb{R}}(X)$ then we consider

$$A_0 + iA_0 \subset C(X).$$

Let ρ^+, ρ^- be real gauges on X and define

$$\rho(x, t) = \begin{cases} \rho^+(x) & \text{for } x \in X \times \{1\} \\ \rho^-(x) & \text{for } x \in X \times \{-1\} \\ +\infty & \text{elsewhere.} \end{cases}$$

Then

$$U = \{f \in C(X): -\rho^- \leq \operatorname{re} f \leq \rho^+\}$$

so that the ρ-interpolation problem for A, E reduces to interpolating by functions in A_0 dominated below and above by $-\rho^-$ and ρ^+ respectively.

We next elaborate on the complex state space discussed in Section 0 by constructing gauges which induce a natural identification of A with real order unit spaces.

We give a partial ordering to the complex numbers $\mathbb{C}$ (considered as a 2-dimensional real space) by letting P be any proper closed convex cone with non-empty interior. Then the dual cone

$$P^* = \{z \in \mathbb{C}: \operatorname{re} az \geq 0 \text{ for all } a \in P\}$$

is also proper with non-empty interior. Let I denote a compact base for P^* with extreme points α and β of modulus 1. Clearly α and β may be any two elements of T (depending on P) such that $|\operatorname{re} \bar{\alpha}\beta| < 1$. Then each $z \in \mathbb{C}$ has a unique representation as a real linear combination of α and β with

$$P^* = \{z \in \mathbb{C}: z = \lambda\alpha + \mu\beta;\ \lambda, \mu \geq 0\}.$$

Let u denote the element in $\operatorname{int} P$ such that $\operatorname{re} \alpha u = \operatorname{re} \beta u = 1$. We note that the base norm (L-norm) on $(\mathbb{C}, P^*)$ is given by

$$|z|_u^* = |\lambda| + |\mu|; \qquad z = \lambda\alpha + \mu\beta.$$

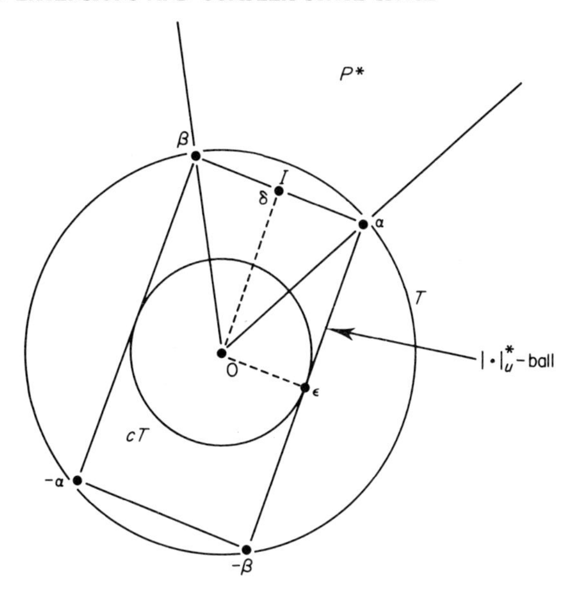

$|\delta| = |\text{re}(\bar{\alpha}\beta)^{1/2}|$, $|\epsilon| = |\text{im}(\bar{\alpha}\beta)^{1/2}|$

One sees from elementary calculus that the minimum of $|z|$ for $|z|_u^* = 1$ occurs midway between α and β so that

$$c|z|_u^* \le |z| \le |z|_u^*$$

for

$$c = \min\{|\text{re}\,(\bar{\alpha}\beta)^{1/2}|, |\text{im}\,(\bar{\alpha}\beta)^{1/2}|\}.$$

Then the order unit (M-norm) on $(\mathbb{C}, u)$ is given by

$$|z|_u = \max\{|\text{re}\,\gamma z| \colon \gamma = \alpha, \beta\}$$

and

$$c|z| \le |z|_u \le |z|.$$

Now let

$$\rho_u(x, t) = \begin{cases} 1 & \text{for } t = \pm\gamma \\ 1/c & \text{for } t \ne \pm\gamma \end{cases}, \qquad \gamma = \alpha, \beta.$$

We denote the gauge induced on $C(X)$ by $\|\cdot\|_u$ so that

$$\|f\|_u = \sup\{|\text{re}\,\gamma f(x)| \colon x \in X, \gamma = \alpha, \beta\} = \sup\{|f(x)|_u \colon x \in X\}.$$

We show next that ρ_u induces an order unit structure on $C(X)$ which allows an identification with real functions on

$$X_0 = X \times \{\alpha, \beta\}.$$

We shall think of X_0 as a disjoint union of X_γ's; $\gamma = \alpha, \beta$ where

$$X_\gamma = X \times \{\gamma\}$$

is identified with X.

The choice $\alpha = 1$, $\beta = -i$, yields the structure introduced in Section 0.

Let A be any separating subspace of $C_{\mathbb{C}}(X)$ with constants and define

$$\Psi\colon C(X) \to C_{\mathbb{R}}(X_0); \qquad \Psi f(x, \gamma) = \operatorname{re} \gamma f(x); \qquad \gamma = \alpha, \beta.$$

and let $\theta = \Psi|_A$. Let $\mathbf{u}$ denote the constant function identically u.

Let $\mu_\gamma(\gamma = \alpha, \beta)$ be measures on X and define $\tilde{\mu}$ on X_0 by $\tilde{\mu} = \mu_\alpha \oplus \mu_\beta$ with $\mu_\gamma = \tilde{\mu}|_{X_\gamma}$. Then

$$\operatorname{re}(f, \Psi^* \tilde{\mu}) = (\Psi f, \tilde{\mu}) = \int_{X_\alpha} \operatorname{re} \alpha f(x)\, \mathrm{d}\mu_\alpha(x) + \int_{X_\beta} \operatorname{re} \beta f(x)\, \mathrm{d}\mu_\beta(x)$$

$$= \operatorname{re} \int_X (\alpha f\, \mathrm{d}\mu_\alpha + \beta f\, \mathrm{d}\mu_\beta) = \operatorname{re}(f, \alpha\mu_\alpha + \beta\mu_\beta)$$

so that

$$\Psi^* \tilde{\mu} = \alpha\mu_\alpha + \beta\mu_\beta.$$

This provides a correspondence between non-negative measures on X_0 and complex measures on X analogous to the Hustad map H of Section 1.

3.1. THEOREM. *The map Ψ is a real Banach space isomorphism into $C_{\mathbb{R}}(X_0)$ such that if $\|\cdot\|_u$ is the Ψ-induced norm given by*

$$\|f\|_u = \|\Psi f\| = \sup \{|\operatorname{re} \gamma f(x)| \colon x \in X; \gamma = \alpha, \beta\}$$

and

$$\mathbf{P} = \{f \colon \Psi f \geq 0\} = \{f \colon f(X) \subset P\}$$

then each $(A, \mathbf{P}, \|\cdot\|_u, \mathbf{u})$ is real isometrically order isomorphic (under θ) to a closed (separating with constants) subspace B of $C_{\mathbb{R}}(X_0)$, with $\theta\mathbf{u} = 1$.

Further,

(i) *The state space, denoted Z_A, of $(A, \mathbf{u})^*$ is given by*

$$Z_A = \operatorname{co}(\alpha S_A \cup \beta S_A)$$

with αS_A, βS_A closed disjoint faces of Z_A.

(ii) *In case $A = C(X)$ then the state space*

$$Z = \operatorname{co}(\alpha S \cup \beta S) \qquad (S = M_1^+(X))$$

where γS $(\gamma = \alpha, \beta)$ are closed complementary split faces of Z.

(iii) *Let μ_γ $(\gamma = \alpha, \beta)$ be probability measures on X and $\tilde{\mu} = \lambda\mu_\alpha \oplus$*

$(1-\lambda)\mu_\beta$ $(0 \le \lambda \le 1)$. *Then*

$$\mu = \Psi^* \tilde{\mu} = \lambda\alpha\mu_\alpha + (1-\lambda)\beta\mu_\beta$$

is a boundary measure for A *if and only if the* μ_γ *are maximal measures on* $X \subset S_A$ *if and only if* $\tilde{\mu}$ *on* $X_0 \subset \alpha X \cup \beta X$ *is maximal on* Z_A.

Furthermore, if $\tilde{\mu} = \mu_\alpha \oplus \mu_\beta$ *for* $\mu_\gamma \in M_{\mathbb{R}}(X)$ *then* $\tilde{\mu} \in A(Z_A)^\perp$ *if and only if* $\mu = \Psi^* \tilde{\mu} \in A^\perp$.

(iv) *For each* $L \in Z_A$ *there is a boundary measure* $\mu \in Z$ *representing* L *and for each* $L \in A^*$ *there is a boundary measure* $\mu \sim L$ *such that* $\|\mu\|_u = \|L\|_u$.

Proof. Clearly Ψ is a real linear map, and in view of the preceding remarks,

$$c\|f\| \le \|f\|_u \le \|f\|$$

where c is as above. Since $\Psi\mathbf{u}(x, \gamma) = \text{re } \gamma u = 1$, $\Psi\mathbf{u} = 1$. Also $\Psi f \ge 0$ if and only if re $\gamma f(x) \ge 0$ $(\gamma = \alpha, \beta)$ for each x, if and only if $f(x) \in P$.

Now let B be the image of A under θ in $C_{\mathbb{R}}(X_0)$. Then B separates points in each X_γ and the function $\theta\bar{\alpha}$ is identically one on X_α and re $\bar{\alpha}\beta$ on X_β. This shows the evaluation map on X_0 is one-to-one and that

$$S_B = \overline{\text{co}}\, X_0 = \text{co}\,(S_\alpha \cup S_\beta)$$

where $S_\gamma = \overline{\text{co}}\, X_\gamma$ is a closed face of S_B. Since, for $f \in A$,

$$\text{re}\,(f, \theta^*(x, \gamma)) = (\theta f, (x, \gamma)) = \text{re}\,(f, \gamma x)$$

we have

$$\theta^*(x, \gamma) = \gamma x, \qquad \theta^* S_\gamma = \gamma S_A, \qquad \theta^* S_B = Z_A$$

and (i) is shown.

Let $A = C(X)$ and let $\mu \in M(X)$. Then, since each $z \in \mathbb{C}$ is a unique real linear combination of α and β, the real measures given by

$$\mu(f) = \alpha\mu_\alpha(f) + \beta\mu_\beta(f); \qquad f \ge 0$$

are uniquely defined. In particular, for $\mu \in Z$, the expression

$$\mu = \lambda\alpha\mu_\alpha + (1-\lambda)\beta\mu_\beta \qquad (0 \le \lambda \le 1, \mu_\gamma \in S)$$

is unique. Hence each S_γ is split with the other as complement.

For (iii) and (iv) we note first that if $\mu \in M(X)$ and $\mu \sim L \in A^*$ and

$$\mu = \alpha\mu_\alpha + \beta\mu_\beta = \Psi^* \tilde{\mu} \qquad (\tilde{\mu} = \mu_\alpha \oplus \mu_\beta \quad \text{on } X_0 = \alpha X \cup \beta X \subset Z_A)$$

then, using the remark preceding the theorem,

$$\int_{X_0} (\theta f)\, d\tilde{\mu} = \text{re} \int_X f\, d\mu = \text{re}\,(f, L)$$

so that $\tilde{\mu}$ represents L with respect to $A(Z_A)=\theta A$, and conversely. Thus $\tilde{\mu}\in A(Z_A)^\perp$ if and only if $\mu\in A^\perp$.

If $L\in Z_A$ then

$$L=\gamma\alpha L_\alpha+(1-\lambda)\beta L_\beta;\qquad L_\alpha, L_\beta\in S_A.$$

Then the Choquet–Bishop–de Leeuw Theorem yields maximal $\mu_\gamma\in S$ with $\mu_\gamma\sim L_\gamma$.

Then if $\mu=\lambda\alpha\mu_\alpha+(1-\lambda)\beta\mu_\beta$ we have $\mu\sim L$ and

$$c(\lambda\mu_\alpha+(1-\lambda)\mu_\beta)\leq|\mu|\leq\lambda\mu_\alpha+(1-\lambda)\mu_\beta$$

so that μ is a boundary measure if and only if each μ_γ is maximal. Since γS_A is a face of Z_A, $\tilde{\mu}$ is maximal on Z_A if and only if μ_γ is maximal on S_A. Since the $\|\cdot\|_u^*$ unit ball is

$$\text{co}\,(Z_A\cup -Z_A)$$

the rest of (iv) follows.

We observe that Theorem 3.1(iv) can be phrased as saying Z_A and $(A, \mathbf{u})_1^*=\text{co}\,(Z_A\cup -Z_A)$ are stable in A^*.

4. PEAK SETS AND M-SETS

The realization of the closed subspace $1\in A\subset C(X)$ as $A(Z_A)$ enables us to apply the techniques of ordered Banach space theory in Chapter 2, Section 8. This allows for a geometrically appealing development of peak set and M-set (defined below) criteria.

A closed subset E of the compact Hausdorff space X is called a *peak set* for the (closed, separating, with constants) subspace A of $C(X)$ if there exists an $f\in A$ with

$$f\equiv 1\quad\text{on } E\quad\text{and}\quad |f|<1\quad\text{on } X\backslash E.$$

In particular peak sets are G_δ's. We say E is a *generalized peak set* for A if E is an intersection of peak sets. Thus E is a generalized peak set if and only if for each compact K disjoint from E there is an $f_K\in A$ with

$$\|f_K\|=1,\qquad f_K\equiv 1\quad\text{on } E\quad\text{and}\quad |f_K|<1\quad\text{on } K.$$

If E is a generalized peak set and a G_δ then E is a peak set (use $f=\sum_{n=1}^\infty 2^{-n}f_n$).

In the following we identify (A, u) with the real affine function space $A(Z_A)$ in accordance with Theorem 3.1(i) where, for $f\in A$, $z\in Z_A$, we have $\theta f(z)=\text{re}\,(f, z)$.

Let E be a closed subset of X and let J be the image under θ of $E^\perp$ so that $J=\{a\in A(Z_A)\colon a\equiv 0 \text{ on } \alpha E\cup\beta E\subset Z_A\}$. Let

$$Z(E)=J^0\cap Z_A.$$

Then $Z(E)$ is a section of Z_A by J.

4.1. THEOREM. *If for some $\alpha,\beta\in T$ with $|\text{re }\bar{\alpha}\beta|<1$, $Z(E)$ is a positively decomposable section of Z_A by J then $h(E)$ is a generalized peak set.*

Proof. We have $J^0=E^{\perp\perp}$ and $h(E)=J^0\cap X\subset S_A$ so that if $y\in X\backslash h(E)$ then $y\notin E^{\perp\perp}\cap X$ and hence $\alpha y,\beta y\in Z_A\backslash Z(E)$. Let K be a compact subset of X disjoint from $h(E)$. Now Theorem 2.8.13 shows there is an $a\in A(Z)$ such that

$$a\geq 0,\qquad a\equiv 0\quad\text{on } Z(E)\qquad\text{and}\qquad a>0\quad\text{on } \alpha K\cup\beta K \text{ in } Z_A.$$

Now $a=\text{re } f$, with $f\in\mathbf{P}$. Thus $f(X)\subset P$. Since for $x\in h(E)$, $\text{re }(f,\alpha x)=0=\text{re }(f,\beta x)$ we have $f\equiv 0$ on $h(E)$. For $y\in K$, $\text{re }(f,\alpha y)$, $\text{re }(f,\beta y)>0$ so that $f(y)\in P\backslash\{0\}$. Since P is proper there is an $s>0$ such that

$$u-P_s\subset D(0,|u|).$$

By multiplying f by a positive scalar, we can assume $f(X)\subset P_r$, $0<r<s$. Thus, if $g=1-(1/u)f$, then

$$g\equiv 1\quad\text{on } h(E),\qquad |g|\leq 1\quad\text{on } X\qquad\text{and}\qquad |g|<1\quad\text{on } K.$$

We now consider two types of peak set conditions that have arisen in the study of function spaces, and show that both are instances of Theorem 4.1. Further measure theoretic conditions will be discussed in Section 6.

For the first, we define E to be a *strong hull* for A if there is a constant $k>0$ such that for each neighbourhood V of E there exists $f_V\in A$ with $f_V\in E^\perp$, $\|f_V\|\leq k$ and $\text{re } f_V\geq 1$ on $X\backslash V$. We say E is a *weak hull* if for some constant $k>0$, for each neighbourhood V of E and $\varepsilon>0$ there exists $f_{V,\varepsilon}$ such that

$$|f_{V,\varepsilon}|\leq\varepsilon\quad\text{on } E,\qquad \|f_{V,\varepsilon}\|\leq k\qquad\text{and}\qquad \text{re } f_{V,\varepsilon}\geq 1\quad\text{on } X\backslash V.$$

If E is a weak or strong hull with constant k then choose P symmetric about the positive real axis such that

$$D(0;k)\cap\{z\in\mathbb{C}\colon \text{re } z\geq 1\}\subset\tfrac{1}{2}+P.$$

Then u is real and we let α, $\bar{\alpha}$ denote the generators of P^*. Let

$$F=\overline{\text{co}}\, E\subset S_A\quad\text{and}\quad \tilde{E}=\text{co}\,(\alpha F\cup\bar{\alpha}F).$$

4.2. THEOREM. *Let E be either a strong hull or a weak hull and interpolation set and let P be as above. Then $Z(E)=\tilde{E}$ and is a positively* *decomposable section of Z_A by J. Furthermore, $E = h(E)$ is a generalized peak set.*

Proof. Assume first that E is a strong hull and let $h \in (A^{**})_k$ be a w^{**}-limit point of the net $\{f_V\}_{E \subset V}$. Since each $f_V \in E^{\perp}$, $f_V \equiv 0$ on $E^{\perp\perp}$ and hence so is h. Now let $z \in Z_A \backslash \tilde{E}$ and let μ be a probability measure on Z_A representing z with supp $\mu \subset \partial \overline{Z} \subset \alpha X \cup \bar{\alpha} X$. We show first that if $\mu(\alpha E \cup \bar{\alpha} E) = 0$ then re $h(z) \geq \frac{1}{2}$. Let W be a neighbourhood of $\alpha E \cup \bar{\alpha} E$ in Z_A with $\mu(W) < \varepsilon$. Let V be any neighbourhood of E such that $\alpha V \cup \bar{\alpha} V \subset W$. Then for $y \in X \backslash V$, re $f_V(y) \geq 1$ so that

$$f_V(y) \in \tfrac{1}{2} + P.$$

Hence re $\alpha f_V(y)$, re $\bar{\alpha} f_V(y) \geq \frac{1}{2}$. Thus, if $B = \alpha V \cup \bar{\alpha} V$ and $C = (\alpha X \cup \bar{\alpha} X) \backslash B$, then

$$\begin{aligned} \theta f_V(z) &= \int_B \theta f_V \, \mathrm{d}\mu + \int_C \theta f_V \, \mathrm{d}\mu \\ &\geq \tfrac{1}{2}\mu(C) - k\mu(B) \geq \tfrac{1}{2}(1-\varepsilon) - k\varepsilon = \tfrac{1}{2} - (k + \tfrac{1}{2})\varepsilon. \end{aligned}$$

Thus re $h(z) \geq \frac{1}{2}$.

Now, for any $x \in Z_A$ and representing measure μ we can, by restricting μ to $\alpha E \cup \bar{\alpha} E$ and its complement in $\alpha X \cup \bar{\alpha} X$, write μ as a convex combination of μ_1, μ_2, where

$$\text{supp } \mu_1 \subset \alpha E \cup \bar{\alpha} E \quad \text{and} \quad \mu_2(\alpha E \cup \bar{\alpha} E) = 0.$$

Hence x is a convex combination of $y \in \tilde{E}$ and z with re $h(z) \geq \frac{1}{2}$. Thus Theorem 2.10.3 shows, since re $h(z) \geq \frac{1}{2} q_1$ on Z_A that $\tilde{E}$ is a positively decomposable section by J. Since $y \in X \backslash E$ implies re $h(y) \geq \frac{1}{2}$, $y \notin E^{\perp\perp}$ so that $y \notin h(E)$. Hence $E = h(E)$ is a generalized peak set.

If E is a weak hull and interpolation set then the same argument shows there exists $h \in (A^{**})_k$ with

$$Z_A = \text{co}\,(\tilde{E} \cup \{z \in Z : \text{re } h(z) \geq \tfrac{1}{2}\}).$$

But the fact that E is an interpolation set means $E^{\perp\perp}$ is the algebraic span of $\tilde{E}$. Hence $h \equiv 0$ on $E^{\perp\perp}$ and $\tilde{E}$ is a section of Z by $E^{\perp}$.

As an application we consider the case where A is a function algebra (also called a *uniform algebra*), that is a closed subalgebra of $C(X)$ containing constants and separating points.

4.3. Corollary (Bishop Peak Point Theorem). *Let A be a function algebra and let $x_0 \in \partial_A X = \text{ext } S_A$. Then x_0 is a generalized peak point.*

Proof. We take $\alpha = 1$ and $\beta = -i$. Since $x_0 \in \text{ext } S_A$, $x_0 \in \text{ext } Z_A$. Let V be a neighbourhood of x_0 and let $K = X \setminus V \subset S_A$. Now Lemma 2.10.7 shows there is a $\theta f \in A(Z_A)$ $(f \in A)$ such that

$$-\delta \leq \theta f \quad \text{on } Z_A$$
$$\theta f(x_0) = 0$$
$$k \leq \theta f \quad \text{on } K$$

where $\delta > 0$ and k large are chosen as follows: let $g = \mathrm{e}^{-f} \in A$ so that $|g(x)| = \mathrm{e}^{-\mathrm{re}\, f(x)}$ and hence

$$|g(x_0)| = 1$$
$$\|g\| \leq \mathrm{e}^{\delta}$$
$$|g| \leq \mathrm{e}^{-k} \quad \text{on } K.$$

Let $h = 1 - \overline{g(x_0)}g$. Then

$$h(x_0) = 0$$
$$\|h\| \leq 1 + \mathrm{e}^{\delta}$$
$$\mathrm{re}\, h \geq 1 - \mathrm{e}^{-k} \quad \text{on } K.$$

Thus k and δ can be chosen so that, say, $2h = f_V$ as in the definition of strong hull. It follows that $\{x_0\}$ is a strong hull and hence x_0 is a peak point.

If $Z(E)$ is a split face of Z_A then it is in particular a positively decomposable section so that Theorem 4.1 again applies. This will be the case precisely when the natural splitting projection of $M(X)$ onto $M(E)$ (by restriction) can be carried down to A^* by the quotient map $\mu \to \hat{\mu}$. This is the basis for the second type of peak set criteria, based on properties of $A^{\perp}$ and $\partial A^{\perp}$.

In general a closed subspace N of a Banach space B is called an *L-ideal* if the unit ball B_1 is split by N. Recall from Chapter 2, Section 9 this means there is a projection π of B onto N with complementary projection σ and

$$\|x\| = \|\pi x\| + \|\sigma x\|, \qquad x \in B.$$

Thus we have that $M(E)$ is an L-ideal in $M(X)$ under

$$\pi_1 \mu = \mu|_E.$$

We denote the complementary projection $\pi_2 \mu = \mu|_{X \setminus E}$.

We say E is an *M-set* for A if

(i) $M(E)$ is A-stable (see Proposition 2.7, for example)

(ii) $m \in \partial_A A^{\perp}$ implies $\pi_1 m = m|_E \in A^{\perp}$.

It follows from Corollary 2.4 (where (ii) holds with $r = 0$) that if E is an M-set then E is an interpolation set. Choose any $\alpha, \beta \in T$ with $|\mathrm{re}\, \bar{\alpha}\beta| < 1$.

We continue with the representation of A as $A(Z_A)$ and let $\hat{}$ denote images in A^* under the quotient map from $M(X)$. As previously we denote

$$F = \overline{\text{co}}\, E \subset S_A, \qquad \tilde{E} = \text{co}\,(\alpha F \cup \beta F).$$

Let $S_E = M(E)_1^+$ and $Z_E = \text{co}\,(\alpha S_E \cup \beta S_E) \subset Z \subset M(X)$. Then

$$Z_A = \hat{Z} \quad \text{and} \quad \tilde{E} = \hat{Z}_E.$$

We let $\mathbf{P}^*$ and $\mathbf{Q}^*$ be the cones in $M(X)$ spanned by Z and Z_E.

4.4. THEOREM. *The following are equivalent, where N denotes $M(E)$:*
(i) *E is an M-set for A;*
(ii) *$\hat{Z}_E$ is a split face of Z_A in A^*;*
(iii) *$\hat{\mathbf{Q}}^* = \hat{N} \cap \hat{\mathbf{P}}^*$ is a complemented subcone in $\hat{\mathbf{P}}^* \subset A^*$;*
(iv) *$\hat{N}$ is an L-ideal in A^* with $\hat{N} \cap S_A = F$.*

Proof. If (i) holds then the map

$$\hat{\pi}_1 = q \circ \pi_1 \circ q^{-1} \quad \text{for } q|_{\partial M(X)}$$

is a well-defined projection on A^* with range contained in $\hat{N}$. The stability of N shows the range of $\hat{\pi}_1$ equals $\hat{N}$. If $x \in \hat{N} \cap S_A$ and $\nu \in \partial S$ with $\hat{\nu} = x$ then choose any $\mu \in \partial N$ with $\hat{\mu} = x$. Then $\nu - \mu \in \partial A^\perp$ so that $(\pi_1 \nu)\hat{} = x$ and $\pi_1 \nu(1) = x(1) = 1$. Hence $\nu = \pi_1 \nu \in S_E$ so that $x \in \overline{\text{co}}\, E$. This shows $\hat{N} \cap S_A = F = \overline{\text{co}}\, E$. Since $\hat{Z}_E = \text{co}\,(\alpha F \cup \beta F)$ we have $\hat{Z}_E = \hat{N} \cap Z_A$ and $\hat{\mathbf{Q}}^* = \hat{N} \cap \hat{\mathbf{P}}^*$. If $x \in \hat{\mathbf{P}}^*$ then Theorem 3.1(iv) shows there is a $\mu \in \partial \mathbf{P}^*$ with $\hat{\mu} = x$. Hence $\pi_1 \mu \in \partial \mathbf{Q}^*$ and so $\hat{\pi}_1 x \in \hat{\mathbf{Q}}^*$. This shows (i) implies (ii) and (iii), the latter two being clearly equivalent. We show next (ii) implies (i).

Since $\hat{Z}_E = \text{co}\,(\alpha F \cup \beta F)$ is a face of Z_A, F is a face of S_A, so that Proposition 2.7 shows $M(E)$ is A-stable. If $m \in \partial A^\perp$ and

$$m = \alpha \mu_\alpha + \beta \mu_\beta, \quad \mu_\gamma \text{ real} \quad (\gamma = \alpha, \beta)$$

and

$$\tilde{m} = \mu_\alpha \oplus \mu_\beta \quad \text{on } \alpha X \cup \beta X \subset Z_A$$

then Theorem 3.1(iii) shows $\tilde{m} \in \partial A(Z_A)^\perp$. Since $\hat{Z}_E$ is split, Theorem 2.10.7 gives $\tilde{m}|_{\hat{Z}_E} \in A(Z_A)^\perp$. But

$$\tilde{m}|_{\hat{Z}_E} = (\mu_\alpha \oplus \mu_\beta)|_{\alpha E \cup \beta E} = \pi_1 \mu_\alpha \oplus \pi_1 \mu_\beta = (\pi_1 m)\tilde{}$$

and hence $\pi_1 m \in A^\perp$. Thus (i), (ii), (iii) are equivalent. If (i) holds then, as above, the map

$$\hat{\pi}_1 = q \circ \pi_1 \circ q^{-1} \quad \text{for } q|_{\partial M(X)}$$

is a well-defined projection onto $\hat{N}$. The fact that $\hat{U}$ $(U = M(X)_1)$ is split by $\hat{N}$ is then immediate from the A-stability of U^0 (Theorem 1.5(i)). Hence (i) implies (iv). If (iv) holds then let $\hat{\pi}_1$ denote the splitting projection onto $\hat{N}$. We show $\hat{\pi}_1$ takes $\hat{\mathbf{P}}^*$ to $\hat{\mathbf{Q}}^*$, establishing (iii). If $x \in S_A$ then, since A_1^* is split by $\hat{N}$,

$$1 = \|x\| = \|\hat{\pi}_1 x\| + \|\hat{\pi}_2 x\| \geq \hat{\pi}_1 x(1) + \hat{\pi}_2 x(1) = x(1) = 1.$$

Thus

$$\hat{\pi}_i x(1) = \|\hat{\pi}_i x\|.$$

Hence $\hat{\pi}_1 x$ belongs to the cone spanned by $S_A \cap \hat{N} = F$. It follows that if $x \in \gamma S_A$ then $\hat{\pi}_1 x$ belongs to $\hat{\mathbf{Q}}^*$ so that $\hat{\pi}_1 \hat{\mathbf{P}}^* = \hat{\mathbf{Q}}^*$.

4.5. COROLLARY. *Let E be an M-set for A. Then*

(i) *E is an interpolation set for A, and*

(ii) *$k(E)$ is an M-set and generalized peak set for A with $(M(k(E)))\hat{} = \hat{N}$.*

Proof. As noted above it follows from Corollary 2.4 that E (and hence $k(E)$) is an interpolation set. Thus $\hat{N}$ is w^*-closed and $\hat{Z}(E) = \hat{N} \cap Z_A = \hat{Z}_E$. It now follows that $h(E) = k(E)$ is a generalized peak set from Theorems 4.4(ii) and 4.1.

Notes

The idea of gauge-dominated extensions introduced in Section 3 goes back to Bishop [**39**] in connection with his generalization of a theorem of Rudin [**186**] and Carleson [**59**]. We shall have more to say on this in the next sections. The idea of T-dependent dominators is due to Roth [**180**]. The extension constant briefly alluded to in Section 3 was introduced in Gamelin [**112**]. This will also be developed further in Section 6.

The Peak Point Theorem for function algebras (Corollary 4.3) was given in 1959 by Bishop [**37**] where the notion of strong hull is introduced in terms of the celebrated "$\frac{1}{4}$–$\frac{3}{4}$ condition". The 2^{-n} iteration used by Bishop and which is generic to results of this type is, in our development, incorporated in the geometric condition of positive decomposability via the Extension Theorem of Chapter 2, Section 10 and ultimately, the Gauge Lemma of Chapter 1, Section 3. The geometric proof of peak point theorems for complex functions spaces appears in Asimow [**24**] and the general framework of Theorems 3.1, 4.1 and 4.2 in Asimow [**26**].

Various *hull* conditions have been given in connection with peak-set criteria. Their relation to approximate identities in general Banach algebras is discussed in Curtis and Figa-Talamanca [**70**].

The measure condition (ii) in the definition of M-set, but without reference to boundary measures, (i.e. condition (iii) of Corollary 2.2 with $r = 0$) appears in 1962 in Bishop [**39**] and Glicksberg [**118**]. The latter shows that for A a function algebra (details in Section 7) the measure condition implies E is a strong hull and hence (using a version of Bishop's argument in [**37**]) a generalized peak set. This is essentially Corollary 4.5 with the more restrictive hypotheses. The Bishop Peak Point Theorem can also be viewed as an instance of Theorem 4.4 since we show in Section 7 that $x_0 \in \partial_A X$ implies the line segment co $(\{x_0, -ix_0\})$ is a split face of Z_A.

As discussed in the notes of Section 2, the definition of M-set originates in Alfsen–Hirsberg [**9**] and Hirsberg [**125**] where it is introduced in analogy with the real $A(K)$ characterization of split faces in Alfsen and Andersen [**6**] (Chapter 2.10.7). The notion of L-ideals (as dual to M-ideals) is introduced by Alfsen and Effros [**18**] in the real case and Hirsberg [**126**] in the present context. The results of Theorem 4.4 are in [**126**]. Similar results for function algebras are given in Ellis [**102**].

5. DOMINATED INTERPOLATION

Consider now the gauge interpolation problem introduced in Section 3, with

$$U = \{f \in C(X): \text{re } tf(x) \leq \rho(x, t); (x, t) \in X \times T\}.$$

We wish to interpret conditions for solutions in terms of measures in $\partial_A A^\perp$ and so the properties of the dual U^0 become important. In particular the A-stability of U^0 is crucial. For this, we require that ρ be A-*superharmonic*, meaning that for each $x \in X$ and probability measure μ on X representing $x \in S_A$ we have

$$\rho(x, t) \geq \int_X \rho(x, t) \, \mathrm{d}\mu(x).$$

We shall assume now that the gauge ρ is bounded above. This minimizes some technical complications in Theorem 5.1(v) below and does not entail a serious loss since if E is an interpolation set ρ can be truncated in accordance with the extension constant $e(A, E)$ and the particular $g \in A|_E$ in question. Thus we take

$$0 < \delta \leq \rho \leq k$$

and let

$$X_0 = \{(x, t, s) \in X \times T \times \mathbb{R}: \rho(x, t) \leq s \leq k\}.$$

Then X_0 is compact since ρ is l.s.c. and

$$Y_0 = \text{graph } \rho = \{(x, t, s): s = \rho(x, t)\}$$

is a G_δ set in X_0.

Take E to be a closed set in X and $\pi_1\mu = \mu|_E$, the projection of $M(X)$ to $N = M(E)$.

Let

$$V=\{g\in C(E)\colon \operatorname{re} tg(x)\le \rho(x,t);\ (x,t)\in E\times T\}.$$

Finally, we define the gauge evaluation

$$\phi_0\colon X_0\to M(X);\qquad (f,\phi_0(x,t,s))=tf(x)/s.$$

5.1. THEOREM.

(i) $U^0=w^*\text{-}\overline{\text{co}}\,(\phi_0(Y_0))$.

(ii) *If h is a bounded Borel function on X with* re $th(x)\le\rho(x,t)$ *on* $X\times T$ *then $h\in U^{00}$ in $C(X)^{**}$, that is,*

$$\operatorname{re}\int_X h\,\mathrm{d}\mu\le 1\quad \textit{for all } \mu\in U^0.$$

(iii) *U^0 is split in $M(X)$ by $N=M(E)$.*

(iv) *$U|_E=V$.*

(v) *If ρ is A-superharmonic then U^0 is A-stable.*

Proof.

(i) If $f\in U$ then

$$\operatorname{re}(f,\phi_0(x,t,s))=\operatorname{re} tf(x)/s\le \operatorname{re} tf(x)/\rho(x,t)\le 1$$

so

$$W\triangleq w^*\text{-}\overline{\text{co}}\,(\phi_0(Y_0))\subset U^0.$$

If $f\in W^0$ then re $tf(x)/\rho(x,t)\le 1$ for all $(x,t)\in X\times T$ so that $f\in U$. Hence $U^0\subset W^{00}=W$.

(ii) Let $\mu\in U^0$ and $\varepsilon>0$ be given. By Lusin's Theorem (see Rudin [**185**, Theorem 2.23]) there is a $g\in C(X)$ and F compact in X such that

$$g|_F=h|_F,\qquad |\mu|(X\backslash F)<\varepsilon/(2\|h\|_\infty),\qquad |g|\le|h|\quad \text{on } X.$$

For each $x\in X\backslash F$ choose $\Psi_x\in C_{\mathbb{R}}(X)$ such that

$$0\le\Psi_x\le 1,\qquad \Psi_x\equiv 1\quad\text{on } F\qquad\text{and}\qquad \Psi_x(x)=0.$$

Let $V_x=\{y\in X\colon \operatorname{re} t\Psi_x(y)g(y)<\rho(y,t)+\varepsilon \text{ for all } t\in T\}$.

Then V_x is an open neighbourhood of $F\cup\{x\}$ and hence there exists $x_1,\ldots,x_n$ and corresponding $\Psi_1,\ldots,\Psi_n$ and $V_1,\ldots,V_n$ such that X is covered by $V_1,\ldots,V_n$. Let

$$\Psi=\Psi_1\wedge\cdots\wedge\Psi_n\quad\text{and}\quad f=g.$$

Then either

$$\operatorname{re} tf(x)\le 0<\rho(x,t)$$

or, for $x\in V_i$

$$0\le \operatorname{re} tf(x)=\Psi(x)\operatorname{re} tg(x)\le\Psi_i(x)\operatorname{re} tg(x)<\rho(x,t)+\varepsilon$$

for all $t \in T$. Hence $f \in (1+\varepsilon)U$ and

$$\text{re } \mu(h) = \text{re } \mu(f) + \text{re} \int_{X \setminus F} (h-f) \, d\mu \leq (1+\varepsilon) + \varepsilon.$$

Thus $h \in U^{00}$.

(iii) If $\pi_1\mu = \mu|_E \in N$ and $\pi_2\mu = \mu|_{X\setminus E}$ then choose $f_i \in U$ $(i = 1, 2)$ such that

$$\text{re } (f_i, \pi_i\mu) \geq \|\pi_i\mu\|_\rho - \varepsilon.$$

Let

$$h = f_1\chi_E + f_2\chi_{X\setminus E}.$$

Then, using (ii)

$$\|\mu\|_\rho \geq \text{re } (h, \mu) \geq \|\pi_1\mu\|_\rho + \|\pi_2\mu\|_\rho - 2\varepsilon.$$

(iv) Clearly $U|_E \subset V$ and hence

$$V^0 \subset U^0 \cap N.$$

Conversely, if $\mu \in U^0 \cap N$ and $g \in V$ let h be the extension of g identically zero on $X\setminus E$. By (ii)

$$\text{re} \int_E g \, d\mu = \text{re} \int_X h \, d\mu \leq 1$$

so that $\mu \in V^0$. But

$$V^0 = U^0 \cap N$$

implies

$$(U|_E)^- = V.$$

But (iii), together with Andô's Theorem (2.9.4) shows $U + M$ is closed in $C(X)$, where

$$M = \{f \in C(X): f \equiv 0 \quad \text{on } E\}.$$

Hence $U|_E$ is closed in $C(E)$ and equals V.

To prove (v) we first demonstrate a special case and then combine this with Hustad's Theorem for the general result. For this we take ρ to be l.s.c. on X with $0 < \delta \leq \rho \leq k$ and A a closed separating (do not assume $1 \in A$) subspace of $C(X)$, such that $X \subset \{x \in A^*: \|x\| = 1\}$ (under evaluation). Let

$$U = \{f \in C(X): \text{re } f \leq \rho\}$$

$$X_0 = \{(x, s) \in X \times \mathbb{R}: \rho(x) \leq s \leq k\}$$

$$Y_0 = \{(x, s) \in X_0: s = \rho(x)\}.$$

We now take

$$\Psi: C(X) \to C(X_0); \qquad \Psi f(x, s) = f(x)/s$$

and $\theta = \Psi|_A$ with closed range $B \subset C(X_0)$. Let $(x_1, s_1) \neq (x_2, s_2)$ in X_0. The Separation Theorem shows there is an $f \in A$ such that $(f, x_1/s_1) \neq (f, x_2/s_2)$ in (A, A^*). Hence B separates points in X_0 so that we can consider X_0 embedded by evaluation in B_1^* and we denote

$$\hat{V} = w^*\text{-}\overline{\text{co}}\, X_0 \subset B_1^*.$$

5.2. PROPOSITION. *In the above set-up, if* $\rho(x) \geq \int_X \rho \, \mathrm{d}\mu$ *whenever* $\mu \in S$ *and* μ *represents* x *with respect to* A *then*

(i) $\theta^* \hat{V} = \hat{U}$;

(ii) *If* ν *is a maximal probability measure on* $\hat{V}$ *then* $\nu(Y_0) = 1$ *and if* $\mu = \Psi^* \nu$ *then*

$$\int_X h \, \mathrm{d}\mu = \int_{Y_0} (h(x)/\rho(x)) \, \mathrm{d}\nu(x, s)$$

for any bounded Borel function h *on* X.

In particular $\mu \in U^0$.

(iii) *Let* $\sigma_\rho : X \to X_0$; $\sigma_\rho(x) = (x, \rho(x))$ *and* $\pi : X_0 \to X$; $\pi(x, s) = s$. *If* μ_0 *is a probability measure on* X *representing* $x_0 \in X \subset S_A$ *and*

$$\mathrm{d}\nu_0 = (\pi/\rho(x_0)) \, \mathrm{d}(\mu_0 \circ \sigma_\rho^{-1}) \quad \text{on } X_0$$

then

$$0 \leq \nu_0 \leq 1, \quad \nu_0 \text{ represents } (x_0, \rho(x_0)) \text{ in } \hat{V} \text{ with respect to } B,$$

and $\Psi^* \nu_0 = \mu_0/\rho(x_0)$.

(iv) *If* ν *is maximal on* $\hat{V}$ *then* $\mu = \Psi^* \nu$ *is maximal on* $K = \overline{\text{co}}\, X \subset A^*$.

(v) *If* $L \in \hat{U}$ *then there is a* $\mu \in U^0$ *maximal on* K *such that* μ *represents* L *with respect to* A.

Proof.

(i) Since $\theta^*(x, s) = x/s$ and $\hat{U} = w^*\text{-}\overline{\text{co}}\,\{x/s : (x, s) \in X_0\}$ we have $\theta^* \hat{V} = \hat{U}$.

(ii) Let $p = 1 - \chi_{\{0\}}$ on $\hat{V}$ so that the lower envelope $\check{p}$ is the Minkowski functional of $\hat{V}$. Since ν is maximal, $\operatorname{supp} \nu \subset X_0$ and

$$1 = \nu[\{(x, s) : p(x, s) = \check{p}(x, s)\}] = \nu[\{(x, s) : \check{p}(x, s) = 1\}]$$

where the last equality follows from the fact that $0 \notin X_0 \subset \hat{V}$.

If $\rho(x) \leq s \leq k$ and $f \in A$ then

$$\theta f(x, s) = f(x)/s = (f(x)/\rho(x))(\rho(x)/s) = (\rho(x)/s)\theta f(x, \rho(x))$$

so that

$$\check{p}(x, s) = (\rho(x)/s)\check{p}(x, \rho(s)) \leq \check{p}(x, \rho(s)).$$

Hence $\check{p}(x, s) = 1$ implies $\rho(x) = s$ and the statements in (ii) follow.

(iii) The definition of ν_0 shows that for any bounded Borel function g on X_0

$$\int_{X_0} g \, d\nu_0 = (1/\rho(x_0)) \int_X \rho(x) g(x, \rho(x)) \, d\mu_0.$$

Thus the super-harmonic property of ρ gives $0 \leq \nu_0 \leq 1$ on X_0. If $f \in C(X)$ then

$$\begin{aligned}(f, \Psi^* \nu_0) = (\Psi f, \nu_0) &= \int_{X_0} (f(x)/s) \, d\nu_0(x, s) \\ &= (1/\rho(x_0)) \int_X \rho(x) f(x)/\rho(x) \, d\mu_0(x) = (1/\rho(x_0))(f, \mu_0).\end{aligned} \qquad (*)$$

Thus $\Psi^* \nu_0 = \mu_0/\rho(x_0)$ and if $f \in A$

$$(f, \Psi^* \nu_0) = f(x_0)/\rho(x_0) = (\theta f)(x, \rho(x_0)) = (\theta f, \nu_0)$$

so that ν_0 represents $(x_0, \rho(x_0))$.

(iv) Let f be a continuous convex function on K and let $g = \Psi(f|_X)$. Then $g \in C(X_0)$ and so $\hat{g}$ is well-defined on $\hat{V}$. Now if $x_0 \in X$ then from Theorem 1.6.2

$$\hat{f}(x_0) = \sup\left\{\int_X f \, d\mu_0 \colon \mu_0 \in S \quad \text{and} \quad \mu_0 \sim x_0\right\}$$

and so by (iii),

$$\hat{g}(x_0, \rho(x_0)) \geq \sup\left\{\int_{X_0} g \, d\nu_0 \colon \mu_0 \in S \quad \text{and} \quad \mu_0 \sim x_0\right\}$$

where ν_0 is related to μ_0 as in (iii).

But then, by (*),

$$\int_{X_0} g \, d\nu_0 = \int_{X_0} (\Psi f) \, d\nu_0 = (1/\rho(x_0)) \int_X f \, d\mu_0$$

so that

$$\hat{g}(x_0, \rho(x_0)) \geq \hat{f}(x_0)/\rho(x_0).$$

Hence, using (ii),

$$\int_X (\hat{f} - f) \, d\mu = \int_{Y_0} ((\hat{f} - f)/\rho) \, d\nu \leq \int_{Y_0} (\hat{g} - g) \, d\nu = 0$$

since ν is maximal. Thus μ is maximal.

(v) Given $L \in \hat{U}$ we let $\tilde{L} \in \hat{V}$ such that $\theta^* \tilde{L} = L$. Then let ν be a maximal probability measure on $\hat{V}$ representing $\tilde{L}$ and let $\mu = \Psi^* \nu$. Then $\mu \in U^0$ by (ii) and is maximal on K by (iv).

To conclude the proof of Theorem 5.1(v) we consider $C(X)$ embedded as the T-invariant functions in $C(X \times T)$ and take

$$\Psi f = \phi f, \qquad \phi(x, t) = t.$$

let

$$U = \{f \in C(X \times T): \text{re } tf(x, t)/\rho(x, t) \leq 1\}$$

$$U^1 = \{f \in C(X \times T): \text{re } f(x, t)/\rho(x, t) \leq 1\}$$

and note that $\Psi U = U^1$. Thus if $\theta = \Psi|_A$ with range B,

$$\hat{U} = \theta^*(U^1)\hat{}; \qquad (U^1)\hat{} \text{ in } B^*.$$

We will apply the set-up of Proposition 5.2 to the subspace B with respect to the set U^1. Clearly $X \times T$ is embedded in the surface of B_1^* by evaluation such that $K = \overline{\text{co}}\,(X \times T) = B_1^*$. We observe next that the A-superharmonicity of ρ yields the hypothesis of Proposition 5.2. For, if $g \in B$ then $g(x, t) = tg(x, 1)$ and

$$S_A = w^*\text{-}\overline{\text{co}}\,(X \times \{1\}) \subset B_1^*$$

is a face of B_1^*. Thus, if μ is a non-negative measure on $X \times T$ representing $(x, 1)$ then $\mu(X \times \{1\}) = 1$. If α_t denotes the homeomorphism of $X \times T$ given by

$$\alpha_t(y, s) = (y, ts)$$

and μ^1 is given by $\mu^1 = \mu \circ \alpha_{\bar{t}}$ then for $g = \phi f$, $f \in A$,

$$\mu^1(g) = \int g \circ \alpha_t \, d\mu = \int_{X \times \{1\}} g(y, t) \, d\mu(y, 1)$$

$$= t \int_{X \times \{1\}} f(y, 1) \, d\mu(y, 1) = g(x, t)$$

so that μ^1 represents (x, t) and has support in $X \times \{t\}$. Conversely, by applying α_t, each μ^1 representing (x, t) is of this form. Hence

$$\int_{X \times T} \rho(y, s) \, d\mu^1(y, s) = \int_{X \times \{1\}} \rho(y, t) \, d\mu(y, 1) \leq \rho(x, t).$$

Now, given $L \in \hat{U}$ in A^* let $L = \theta^* \tilde{L}$, $\tilde{L} \in (U^1)\hat{} \in B^*$. Proposition 5.2(v) yields $\tilde{\mu} \in (U^1)^0$ such that $\tilde{\mu}$ is maximal on B_1^* and represents $\tilde{L}$ with respect to B. It follows from Theorem 1.5 that $\mu = H\tilde{\mu} = \Psi^* \tilde{\mu}|_M$ is a boundary measure for A on X and by the above μ represents L and belongs to U^0.

We now characterize, in measure-theoretic terms, approximate ρ-interpolation sets for a closed, separating, with constants, subspace A. The results here are refinements of conditions in Section 2, using the gauges $\|\cdot\|_\rho$ in place of the standard norms in $C(X)$, $M(X)$. We take E to be a closed subset of X with

$$M=\{f\in C(X): f|_E=0\}$$

and $N=M(E)\subset M(X)$. We continue to denote quotient map images in A^* by $\hat{}$, and

$$U=\{f\in C(X): \operatorname{re} tf(x)/\rho(x,t)\leq 1\}.$$

5.3. THEOREM. *The following are equivalent for a bounded dominator, ρ, and imply that E is an interpolation set*:

(i) *E is an approximate ρ-interpolation set for A;*

(ii) *$A+M$ is closed in $C(X)$ and*

$$(A+M)\cap(U+M)=(A\cap U+M)^-;$$

(iii) $\hat{U}\cap\hat{N}=(U^0\cap N)\hat{}$;

(iv) $\|\mu+A^\perp\cap N\|_\rho=\|\mu+A^\perp\|_\rho$ *for all* $\mu\in N$;

(v) $\|\pi_1 m+A^\perp\cap N\|_\rho\leq\|-\pi_2 m\|_\rho$ *for all* $m\in A^\perp$.

Proof. If (i) holds then E is an interpolation set so that $A+M$ is closed in $C(X)$. Since U^0 is split (Theorem 5.1(iii)) by N, Andô's Theorem shows $U+M$ is closed. Thus, the fact that E is an approximate ρ-interpolation set translates into (ii). The polar of the statement in (ii) is

$$(A^\perp+U^0)\cap N=A^\perp\cap N+U^0\cap N \tag{$*$}$$

or, equivalently,

$$(U^0+A^\perp)\cap(N+A^\perp)=U^0\cap N+A^\perp.$$

Then applying the quotient map shows this is equivalent to (iii) and also (iv). The equivalence of (iv) and (v) is the same computation as in Corollary 2.2 since $(A|_E)^\perp=A^\perp\cap N$ and

$$\|\mu\|_\rho=\|\pi_1\mu\|_\rho+\|\pi_2\mu\|_\rho,$$

keeping in mind the presence of the minus sign in (v), reflecting the non-homogeneity of $\|\cdot\|_\rho$. Finally, if $(*)$ holds then the polar statement is

$$(A+M)^-\cap(U+M)=(A\cap U+M)^-.$$

If ρ is bounded the equivalence of $\|\cdot\|$ and $\|\cdot\|_\rho$ shows, using Corollary 2.2, that (v) implies E is an interpolation set so that $A+M$ is closed.

5.4. COROLLARY. *Let $N=M(E)$ and U^0 be A-stable sets. Then the following are equivalent for bounded dominators, ρ:*

(i) *E is an approximate ρ-interpolation set for A;*
(ii) $\|\mu + A^{\perp} \cap N\|_{\rho} = \|\mu + \partial A^{\perp}\|_{\rho}$ *for all* $\mu \in \partial N$;
(iii) $\|\pi_1 m + A^{\perp} \cap N\|_{\rho} \leq \|-\pi_2 m\|_{\rho}$ *for all* $m \in \partial A^{\perp}$.

Proof. Note that if μ is a boundary measure and U^0 is A-stable then (as in Corollary 2.4)

$$\|\mu + A^{\perp}\|_{\rho} = \|\mu + \partial A^{\perp}\|_{\rho} \quad \text{for } \mu \in \partial N. \tag{$*$}$$

Thus (i) implies (ii) implies and implied by (iii) follows as in Theorem 5.3 and Corollary 2.2. To show (ii) implies (i) it suffices, by 5.3, to show

$$\hat{U} \cap \hat{N} \subset (U^0 \cap N)\hat{}.$$

If $x \in \hat{U} \cap \hat{N}$ then choose $\mu \in \partial N$ with $\hat{\mu} = x$ and

$$\mu \in U^0 + A^{\perp}.$$

Then (ii), together with ($*$) shows

$$\|\mu + A^{\perp} \cap N\|_{\rho} = \|\mu + \partial A^{\perp}\|_{\rho} = \|\mu + A^{\perp}\|_{\rho} \leq 1$$

so that

$$\mu = \nu + m; \qquad \nu \in U^0 \quad \text{and} \quad m \in A^{\perp} \cap N.$$

Hence $\nu \in N$ and

$$\hat{\mu} = x = \hat{\nu} \in (U^0 \cap N)\hat{}.$$

5.5. COROLLARY. *Let E be a closed subset of X such that either*
(a) $E \subset \partial_A X$, *or*
(b) $E = F \cap X$, *F a closed face of S_A,*
and let ρ be a bounded A-superharmonic dominator. Then the following are equivalent:
(i) *E is an approximate ρ-interpolation set for A;*
(ii) $\|\pi_1 m + A^{\perp} \cap N\|_{\rho} \leq \|-\pi_2 m\|_{\rho}$ *for all* $m \in \partial A^{\perp}$.

Proof. If (a) or (b) and (i) or (ii) holds for ρ bounded then E is an interpolation set and Proposition 2.7 shows N is A-stable. Theorem 5.1(v) shows U^0 is A-stable so that Corollary 5.4 applies.

We show next that if E is an M-set for A then we can conclude exact solutions for all A-superharmonic dominators. This is a special case of more general criteria presented in the next section. Exactness is tantamount to the closedness of the set $A \cap U + M$ in $C(X)$.

5.6. THEOREM. *The following are equivalent for a bounded dominator, ρ:*

(i) *E is an exact ρ-interpolation set for A;*

(ii) *$A+M$ is closed, $A\cap U+M$ is closed, and*

$$(A+M)\cap(U+M)=A\cap U+M;$$

(iii) *$A\cap U+E^{\perp}$ is closed in A, where $E^{\perp}=M\cap A$, and*

$$A\cap(U+M)=A\cap U+E^{\perp}.$$

Proof. If (i) holds then E is an interpolation set and $A+M$ is closed. If $g\in(A\cap U+M)^{-}$ then $g\in A|_E$ and Theorem 5.1(iv) shows $\|g\|_\rho\leq 1$ so that (i) implies $g\in A\cap U+M$. If (ii) holds then $A\cap(A\cap U+M)=A\cap U+E^{\perp}$ is closed and hence $A\cap(U+M)=A\cap U+E^{\perp}$ follows from the analogous equality in (ii). If (iii) holds with ρ bounded and

$$f\in(A+M)^{-}\cap(U+M)$$

then given $\varepsilon>0$,

$$f=a+m+z=u+m';\qquad a\in A,\ m,\ m'\in M,\qquad \|z\|<\varepsilon,\quad \text{and}\quad u\in U.$$

Now $\|\cdot\|_\rho\leq k\|\cdot\|$ in $C(X)$ so that

$$a\in A\cap((1+k\varepsilon)U+M)=(1+k\varepsilon)U\cap A+E^{\perp}.$$

Hence $f\in(A\cap U+M)^{-}$. Therefore

$$(A+M)^{-}\cap(U+M)=(A\cap U+M)^{-}$$

and the equivalence of $\|\cdot\|_\rho$ and $\|\cdot\|$ shows, by Theorem 2.1, that E is an interpolation set so that $A+M$ is closed. The set equality in (iii) then easily yields

$$(A+M)\cap(U+M)=A\cap U+M.$$

We can extend Theorem 4.4 as follows.

5.7. THEOREM. *The following are equivalent:*

(i) *E is an M-set for A;*

(ii) *For any bounded A-superharmonic dominator ρ,*

$$\hat{U} \textit{ is split by } \hat{N} \textit{ in } A^* \textit{ with } \hat{U}\cap\hat{N}=(U^0\cap N)\hat{}.$$

Proof. The fact that E is an M-set shows that

$$\hat{\pi}_1=q\circ\pi_1\circ q^{-1}\quad \text{for } q|_{\partial M(X)}$$

is well-defined (Theorem 4.4) so that the A-stability of U^0 (Theorem 5.1(v)) shows $\hat{U}$ is split by $\hat{N}$ under $\hat{\pi}_1$. Since $\hat{\pi}_1\circ q=q\circ\pi_1$, $\hat{U}\cap\hat{N}=[\partial(U^0\cap N)]\hat{}$.

If (ii) holds, then the case $\rho \equiv 1$ gives Theorem 4.4(iv) and hence E is an M-set.

5.8. COROLLARY. *If E is an M-set for A then E is an exact ρ-interpolation set for any A-superharmonic dominator ρ.*

Proof. If ρ is bounded then Theorem 5.7(ii) shows

$$(U+A^{\perp})\cap(N+A^{\perp})=U^{0}\cap N+A^{\perp}$$

so that

$$(A\cap U+E^{\perp})^{-}=A\cap(U+M).$$

But Andô's Theorem shows $A\cap U+E^{\perp}$ is closed so that Theorem 5.6 gives exact interpolation. Since $\rho \wedge k$ is A-superharmonic for any $k \geq 0$ the conclusion holds in general.

The conditions for exact solutions in these results depend on splitness in A^*. We give some more general criteria for exact solutions based on *decomposability* in the next section.

The utility of boundary measures and superharmonic dominators is made most apparent by considering the situation for real $A(K)$ spaces. The closer the measures come to having their support on the boundary (ext K) the more accurately they portray the geometry of K. To take the simplest possible case let K be the unit interval $[0, 1]$ and let $E=\{0\}$. Then the condition

$$m \in A^{\perp} \text{ implies } \pi_1 m \in A^{\perp}$$

fails (take $m=(\varepsilon_0+\varepsilon_1)/2-\varepsilon_{1/2}$) unless one restricts attention to boundary measures. This phenomena is also reflected in less trivial settings. For example if A is the disc algebra (details in Section 6) on the unit disc $\Delta \subset \mathbb{C}$ then normalized Lebesgue measure σ on T represents 0 so that $\sigma-\varepsilon_0 \in A^{\perp}$. One gets around this by restricting A to T since T equals $\partial_A \Delta$.

Again, considering the case $A(K)$, we noted in Section 3 that

$$U=\{f: -\rho^{-} \leq f \leq \rho^{+}\}.$$

Then ρ is A-superharmonic if and only if (Proposition 1.6.2) $\rho^{+}=(\rho^{+})\hat{}$ and $\rho^{-}=(\rho^{-})\hat{}$, that is, if and only if ρ^{+} and ρ^{-} are concave. For extending affine functions this is clearly a relevant hypothesis (as the above example shows).

Notes

Theorem 5.1 is due to W. Roth [**182**, **183**]. Abstract versions of the (approximate solution) Theorem 5.3 using set inclusion formulations are given in Roth [**182**], Andô [**19**], Asimow [**28**]. These are outgrowths of conditions based on the gauge ρ_r (introduced in Section 3) developed in Gamelin [**13**] and, in the case of $\partial A^{\perp}$, Briem [**50**]. We discuss this in more detail in connection with exact solutions in Section 6.

Theorem 5.7 relating M-sets and exact interpolation for superharmonic gauges is given by Roth [**182**]. Corollary 5.8 is essentially the Alfsen–Hirsberg Theorem [**9**].

6. DECOMPOSABILITY AND EXACT INTERPOLATION

In Theorem 4.1 the positive decomposability of Z_A by $E^{\perp\perp}$ was used, via Theorem 2.8.13, to conclude the positive extension property (PEP) for the subcone $\mathbf{Q}^* = \mathbf{P}^* \cap E^{\perp\perp}$. This in turn is equivalent to $\mathbf{P} + E^{\perp}$ being a closed subset in A. In Corollary 5.8 we used the splitting of polar sets $\hat{U}$ to conclude the closedness of $A \cap U + E^{\perp}$.

It is convenient to introduce now a broader concept of decomposability that leads to a more widely applicable closedness condition.

Let S be a closed convex subset of the (real or complex) Banach space E and let M be a closed subspace of E with $N = M^0$ in E^*. Following the definition for decomposability of cones in Chapter 2, Section 8, we let W be a norm-closed convex neighbourhood of 0 in E^* and define the distance, d_W, to N via W by

$$d_W(x) = p(N + W)(x) = \inf\{r \geq 0 : x \in N + rW\} = p(W)(x + N)$$

where the notation $p(A)$ means the Minkowski functional of A.

Let $S^0 \cap N$ be a section of S^0 by M, i.e. $(S^0 \cap N)^{\perp} = M$ in E. Then we say $S^0 \cap N$ is a *decomposable section* of S^0 by M if there is a positive homogeneous map $\pi_1 : S^0 \to S^0 \cap N$ with

$$\pi_2 x = x - \pi_1 x$$

such that

(i) $\pi_1 x = x$ if $x \in S^0 \cap N$
(ii) $p(S^0) = p(S^0) \circ \pi_1 + p(S^0) \circ \pi_2$
(iii) $\|\pi_2 x\| \leq d_W(x)$ for $x \in S^0$.

If S^0 is a cone then this agrees with the previous definition since $p(S^0)(x) = 0$ for $x \in S^0$ and $+\infty$ otherwise. If π_1 is a bounded linear projection such that (ii) holds, ((i) and (iii) being automatic in this case) then S^0 is split by N.

6.1. THEOREM. *If $S^0 \cap N$ is a decomposable section of S^0 by M in E^* then $S + M$ is closed in E.*

Proof. If we show, for some bounded convex neighbourhood V of 0 in E,

$$(S + M)^- \cap (S + E_r) \subset (S + rV \cap M)^- \quad \text{for all } r > 0 \qquad (*)$$

then the Gauge Lemma (1.3.2) applies with

$$A \leftrightarrow S, \qquad B \leftrightarrow V \cap K, \qquad D \leftrightarrow (S+M)^-$$

to yield $d(=d(A, B)) = \bar{d}$ on $D = (S+M)^-$. But $x \in D$ implies, by (*), that $\bar{d}(x) < \infty$. Hence $d(x) < \infty$ so that $x \in S + rV \cap M$ (some r) $\subset S + M$.

To show (*) it suffices to demonstrate

$$(S + rV \cap M)^0 \subset \overline{\text{co}}\,(S^0 \cap N, (S + E_r)^0). \qquad (*)^0$$

Now, if S^0 is decomposable by N then we can choose a bounded convex neighbourhood V of 0 in E such that $V^0 \subset W$, where $\|\pi_2 x\| \leq d_W(x)$ on S^0. Then the Minkowski functional $p((S + rV \cap M)^0)$ is the support functional $\rho(S + rV \cap M) = \rho(S) + \rho(rV \cap M)$. Hence for $x \in (S + rV \cap M)^0$, we have $x \in S^0$ and

$$\begin{aligned} 1 &\geq p((S + rV \cap M)^0)x = p(S^0)x + p((rV \cap M)^0)x \\ &\geq p(S^0)\pi_1 x + p(S^0)\pi_2 x + r d_W(x) \\ &\geq p(S^0)\pi_1 x + (p(S^0) + r\|\cdot\|)\pi_2 x \\ &= p(S^0)\pi_1 x + p((S + E_r)^0)\pi_2 x. \end{aligned}$$

Hence, if $\lambda = p(S^0)\pi_1 x$, $\mu = p(S + E_r)^0 \pi_2 x$ and $\alpha = p((S + rV \cap M)^0)x$, then $1 \geq \alpha \geq \lambda + \mu$ and

$$x = \alpha[(\lambda/\alpha)(\pi_1 x/\lambda) + (\mu/\alpha)(\pi_2 x/\mu)]$$

where $0/0$ is taken to be 1. Thus $(*)^0$, and therefore (*), holds and the proof is complete.

If E is a closed subset of the compact Hausdorff space X then the space $M(E)$ is a w^*-closed subspace of $M(X)$ and the range of the projection $\pi_1 \mu = \mu|_E$. Various subsets of $M(X)$ are split under π_1, including the positive cone (in the real case), the generalized positive cones of Section 3, the unit ball and more generally the sets U^0 of Theorem 5.1. We discuss next when this is inherited, in the form of decomposability, under the quotient map

$$q: M(X) \to M(X)/A^\perp = A^*.$$

We assume in the following that E is an interpolation set, so that $q(M(E))$ is w^*-closed in A^*.

Let $0 \in S$, a w^*-closed convex set split under π_1 and let W be any w^*-closed convex norm-neighbourhood of 0 such that

$$\pi_2(W) \subset W.$$

Denote images $q\mu$ by $\hat{\mu} \in A^*$ and assume that the image $\hat{S}$ is w^*-closed in A^*. Since E is an interpolation set, $\hat{N}$ (for $N = M(E)$) is w^*-closed. We note that

$$p(\hat{S})\hat{\mu} = p(S + A^{\perp})\mu \leq p(S)\mu.$$

6.2. PROPOSITION. *If $\mu \in S$ with $p(S)\mu = p(\hat{S})\hat{\mu}$ then*

$$p(S)\pi_i\mu = p(\hat{S})(\pi_i\mu)\hat{} \qquad (i = 1, 2).$$

Proof.

$$\begin{aligned} p(\hat{S})(\pi_1\mu)\hat{} &= p(S + A^{\perp})\pi_1\mu = p(S + A^{\perp})\mu - p(S + A^{\perp})\pi_2\mu \\ &\geq p(S)\mu - p(S)\pi_2\mu = p(S)\pi_1\mu \geq p(\hat{S})(\pi_1\mu)\hat{}. \end{aligned}$$

6.3. PROPOSITION. *The Minkowski functional*

$$p(W + N + A^{\perp}) = p(W + \pi_2 A^{\perp}) \circ \pi_2.$$

Proof. If $\mu = r\nu + \eta + \xi$ with $\nu \in W$, $\eta \in N$, $\xi \in A^{\perp}$ and $r \geq 0$ then

$$\pi_2\mu = r\pi_2\nu + \pi_2\xi \in r\pi_2 W + \pi_2 A^{\perp} \subset rW + \pi_2 A^{\perp}.$$

Conversely, if $\pi_2\mu = r\nu + \pi_2\xi$ then

$$\mu = r\nu + (\pi_1\mu - \pi_1\xi) + \xi \in rW + N + A^{\perp}.$$

6.4. PROPOSITION. *If W and N are A-stable sets and $\mu \in \partial_A M(X)$ then*

$$p(W + N + A^{\perp})\mu = p(W + \pi_2\partial_A A^{\perp}) \circ \pi_2\mu.$$

Proof. If $\mu = r\nu + \eta + \xi \in rW + N + A^{\perp}$ then

$$\mu = r\nu' + \eta' + \xi'; \qquad \nu' \in \partial W, \qquad \nu' \in \partial N, \qquad \xi' \in A^{\perp}.$$

Since $\mu \in \partial M(X)$, $\xi' \in \partial A^{\perp}$ so that

$$p(W + N + A^{\perp})\mu = p(W + \partial A^{\perp})\mu.$$

But it follows, as in Proposition 6.3, that

$$p(W + N + \partial A^{\perp}) = p(W + \pi_2\partial A^{\perp}) \circ \pi_2.$$

6.5. LEMMA (DECOMPOSABILITY LEMMA). *If for each $x \in \hat{S}$ there exists a $\mu \in S$ satisfying for some $k > 0$ and W as above*

(i) *$\hat{\mu} = x$ and $p(\hat{S})\hat{\mu} = p(S)\mu$*
(ii) *$p(W)\pi_2\mu \leq kp(W + \pi_2 A^{\perp})\pi_2\mu$*

then

$$\hat{S} \cap \hat{N} = (S \cap N)\hat{} \quad \textit{is a decomposable section of } \hat{S}.$$

If W and N are A-stàble and μ can be chosen in ∂S satisfying (i) *and*

(ii)′ $p(W)\pi_2\mu \le kp(W+\pi_2\partial A^{\perp})\pi_2\mu$
then the same conclusion holds.

Proof. Assume first $p(\hat{S})x = 1$. Choose μ according to the hypothesis and define

$$\hat{\pi}_1 x = (\pi_1\mu)\hat{}\in (S\cap N)\hat{}$$

and

$$\hat{\pi}_2 x = x - \hat{\pi}_1 x = (\pi_2\mu)\hat{}.$$

If $p(\hat{S})x = 0$ and $p(\hat{W})x = 1$ then again choose μ and define $\hat{\pi}_1 x$ as above. Since $p(\hat{S})x = p(S)\mu$ we can extend $\hat{\pi}_1$ to be a positive homogeneous function on $\hat{S}$. Then

$$p(\hat{S})x = p(S)\mu = p(S)\pi_1\mu + p(S)\pi_2\mu \ge p(\hat{S})\hat{\pi}_1 x + p(\hat{S})\hat{\pi}_2 x \ge p(\hat{S})x$$

and, by Proposition 6.3

$$p(\hat{W})\hat{\pi}_2 x \le p(W)\pi_2\mu \le kp(W+\pi_2 A^{\perp})\pi_2\mu = kp(W+N+A^{\perp})\mu$$
$$= kp(\hat{W}+\hat{N})x = kd_{\hat{W}}(x).$$

If $x\in \hat{S}\cap\hat{N}$ then $d_{\hat{W}}(x)=0$ so that $p(\hat{W})\hat{\pi}_2 x = 0$. Hence $0=\hat{\pi}_2 x$ and $x = \hat{\pi}_1 x \in (S\cap N)\hat{}$. This shows $\hat{S}\cap\hat{N} = (S\cap N)\hat{}$ is a decomposable section of $\hat{S}$. The proof using (ii)′ is the same with Proposition 6.4 in place of Proposition 6.3.

As an application of Lemma 6.5 we develop another, measure theoretic, peak set criterion based on the set-up of Sections 3 and 4.

Let A be a closed separating subspace of $C(X)$ containing constants and let E be a closed interpolating (with respect to A) subset of X. We take, P, P^* as in Theorem 3.1 with generators α, $\bar{\alpha}$ for P^* and $u = (\mathrm{re}\ \alpha)^{-1}$. As in Section 4, we take

$$Z = \mathrm{co}\ (\alpha S \cup \bar{\alpha} S) \subset M(X)$$
$$Z_E = \mathrm{co}\ (\alpha S_E \cup \bar{\alpha} S_E) \subset M(E)$$
$$F = \overline{\mathrm{co}}\ E \subset S_A$$
$$\hat{Z}_E = \mathrm{co}\ (\alpha F \cup \bar{\alpha} F).$$

6.6. THEOREM. *Let E be a closed interpolation set for A. If there is an s, $0\le s<1$, such that for each $m\in A^{\perp}$*

$$|m(E)| \le s\|\pi_2 m\|$$

then $\hat{Z}_E$ is a positively decomposable section of Z_A and $k(E)$ is a generalized peak set.

Proof. Choose the generators $\alpha, \bar{\alpha}$ of $P^* \subset \mathbb{C}$ such that $|\alpha| = 1$ and $s = \text{re}\,\alpha = 1/u$. If $\mu \in Z$ then

$$\|\mu\|_u = \|\hat{\mu}\|_u = \text{re}\,\mu(\mathbf{u}) = u\ \text{re}\,\mu(X)$$

and generally,

$$c\|\mu\|_u \leq \|\mu\| \leq \|\mu\|_u.$$

We apply Lemma 6.5 with S the cone spanned by Z and U the $\|\cdot\|_u$ unit ball in $M(X)$. Let $x \in Z_A = \hat{Z}$ and $\mu \in Z$ with $\hat{\mu} = x$.

Then, given $m \in A^\perp$,

$$\begin{aligned} s\|\pi_2\mu + \pi_2 m\|_u &\geq s\|\pi_2 m\|_u - s\|\pi_2\mu\|_u \geq s\|\pi_2 m\| - s\|\pi_2\mu\|_u \\ &\geq |m(E)| - s\ \text{re}\,\pi_2\mu(\mathbf{u}) \geq s\ \text{re}\,\pi_1 m(\mathbf{u}) - s\ \text{re}\,\pi_2\mu(\mathbf{u}) \\ &= -s\ \text{re}\,\pi_2 m(\mathbf{u}) - s\ \text{re}\,\pi_2\mu(\mathbf{u}) \\ &= (1-s)\ \text{re}\,\pi_2\mu(\mathbf{u}) - (s\ \text{re}\,\pi_2 m(\mathbf{u}) + \text{re}\,\pi_2\mu(\mathbf{u})) \\ &\geq (1-s)\|\pi_2\mu\|_u - \|s\pi_2 m + \pi_2\mu\|_u. \end{aligned}$$

Hence

$$\|\pi_2\mu\|_u \leq ((1+s)/(1-s))\|s\pi_2 m + \pi_2\mu\|_u.$$

Since $m \in A^\perp$ was arbitrary this establishes (ii) of Lemma 6.5 and therefore

$$\hat{Z}_E = Z_A \cap \hat{N}$$

is a decomposable section of Z_A.

We show next that $\hat{Z}_E$ is a positively decomposable section. It suffices to show the positive cone in $A^*/\hat{N}$ is α-additive for some $\alpha > 0$ under the quotient norm induced by

$$\hat{q}: A^* \to A^*/\hat{N}.$$

Thus, given $\mu_i \geq 0(P^*)$ in $M(X)$ $(i = 1, \ldots, n)$, we have

$$\begin{aligned} \left\|\hat{q}\left(\sum_{i=1}^n \hat{\mu}_i\right)\right\|_u &= \|\textstyle\sum \mu_i + A^\perp + N\|_u = \|\textstyle\sum \pi_2\mu_i + \pi_2 A^\perp\|_u \\ &\geq ((1-s)/(1+s))\|\textstyle\sum \pi_2\mu_i\|_u = ((1-s)/(1+s)) \textstyle\sum \|\pi_2\mu_i\|_u \\ &\geq ((1-s)/(1+s)) \textstyle\sum \|\hat{q}\hat{\mu}_i\|_u. \end{aligned}$$

Finally, we note that $(M_1^+(E))\hat{} = \overline{\text{co}}\,E$, so that

$$\hat{Z}_E = (\alpha\,\overline{\text{co}}\,E \cup \bar{\alpha}\,\overline{\text{co}}\,E) = \hat{N} \cap Z_A.$$

Thus $h(E) = k(E)$ is a generalized peak set, by Theorem 4.1.

6.7. Corollary. *Let E be a closed interpolation set for A such that $N + M(E)$ is A-stable. Then if there is an s, $0 \leq s < 1$, such that for each*

$m \in \partial A^{\perp}$

$$|m(E)| \leq s\|\pi_2 m\| \tag{*}$$

then $\hat{Z}_E$ is a positively decomposable section of Z_A and $k(E)$ is a generalized peak set.

Proof. Theorem 3.1(iv) shows the $\|\cdot\|_u$ ball is A-stable and that the measure μ such that $\hat{\mu} = x$ can be chosen to be a boundary measure. Thus the same computation as in Theorem 6.6 shows, with Proposition 6.4, that

$$\|\pi_2\mu\|_u \leq ((1+s)/(1-s))\|\pi_2\mu + \pi_2\partial A^{\perp}\|_u.$$

Then Lemma 6.5(i) and (ii)′ yield the same conclusion. The positive decomposability follows in exactly the same fashion.

We turn now to some general measure-theoretic conditions sufficient for exact ρ-interpolation. We shall apply Theorem 5.6, employing the closedness condition of Theorem 6.1.

6.8. COROLLARY. *Let ρ be a bounded dominator or let E be an interpolation set with ρ any dominator. Then, if*

$$\hat{U} \cap \hat{N} = (U^0 \cap N)\hat{}$$

is a decomposable section of $\hat{U}$ in A^*, E is an exact ρ-interpolation set for A.

Proof. In either case N is w^*-closed in A^* so that Theorem 6.1 yields Theorem 5.6(iii).

6.9. THEOREM. *Let E be an approximate ρ-interpolation set for A with ρ a bounded dominator such that for some x, $0 \leq s < 1$,*

$$\|\pi_1 m + A^{\perp}\|_\rho \leq s\|-\pi_2 m\|_\rho \quad \textit{for all } m \in A^{\perp}. \tag{*}$$

Then E is an exact ρ-interpolation set for A. If, in addition, U^0 and N are A-stable and $(*)$ holds with $m \in \partial A^{\perp}$ only, then E is an exact ρ-interpolation set. If $\rho \leq 1$ and $\rho \equiv 1$ on E then $(*)$ also implies $k(E)$ is a generalized peak set.

Proof. Theorem 5.3 shows $\hat{U} \cap \hat{N} = (U^0 \cap N)\hat{}$ so, by Corollary 6.8, it remains only to show the decomposability of $\hat{U}$ in A^*. We apply Lemma 6.5 with $S = U^0 = W$. If $\|x\|_\rho = 1$ in A^* then we can choose μ with $\hat{\mu} = x$ and $\|\mu\|_\rho = 1$. If U^0 is A-stable then we can choose $\mu \in \partial_A U^0$. Since U^0 is split, Proposition 6.2 shows $\|\pi_i\mu\|_\rho = \|\pi_i\mu + A^{\perp}\|_\rho$ $(i = 1, 2)$. Given $m \in A^{\perp}(\partial A^{\perp})$ choose $n \in A^{\perp}$ with

$$\|-\pi_1 m + n\|_\rho \leq s\|\pi_2 m\|_\rho.$$

Since $\|\cdot\|_\rho$ and $\|\cdot\|$ are equivalent there is a k such that for $\nu \in M(X)$,

$$\|-\nu\|_\rho \leq k\|\nu\|_\rho.$$

Thus

$$s\|\pi_2 m-\pi_2\mu\|_\rho \geq s\|\pi_2 m\|_\rho - s\|\pi_2\mu\|_\rho \geq \|-\pi_1 m+n\|_\rho - s\|\pi_2\mu\|_\rho$$
$$= \|\pi_2 m - m + n\|_\rho - s\|\pi_2\mu\|_\rho$$
$$= \|(\pi_2\mu + n - m) - (\pi_2\mu - \pi_2 m)\|_\rho - s\|\pi_2\mu\|_\rho$$
$$\geq \|\pi_2\mu\|_\rho - \|\pi_2\mu - \pi_2 m\|_\rho - s\|\pi_2\mu\|_\rho.$$

Hence

$$\|\pi_2\mu\|_\rho \leq (1+ks)/(1-s)\|\pi_2\mu - \pi_2 m\|_\rho$$

which shows Lemma 6.5(ii), or 6.5(ii′) in case U^0 and N are A-stable. If $\rho \leq 1$ and $\rho \equiv 1$ on E then

$$\|\pi_1 m + A^\perp\|_\rho \geq \|\pi_1 m(1)\|_\rho = |m(E)|$$

and

$$\|-\pi_2 m\|_\rho \leq \|-\pi_2 m\| = \|\pi_2 m\|$$

so that Theorem 6.6 or Corollary 6.7 applies to show $k(E)$ is a generalized peak set.

6.10. COROLLARY. *Let E be a closed set and let ρ be a bounded dominator such that for some s,* $0 \leq s < 1$,

$$\|\pi_1 m + A^\perp \cap N\|_\rho \leq s\|-\pi_2 m\|_\rho \quad \text{for all } m \in A^\perp. \tag{$*$}$$

Then E is an exact ρ-interpolation set for A.

If ρ is A-superharmonic and E satisfies either

(i) $E \subset \partial_A X$, *or*

(ii) $E = F \cap X$ *for F a closed face of* S_A,

and if (∗) *holds for* $m \in \partial A^\perp$ *only, then E is an exact interpolation set for A.*

Proof. If (∗) holds for $m \in A^\perp$, $(\partial A^\perp)$, then Theorem 5.3 or Corollary 5.4 shows E is an approximate ρ-interpolation set for A. Since

$$\|\pi_1 m + A^\perp\|_\rho \leq \|\pi_1 m + A^\perp \cap N\|_\rho$$

the rest follows from Theorem 6.9.

We note that if $s = 0$ in (∗) then E is an M-set and (∗) is satisfied for any ρ (superharmonic ρ in the $\partial A^\perp$ case). This is just the conclusion of Corollary 5.8 in this case.

We consider now some examples illustrating the various dual interpolation criteria of the preceding sections. First we present an approximate interpolation set which is not exact.

Let X be the unit ball in the real sequence space l^1 (w^*-topology) and let $\rho^+ = \rho^- = 1$. Then take $A = c_0$, the pre-dual of l^1, so that $\|a\|_\rho = \|a\| = \sup\{|a_n|\}$. Let E be the singleton $\{x^0\}$,

$$x_n^0 = 1/2^n, \qquad n = 1, 2, \ldots .$$

6.11. EXAMPLE. *Let A, E, X, ρ be as above. Then E is an approximate, but not exact, ρ-interpolation set for A.*

Proof. If $a \in c_0$ and $(a, x^0) = \sum_{n=1}^{\infty} a_n/2^n = 1$, then $\|a\|_\rho = 1$ implies each $a_n = 1$. But $a \in c_0$ means that $a_n \to 0$ so exact ρ-interpolation is impossible. It is clear, however, that we can find $a \in c_0$ with $(a, x^0) = 1$ and $\|a\| \leq 1 + \varepsilon$ for any given $\varepsilon > 0$. Thus E is an approximate ρ-interpolation set for A.

The next few examples (as well as some in Section 7) are based on the disc algebra whose basic properties we now sketch.

6.12. The Disc Algebra

Let Δ be the unit disc in the complex plane $\mathbb{C}$ and let

$$A(\Delta) = \{f \in C(\Delta) : f \text{ is holomorphic on int } \Delta\}.$$

Let $\mathrm{d}\sigma$ denote integration on $T \subset \Delta$ using normalized Lebesgue measure and define for $\mu \in M(T)$ the *Fourier coefficients*

$$\hat{\mu}(n) = \int_T \bar{z}^n \, \mathrm{d}\mu(z); \qquad n = 0, \pm 1, \pm 2, \ldots .$$

If $f \in C(T)$ then by $\hat{f}(n)$ we mean $(f\,\mathrm{d}\sigma)\hat{}\,(n)$ so that

$$\hat{f}(n) = \int_T \bar{z}^n f(z) \, \mathrm{d}\sigma(z).$$

If $g \in C(T)$ and $\mu \in M(T)$ then define the *convolution*

$$(g * \mu)(w) = \int_T g(w\bar{z}) \, \mathrm{d}\mu(z); \qquad w \in T.$$

(1) *The polynomials in z are dense in $A(\Delta)$.*

Proof. Take $f_r(z) = f(rz)$ ($f \in A(\Delta)$ and $0 \leq r < 1$) and note that f_r is holomorphic on a neighbourhood of Δ, hence uniformly approximable by a polynomial, and $f_r \to f$ uniformly on Δ.

(2) *The circle T is a boundary for $A(\Delta)$.*

Proof. That the map $f \to f|_T$ is an isometry is immediate from the Maximum Modulus Principle (Rudin [**187**]).

We denote $A(T) \doteq A(\Delta)|_T$.

(3) *Let* $\mu \in M(T)$. *Then* $\mu = 0$ *if and only if* $\hat{\mu}(n) = 0$ *for all integers n. If* μ *is real then* $\mu = 0$ *if and only if* $\hat{\mu}(n) = 0$ *for* $n = 0, 1, 2, \ldots$. *If* $g \in C(T)$ *then* $g * \mu \in C(T)$ *and*

$$(g * \mu)\hat{} = (\hat{g})(\hat{\mu}).$$

Proof. The Stone–Weierstrass Theorem shows that polynomials in $z, \bar{z}$ are dense in $C(T)$. Hence $\hat{\mu} = 0$ for all n implies $\mu \in C(T)^{\perp} = \{0\}$. If μ is real then

$$\hat{\mu}(-n) = \overline{\hat{\mu}(n)}$$

so the second statement follows. The continuity of $g * \mu$ is easy to check and the convolution formula follows from the calculation

$$\begin{aligned}(g * \mu)\hat{}(n) &= \int_T \bar{w}^n (g * \mu)(w)\, \mathrm{d}\sigma(w) = \int_T \int_T \bar{w}^n g(w\bar{z})\, \mathrm{d}\mu(z)\, \mathrm{d}\sigma(w) \\ &= \int_T \int_T \bar{u}^n \bar{z}^n g(u)\, \mathrm{d}\sigma(uz)\, \mathrm{d}\mu(z) \\ &= \left(\int_T \bar{u}^n g(u)\, \mathrm{d}\sigma(u)\right)\left(\int_T \bar{z}^n\, \mathrm{d}\mu(z)\right) = \hat{g}(n)\hat{\mu}(n)\end{aligned}$$

where we have used the change of variable $w = uz$, Fubini's Theorem and the translation invariance of $\mathrm{d}\sigma$ on T.

(4) $A(T) = \{f \in C(T) \colon \hat{f}(n) = 0;\ n = -1, -2, \ldots\}$.

Proof. From (1) and the fact that $\hat{\sigma}(-n) = \int_T z^n\, \mathrm{d}\sigma(z) = 0$ for $n = 1, 2, \ldots$ we see that $A \subset B$, where B is the set on the right. We show next $A^{\perp} \subset B^{\perp}$. Let $\nu \in A^{\perp}$ and define

$$\mu(f) = \int_T f(\bar{z})\, \mathrm{d}\nu(z); \qquad f \in C(T).$$

Then

$$\hat{\mu}(n) = \int_T \bar{z}^n\, \mathrm{d}\mu(z) = \int_T z^n\, \mathrm{d}\nu(z) = 0; \qquad n = 0, 1, 2, \ldots.$$

If $g \in B$ then

$$(g * \mu)\hat{}(n) = \hat{g}(n)\hat{\mu}(n) = 0 \quad \text{for all } n$$

so that $g * \mu = 0$. But

$$0 = g * \mu(1) = \int_T g(\bar{z})\, d\mu = \int_T g\, d\nu$$

so that $\nu \in B^\perp$.

(5) (*Wermer Maximality Theorem*). *Let $A(T) \subset B \subset C(T)$ where B is a closed sub-algebra of $C(T)$. Then either $B = A(T)$ or $B = C(T)$.*

Proof. If $\bar{z} = z^{-1} \in B$ then $B = C(T)$ by the Stone–Weierstrass Theorem. If $z^{-1} \notin B$ then z generates a proper maximal ideal in B so that the Gelfand–Naimark Theorem (see, for example Stout [**204**]) shows there is an algebra homomorphism $\phi \in S_B$ such that $\phi(z) = 0$. Thus, if $\mu \in M_1^+(X)$ and represents ϕ with respect to B then

$$\hat{\mu}(n) = \overline{\phi(z^n)} = 0; \qquad n = 1, 2, \ldots$$

so that $(\mu - \sigma)\hat{}\,(n) = 0$; $n = 0, 1, 2, \ldots$ and $\mu = \sigma$. But then if $g \in B$ and $n = -1, -2, \ldots$ we have

$$\hat{g}(n) = \int_T z^{|n|} g(z)\, d\sigma(z) = \phi(z^{|n|} g(z)) = (\phi(z))^{|n|} \phi(g) = 0.$$

Then (4) shows $g \in A(T)$.

6.13. Example. *Let $A = A(T) \subset C(T)$ and let E be a non-empty proper closed subset of T. Then*

(1) *$A|_E$ is dense in $C(E)$.*

(2) *If E contains an arc then E fails to be an interpolation set for A.*

Proof. Let J be a closed sub-arc of T and let $f \in A$ with $f \equiv 0$ on J. Then $f \equiv 0$ on T, as can be seen by taking a finite number of rotations

$$f_i(z) = f(\alpha_i z); \qquad \alpha_i \in T$$

such that the product $g = f_1 \cdots f_n$ is identically zero on T. But then the holomorphic extension of g is identically zero on Δ implying that one of the factors, and hence f, has uncountably many zeros on the interior of Δ. This means $f \equiv 0$ in $A(\Delta)$, and thus, $f \equiv 0$ on T.

Now let $B = (A|_E)^- \subset C(E)$, $M = \{f \in C(T): f|_E = 0\}$ and let $A_1 = (A + M)^-$ in $C(T)$ so that A_1 is a closed sub-algebra containing A with $A_1|_E = B$. Since E is proper in T there is a closed sub-arc $J \subset T \backslash E$ and $f \in C(T)$ with $f \equiv 0$ on J, $f \equiv 1$ on E and

$$f = 1 + (f - 1) \in A + M.$$

But the above shows $f \notin A$ so that

$$A \subsetneq A_1.$$

Hence 6.12(5) gives $A_1 = C(T)$ so that $(A|_E)^- = A_1|_E = C(E)$.

Similarly, if E contains an arc then there is a $0 \neq f \in C(E)$ such that f is identically zero on a sub-arc of E. But again, such an f is not the restriction of an element of A.

We will discuss $A^\perp$ (A the disc algebra) at the end of Section 7 and use this to relate the exactness criteria of Corollary 5.8 to the original Rudin and Carleson theorems. For now we return to the more general sub*space* exactness criteria.

If E is a closed subset of X with $N = M(\check{E})$ A-stable and ρ is A-superharmonic then we have the following measure theoretic conditions:

(i) 6.10(*) $\quad \|\pi_1 m + A^\perp \cap N\|_\rho \leq s\|-\pi_2 m\|_\rho \quad$ for all $m \in \partial A^\perp$

(ii) 6.9(*) $\quad \|\pi_1 m + A^\perp\|_\rho \leq s\|-\pi_2 m\|_\rho \quad$ for all $m \in \partial A^\perp$

(iii) 6.7(*) $\quad |m(E)| \leq s\|\pi_2 m\| \quad$ for all $m \in \partial A^\perp$

It is clear that (i) implies (ii) and, for $\rho \equiv 1$,

$$|m(E)| = |\pi_1 m(1)| = |\pi_1 m(1) + \eta(1)| \leq \|\pi_1 m + \eta\| \quad (\eta \in A^\perp)$$

so that (ii) implies (iii). Thus any of these conditions for $\rho \equiv 1$ imply, by Corollary 6.7, that $\overline{\text{co}}\, E$ is a *face* of S_A.

Thus the condition (*) of Corollary 6.10, while consonant with the traditional formulations, is geometrically rather restrictive. The next example shows that it may be more productive to combine Lemma 6.5 and Corollary 6.8 to obtain an exactness condition based directly on decomposability.

Let X be a square in $\mathbb{R}^2$ with vertices denoted $\{1, 2, 3, 4\}$. Let $E = \{1, 2\}$ with 1 and 2 diagonally opposite vertices. Take $A = A(X)$ (real affine functions) and $\rho^+ = \rho^- = 1$. One can verify with elementary calculations that E is an exact ρ-interpolation set for A, but it would be reassuring to have this confirmed by a theoretical result from this section. The basis is of course the decomposability of X by the diagonal through E.

6.14. Example. *In the above set-up, Corollary* 5.4 *shows E is an approximate ρ-interpolation set. The condition* (*) *of Theorem* 6.9 *fails. However Lemma* 6.5(i) *and* (ii)′ *hold* (*for* $S = U^0$) *so that the conclusion of Corollary* 6.8 *holds, showing that the diagonal*

$$F = \text{co}\, E$$

is a decomposable section of X and E is an exact ρ-interpolation set for A.

Proof. We note that $\partial_A M(X)$ is the four-dimensional space spanned by the point-masses $\{\varepsilon_i\}_{i=1}^4$ at the vertices. Then $\partial A^\perp$ is the one-dimensional space generated by

$$m = \varepsilon_1 + \varepsilon_2 - \varepsilon_3 - \varepsilon_4.$$

If $\mu = \sum \lambda_i \delta_i$ then $\|\mu\|_\rho = \sum |\lambda_i| = \|\mu\|$. Since $E \subset \partial_A X$, N is stable and so is $U^0 = M_1(X)$ with

$$\partial U^0 = \text{co}\,\{\pm\varepsilon_i\}_{i=1}^4.$$

Clearly $A^\perp \cap N = \{0\}$ since F is a (one-dimensional) simplex. Thus

$$\|\pi_1 m + A^\perp \cap N\| = \|\pi_1 m\| = \|\pi_2 m\|$$

and so Corollary 5.4(iii) shows E is an approximate ρ-interpolation set but Theorem 6.9(*) fails. Let $S = M_1(X)$. To show Lemma 6.5 applies it suffices to demonstrate (ii)′ since the stability requirements and (i) are immediate.

Let $\mu = \sum \lambda_i \varepsilon_i$, then

$$\|\pi_2\mu + \pi_2 \partial A^\perp\| = \inf\{|\lambda_3 - \lambda| + |\lambda_4 - \lambda| : \lambda \in \mathbb{R}\} = |\lambda_4 - \lambda_3|.$$

If λ_3 and λ_4 are opposite in sign then

$$\|\pi_2\mu + \partial A^\perp\| \le \|\pi_2\mu\| = |\lambda_3| + |\lambda_4| = |\lambda_4 - \lambda_3| = \|\pi_2\mu - \pi_2\partial A^\perp\|.$$

If, say, $0 \le \lambda_3 \le \lambda_4$, consider $\nu = \mu + \lambda_3 m$. Then $\hat{\nu} = \hat{\mu}$ and

$$\|\nu\| = \sum|\lambda_i - \lambda_3| \le (|\lambda_1| + |\lambda_2| + 2|\lambda_3|) + |\lambda_4| - |\lambda_3| = \|\mu\|$$

and

$$\|\pi_2\nu + \partial A^\perp\| \le \|\pi_2\nu\| = \lambda_4 - \lambda_3 = \|\pi_2\mu + \pi_2\partial A^\perp\|.$$

Hence in either case Lemma 6.5(ii)′ holds.

6.15. EXAMPLE. *Let X be the square and now $E = \{1, 3\}$ consists of adjacent vertices. Again, Corollary 5.4 shows E is an approximate ρ-interpolation set for $A(X)$ but Theorem 6.9(*) fails. However Corollary 6.7(*) holds showing that the edge*

$$F = \text{co}\, E$$

is a positively decomposable face of X and hence an exact ρ-interpolation set.

Proof. The first part is the same as in 6.14 and to check Corollary 6.7(*) we have

$$|m(E)| = |\delta_1(1) - \delta_3(3)| = 0.$$

Gamelin [**114**] and Briem [**50**] present interpolation results in terms of an extension constant $e(A, E)$. We conclude by illustrating the relationship between this and Corollary 5.4 and Corollary 6.10. We shall deal with the boundary measure case only, noting that similar but somewhat easier arguments apply in the earlier version.

Let X be a compact *metric* space, $1 \in A$ a closed separating subspace of $C(X)$ and E a compact subset of $\partial_A A$. Let $e(A, E)$ denote the smallest number k such that

$$\|\pi_1 m + A^\perp \cap N\| \leq k\|\pi_2 m\| \quad \text{for all } m \in \partial A^\perp.$$

Let $\mathcal{G}$ be the collection of all compact subsets $G \subset \partial_A X \backslash E$ and let

$$\rho(G, k)(x, t) = \begin{cases} 1 & \text{for } (x, t) \in E \times T \\ k & \text{for } (x, t) \in G \times T \\ 1 \vee k & \text{otherwise.} \end{cases}$$

Then ρ is l.s.c. and hence a gauge on X.

6.16. THEOREM.
(1) *The extension constant $e(A, E) \leq k$ if and only if E is an approximate $\rho(G, k)$-interpolation set for all $G \in \mathcal{G}$.*
(2) *If $e(A, E) < 1$ then E is an exact $\rho(G, k)$ interpolation set for all k such that $e(A, E) < k$.*

Proof. Let $\rho = \rho(G, k)$ and note that, by Theorem 5.1(iii)

$$\|\mu\|_\rho = \|\mu|_E\| + k\|\mu|_G\| + (k \vee 1)\|\mu|_Y\| \quad (Y = X \backslash (E \cup G))$$

for all $\mu \in M(X)$. The stability of N and U^0 are immediate. Now we have

$$\|\pi_1 m + A^\perp \cap N\| = \|\pi_1 m + A^\perp \cap N\|_\rho.$$

If $\mu \in \partial_A M(X)$ then, using the metrizability of X,

$$|\mu|(X \backslash E) = |\mu|(\partial_A X \backslash E) = \sup\{|\mu|(G) \colon G \in \mathcal{G}\}.$$

Thus

$$k\|\pi_2 m\| = \sup\{\|\pi_2 m\|_\rho \colon \rho = \rho(G, k) \text{ and } G \in \mathcal{G}\}. \tag{3}$$

Now, if $e(A, E) \leq k$ then

$$\|\pi_1 m + A^\perp \cap N\| \leq k\|\pi_2 m\| \quad \text{for all } m \in \partial A^\perp \tag{4}$$

so that

$$\|\pi_1 m + A^\perp \cap N\|_\rho \leq k(\|m|_G\| + \|m|_Y\|) \leq \|\pi_2 m\|_\rho \tag{5}$$

and E is an approximate $\rho(G, k)$ interpolation set by Corollary 5.4.

Conversely, the inequalities (5) imply, by (3), that (4) holds. If $k_0 = e(A, E) < k \wedge 1$ then for $m \in \partial A^\perp$ and $\rho = \rho(G, k)$

$$\|\pi_1 m + A^\perp \cap N\|_\rho = \|\pi_1 m + A^\perp \cap N\| \le k_0(\|m|_G\| + \|m|_Y\|)$$
$$\le (k_0/k)(k\|m|_G\| + (k \vee 1)\|m|_Y\|) = (k_0/k)\|\pi_2 m\|_\rho$$

so that Corollary 6.10 gives the exactness of solutions to $\rho(G, k)$.

Notes

The usual procedure in the literature for showing E is an exact interpolation set is to show E is a generalized peak set and then use the 2^{-n} iteration of Bishop [**39**] and Glicksberg [**118**]. Thus Corollary 6.10(*) with $s = 0$ is shown in these papers to imply E is a (generalized) peak exact ρ-interpolation set (for continuous T-invariant ρ). Gamelin [**113**] shows that Corollary 6.10(*) with $\rho \equiv 1$ implies that E is a (generalized) peak exact interpolation set. He also relates the smallest value of s to the extension constant of Theorem 6.16 and proves the pre-boundary measure version of Theorem 6.16. The version given in Theorem 6.16 is due to Briem [**50**]. Andô [**19**] gives a general Banach space formulation of Corollary 6.10(*) and introduces the polar geometry which is instrumental in the development herein (cf. Asimow [**28, 29**]). The peak set conditions and the notion of positive decomposability in Theorem 6.6 and Corollary 6.7 are given in Asimow [**29**]. The case $s = 0$ is proved in Alfsen and Hirsberg [**9**] and the fact that Corollary 6.10(*) ($\rho \equiv 1$, $m \in \partial A^\perp$) implies $k(E)$ is a peak set (X metrizable) is given by Briem [**50**]. He also gives the condition in Corollary 6.7 for the case E a singleton. The application of decomposability to exact interpolation criteria is given in [**28, 29**].

7. M-HULLS AND FUNCTION ALGEBRAS

We say the closed subset E of X is an *M-hull* for A ($1 \in A$, a separating subspace of $C(X)$) if

(i) $M(E)$ is A-stable,

(ii) $m \in \partial_A A^\perp$ implies $\pi_1 m = m|_E \in A^\perp$

(iii) $E = k(E)$ ($k(E) = (\overline{\text{co}}\, E) \cap X \subset S_A \subset A^*$)

As in Section 4, if E satisfies (i) and (ii) E is termed an M-set and Corollary 4.5 shows that if E is an M-set then $k(E)$ is an M-hull. It is easy to see that an M-set E may be a proper subset of $k(E)$—for example, if X is the unit interval and $A = A(X)$ then $E = \{0, 1\}$ is an M-set whose hull is X. For A a function *algebra* the natural objects to study are the M-hulls and we shall see first that the characterizations of Theorem 4.4 can be substantially sharpened.

Throughout this section we take the complex state space Z_A to be determined by $\alpha = 1$ and $\beta = -i$ so that

$$Z_A = \text{co}\,(S_A \cup -iS_A).$$

We refer to convex subsets of Z_A as *symmetric* if they are of the form

$$\text{co}\,(F \cup -iF).$$

In particular the set $\hat{Z}_E = \text{co}\,(F \cup -iF)$, $(F = \overline{\text{co}}\, E \subset S_A)$ is symmetric.

7.1. THEOREM. *Let A be a function algebra in $C(X)$ and E a closed subset of X. The following are equivalent:*
- (i) *E is a generalized peak set;*
- (ii) *$m \in A^\perp$ implies $\pi_1 m \in A^\perp$;*
- (iii) *E is an M-hull;*
- (iv) *$E = k(E)$ and $\hat{Z}_E$ is a split face of Z_A;*
- (v) *$E = k(E)$ and $\hat{Z}_E$ is a face of Z_A.*

Proof. Let E be a generalized peak set and let $m \in A_1^\perp$, $g \in A$ and $\varepsilon > 0$ be given. Choose K, a compact subset of X, such that

$$K \cap E = \varnothing \quad \text{and} \quad |m|(X \backslash E) \le |m|(K) + \varepsilon.$$

Choose $f \in A$ such that $\|f\| = 1$ and

$$|f| < 1 \quad \text{on } K, \qquad f \equiv 1 \quad \text{on } E.$$

Since A is an algebra we can assume, by taking f^n for n sufficiently large, that

$$|f| < \varepsilon \quad \text{on } K.$$

Now $\pi_1 m = m|_E$ so that

$$|\pi_1 m(g)| = \left|\int_E g \,\mathrm{d}m\right| = \left|\int_E g \cdot f \,\mathrm{d}m\right| = \left|\int_{X\backslash E} gf \,\mathrm{d}m\right|$$

$$\le \|g\| \left(\int_K |f| \,\mathrm{d}|m| + \varepsilon\right) \le 2\varepsilon \|g\|.$$

Hence $\pi_1 m \in A^\perp$ and (i) implies (ii) is shown.

If (ii) holds and $x \in k(E)$ then there is a probability measure μ supported on E such that μ represents x. Thus

$$\mu - \varepsilon_x \in A^\perp$$

so that

$$(\mu - \varepsilon_x)|_E \in A^\perp.$$

Since $\mu(1) = 1$ it follows that $x \in E$. Thus $E = k(E)$ and, if μ is a maximal measure representing x, then

$$\mu|_E - \varepsilon_x \in A^{\perp}$$

so that $\mu(E) = 1$. This shows E is A-stable and therefore E is an M-hull. Now Theorem 4.4 and Corollary 4.5 show (iii) implies (iv) and (iv) implies (i). Thus (i)–(iv) are equivalent and trivially imply (v).

If (v) holds then we shall show (ii). Thus let $m \in A_1^{\perp}$, $g \in A$ and $\varepsilon > 0$ be given.

Choose K compact in X such that $K \cap E = \emptyset$ and

$$|m|(X \backslash E) \leq |m|(K) + \varepsilon.$$

Now K is a compact subset of Z_A disjoint from the face $\hat{Z}_E$. Thus Proposition 2.10.6 shows there exists $\theta f \in A(Z_A)$ such that

$$0 \leq \theta f \quad \text{on } Z_A$$

$$\theta f \leq \delta \quad \text{on } \hat{Z}_E$$

$$k \leq \theta f \quad \text{on } K$$

where δ and k are chosen as follows: Let $h = e^{-f} \in A$ and note that $|f| \leq \sqrt{2}\delta$ on E so that δ is chosen sufficiently small that

$$|1 - h| \leq \varepsilon \quad \text{on } E.$$

Since $|h| = \exp(-\text{re } f)$ we have $|h| \leq 1$ on X and can choose k so large that $|h| \leq \varepsilon$ on K.

Hence

$$\begin{aligned} |\pi_1 m(g)| &= \left| \int_E (gh + g(1-h)) \, \mathrm{d}m \right| \leq \left| \int_E gh \, \mathrm{d}m \right| + \int_E |g(1-h)| \, \mathrm{d}|m| \\ &= \left| \int_{X \backslash E} gh \, \mathrm{d}m \right| + \int_E |g||1-h| \, \mathrm{d}m \\ &\leq \|g\| [\varepsilon(|m|(K) + 1) + \varepsilon |m|(E)]. \end{aligned}$$

We see that the proof of the last implication is a variation on the argument used to prove the Bishop Peak Point Theorem (4.3) where we show co $(x, -ix)$ is a face of Z_A.

The equivalence (ii) shows that once E is known to be an M-hull the key property (ii) of the definition holds for all $m \in A^{\perp}$, not just the boundary measures.

We note as well that, from Corollary 6.10 ($s = 0$), the M-hulls are exact ρ interpolation sets for any dominator ρ and that the extension constant $e(A, E)$ is either zero or greater than or equal to one. We discuss an example of the later phenomena in Example 7.20.

It follows from Theorem 7.1(iv) that the collection of M-hulls for A is closed under intersections and finite unions. We shall use the notation

$$\partial_A X = \text{ext}(S_A)$$

and

$$\partial_A E \doteq E \cap \partial_A X.$$

Since $F = \text{co}\, E \subset S_A$ is a face, $\partial_A E = \text{ext}\, F$, and, if E_j $(j = 1, 2)$ are M-hulls then $E_1 \subset E_2$ if and only if $\partial_A E_1 \subset \partial_A E_2$.

Statements (iv) and (v) serve to characterize the symmetric split faces of Z_A. We now investigate the properties of A pertaining to general split faces of Z_A, interpreting some standard function algebra concepts from this geometric perspective. This will be in the same spirit as the result in Theorem 0.2.

A useful tool in relating facial structure in Z_A to A is the Šilov decomposition.

Let A be a closed subalgebra of functions in $C(X)$, but do not assume now that A contains constants or separates points. Let $A_0 = \{u \in A : u \text{ is real-valued}\}$. We say E is *antisymmetric* for A if $f \in A$ and $f|_E$ real-valued implies $f|_E$ is constant. If X is antisymmetric for A we say A is an *anti-symmetric* algebra.

7.2. PROPOSITION. *Let A be as above and let $\mu \in \text{ext}(A_1^\perp)$. Then* supp μ *is anti-symmetric. If $1 \in A$ then* supp μ *is not a singleton.*

Proof. Let $E = \text{supp}\, \mu$ and let $f \in A$ with $f|_E$ real. Choose a, b real $(a \neq 0)$ so that if $g = af + b$ then $0 < g < 1$ on E. Then, g need not be in A, but since A is an algebra

$$g\mu \quad \text{and} \quad (1-g)\mu \in A^\perp.$$

Now

$$1 = \|\mu\| = \int_E \mathrm{d}|\mu| = \int_E g \,\mathrm{d}|\mu| + \int_E (1-g)\,\mathrm{d}|\mu| = \|g\mu\| + \|(1-g)\mu\|.$$

Hence μ is a convex combination of $g\mu/\|g\mu\|$ and $(1-g)\mu/\|(1-g)\mu\|$ in $A_1^\perp$ so that $g = \|g\mu\|$ (a.e. $\mathrm{d}|\mu|$). Since g is continuous g is constant on E and thus, so is f.

If $1 \in A$ it is clear that no element of $A^\perp$ can be a non-zero point mass.

We now define the *Šilov Decomposition* of X *with respect to* A to be the collection $\mathscr{S}$ of the maximal sets of constancy for functions in A_0. Thus, for each $x \in X$, if

$$E_x = \{y : u(x) = u(y) \text{ for all } u \in A_0\}$$

then $\mathscr{S}=\{E_x\}_{x\in X}$. The elements of $\mathscr{S}$ form a partition of X into compact subsets.

7.3. THEOREM. *Let $\mathscr{S}$ be the Šilov decomposition of X with respect to the closed algebra of functions A.*

(i) *If $1\in A$ then each $E\in\mathscr{S}$ is a generalized peak set.*

(ii) *If E_0 is antisymmetric for A then $E_0\subset E$ for some $E\in\mathscr{S}$.*

(iii) $A=\{f\in C(X): f|_E\in A|_E, \forall E\in\mathscr{S}\}$.

Proof.

(i) If $x\notin E$ there is a $u\in A_0$ such that $u|_E\equiv 0$, $u(x)\neq 0$ and $\|u\|<1$. If $v=1-u^2$ then $\|v\|=1$, $v(x)<1$ and $E\subset v^{-1}(1)$. This shows E is an intersection of peak sets.

Statement (ii) is by definition.

(iii) Let $f\in C(X)$ and $f|_E\in A|_E$ for all $E\in\mathscr{S}$. If $A^\perp\neq\{0\}$ then choose $\mu\in\operatorname{ext}A_1^\perp$. Then

$$E_0=\operatorname{supp}\mu\subset E\in\mathscr{S}.$$

Thus, if $f|_E=g|_E$ for $g\in A$,

$$\int_X f\,d\mu=\int_E f\,d\mu=\int_E g\,d\mu=\int_X g\,d\mu=0$$

so that $f\in A^{\perp\perp}=A$.

If A is a self-adjoint function algebra (separating and containing constants) then A_0 separates points so that $\mathscr{S}=\{x\}_{x\in X}$. Then (iii) shows $A=C(X)$, which is precisely the complex Stone–Weierstrass Theorem. If K is closed in X we denote by I_K the closed ideal of all functions in $C(X)$ vanishing on K. Clearly

$$(I_K)^\perp=\{\mu\in M(X): \operatorname{supp}\mu\subset K\}.$$

If $K=\varnothing$ then $I_K=C(X)$ and if $K=X$, $I_K=\{0\}$. We observe next that all closed ideals in $C(X)$ are of the form I_K.

7.4. COROLLARY. *Let A be a closed ideal in $C(X)$ and let*

$$K=\{x: f(x)=0\ \forall f\in A\}.$$

Then $A=I_K$.

Proof. Clearly $A\subset I_K$. Let $\mathscr{S}$ be the Šilov decomposition with respect to A. If $x\in K$ then clearly $K\subset E_x\in\mathscr{S}$. If $y\notin K$ then there exists $f\in A$ such that $f(y)=1$. Let

$$g=|f|^2=\bar{f}f\in A_0.$$

Then $g(y)=1$. If $z \in X\backslash\{y\}$ use the Tietze Extension Theorem to choose $h \in C_{\mathbb{R}}(X)$ with

$$h \in I_K, \qquad h(y)=1 \quad \text{and} \quad h(z)=0.$$

Then $u = hg \in A_0$ with $u(y)=1$ and $u(z)=0$. Thus $E_y=\{y\}$ and it follows that $\mathcal{S}$ consists precisely of K (if $K \neq \varnothing$) and the singletons $\{y\}$ for $y \in K$ (if $K \neq X$). Thus $f \in I_K$ implies

$$f|_E \in A|_E \ \forall E \in \mathcal{S}$$

so that $I_K \subset A$.

We now take A to be a function algebra so that $1 \in A$ and A separates points. The largest closed ideal of $C(X)$ which is contained in A is called the *essential ideal* of A. This is well-defined since if I_1 and I_2 are ideals in A then so is $(I_1+I_2)^-$. Then Corollary 7.4 shows the essential ideal is I_K for a closed subset K of X which we call the *essential set*. If $K = X$ (equivalently $I_K=\{0\}$) then we say A is *essential*.

7.5. THEOREM. *Let A be a function algebra and let K be the essential set for A.*

(i) *K is a generalized peak set (equivalently an M-hull).*
(ii) *$K \subset E$ (E closed) if and only if* supp $\mu \subset E$ *for all $\mu \in A^{\perp}$.*
(iii) *If $x \in K$ then $E_x \subset K$ ($E_x \in \mathcal{S}$).*
(iv) *If $x \notin K$ then $E_x = \{x\}$.*
(v) *$A = \{f \in C(X): f|_K \in A|_K\}$.*
(vi) *$A|_K$ is essential in $C(K)$.*

Proof. (i) follows from the Tietze Theorem and the fact that $1 \in A$. For (ii), $K \subset E$ if and only if $I_E \subset I_K$ if and only if $A^{\perp} \subset (I_E)^{\perp}$. Parts (iii) and (iv) are straightforward applications of the Tietze Theorem. In (v), if $f|_K = g|_K$ for $g \in A$ then $f-g \in I_K \subset A$ so that $f \in A$. For (vi) let $K_0 \subset K$ and suppose $I_{K_0}|_K \subset A|_K$. If $g \in I_{K_0} \subset C(X)$ then $g|_K \in I_{K_0}|_K \subset A|_K$ so that (v) gives $g \in A$. Thus $I_{K_0} \subset A$, implying $K = K_0$.

We now return to the representation $\theta A = A(Z_A)$ and identify the split faces of Z_A.

7.6. THEOREM. *Let $F_j (j=1,2)$ be closed and convex in S_A such that*

$$F = \text{co}\,(F_1 \cup -iF_2)$$

is split in Z_A. Then $E_j = F_j \cap X$ is an M-hull.

Proof. Let G be the complementary face of F in Z_A. Then

$$G = \text{co}\,(G_1 \cup -iG_2)$$

where

$$G_1 = G \cap S_A \quad \text{and} \quad G_2 = i(G \cap (-iS_A)).$$

Let h be the affine function on Z_A such that

$$h|_F = \theta 1|_F \quad \text{and} \quad h|_G \equiv 0.$$

Then h is a well-defined u.s.c. affine function such that $0 \leq h \leq 1$ and $h|_{F_1} \equiv 1$. Furthermore $h|_{-iF_2} \equiv 0$ and $h|_{-iG_2} \equiv 0$ so that $h|_{-iS_A} \equiv 0$. Also $h < 1$ on $S_A \backslash F_1$. Now Theorem 1.6.1(ix) shows there is a net $(f_\alpha)_{\alpha \in J}$ in A such that $\theta f_\alpha \downarrow$ pointwise to h. Then $\theta((1+i)f_\alpha)$ converges pointwise to an affine function h_0 on Z_A such that

$$h_0(x) = h(x) - h(-ix) \quad \text{for } x \in S_A$$

$$h_0(-ix) = h(x) + h(-ix) \quad \text{for } -ix \in -iS_A.$$

Hence

$$0 \leq h_0 \leq 1 \quad \text{on } Z_A$$

$$h_0 \equiv 1 \quad \text{on} \quad \text{co}\,(F_1 \cup -iF_1) \qquad \text{and} \qquad h_0 < 1 \quad \text{on} \quad Z_A \backslash \text{co}\,(F_1 \cup -iF_1).$$

Therefore co $(F_1 \cup -iF_1)$ is a face of Z_A and the conclusion for E_1 follows from Theorem 7.1(v). The argument for E_2 is the same.

If E_1 and E_2 are closed sets in X we say $E_1 \cup -iE_2$ is an *A-interpolation pair* if $E_1 \cup -iE_2$ is an interpolation set for $A(Z_A)$ as a real subspace of $C_{\mathbb{R}}(X_0)$ as in Theorem 3.1. As usual this just amounts to the real linear span, $\hat{N}$, of $F_1 \cup -iF_2$ ($F_j = \overline{\text{co}}\, E_j \subset S_A$) being w^*-closed in A^*. Here

$$N = \{\mu \in M(X) \colon \mu = \mu_1 - i\mu_2;\ \mu_i \in M_{\mathbb{R}}(E_j), j = 1, 2\}$$

so that $\hat{N}$ is indeed the quotient image of N.

7.7. Proposition. *The pair $E_1 \cup -iE_2$ is an A-interpolation pair if and only if $N + A^\perp$ is w^*-closed in $M(X)$.*

Proof. We have $E_1 \cup -iE_2$ an A-interpolation pair if and only if $\hat{N}$ is w^*-closed in A^* if and only if $N + A^\perp$ is w^*-closed in $M(X)$.

We let

$$I = \{f = u_1 + iu_2 \in C(X);\ u_j \text{ real and } u_j|E_j \equiv 0, j = 1, 2\}.$$

We have I is a real subspace in $C(X)$ and $I^0 = N$. Thus, if $B = I \cap A$ then

$$B^0 = (I^0 + A^0)^- = (N + A^\perp)^- \quad (w^*\text{-closure in } M(X)).$$

7.8. THEOREM. *Let E_j $(j=1,2)$ be M-hulls for the function algebra A such that $X=E_1\cup E_2$. If $E_1\cup -iE_2$ is an A-interpolation pair then the essential set $K\subset E_1\cap E_2$ so that*

$$I_{E_1\cap E_2}\subset A.$$

Proof. Let

$$B_j=\{f\in A|_{E_j}: f\equiv 0 \quad \text{on} \quad E_1\cap E_2\} \quad (j=1,2)$$

and let I be as above. We note that

$$(I\cap A)|_{E_2}\subset (B_2)_0 \quad \text{(the real-valued functions in } B_2) \qquad (*)$$
$$(I\cap A)|_{E_1}\subset i(B_1)_0.$$

We show that the Šilov decomposition $\mathscr{S}(B_j)$ of B_j as a subspace of $C(E_j)$ consists of the set $E_1\cap E_2$ and singletons in $E_j\backslash(E_1\cap E_2)$. If $x\in E_1\cap E_2$ then clearly $E_1\cap E_2\subset E_x$ $(j=1,2)$. Given $x\in E_j$, let

$$K_x=\{y\in E_j: f(y)=f(x)\ \forall f\in I\cap A\}.$$

Then $(*)$ shows $E_x\subset K_x$, and it suffices to show K_x is a singleton if $x\notin E_1\cap E_2$. We observe first that K_x is a generalized peak set for $A|_{E_j}$, since $z\in E_j\backslash K_x$ implies there is an $f\in I\cap A$ with $f|_{K_x}$ constant and $f(z)\neq f|_{K_x}$. If $j=1$ then $(*)$ shows

$$f|_{E_1}=iu_2$$

and if $j=2$ then

$$f|_{E_2}=u_1 \quad (u_j \text{ real}, j=1,2).$$

Thus for $\varepsilon>0$ sufficiently small

$$g=1\pm\varepsilon(f-f(x))^2 \quad (+ \text{ for } j=1 \text{ and } - \text{ for } j=2)$$

satisfies

$$0\le g\le 1, \qquad g|_{K_x}=1 \quad \text{and} \quad g(z)<1.$$

Thus K_x is a generalized peak set for $A|_{E_j}$, hence an M-hull for $A|_{E_j}$, and therefore (since E_j an M-hull for A), K_x is an M-hull for A.

Hence $K_x\subset E_1\cap E_2$ if and only if $\partial_A K_x\subset E_1\cap E_2$. Thus one can assume $x\in\partial_A K_x\backslash(E_1\cap E_2)$. Suppose $y\in K_x$ and $y\neq x$. Then, since $x\in\partial_A X$, Corollary 4.3 shows there is a peak function f in A such that $f(x)=1$ and $|f|<1$ on $(E_1\cap E_2)\cup\{y\}$. Since x, $y\in K_x$ we have

$$\varepsilon_x-\varepsilon_y\in(I\cap A)^0=N+A^\perp$$

which is w^*-closed by Proposition 7.7. Thus

$$\varepsilon_x-\varepsilon_y=\mu+(\nu_1-i\nu_2); \qquad \mu\in A^\perp \qquad \nu_j \text{ real and supp } \nu_j\subset E_j \quad (j=1,2).$$

If f_0 is the point-wise limit of f^n as $n \to \infty$ then f_0 is a characteristic function and evaluating at f_0 gives the contradiction

$$1 = i\nu_2(f_0).$$

Hence y must equal x and $K_x = \{x\}$ if $x \in E_j \backslash (E_1 \cap E_2)$ $(j = 1, 2)$. This shows $\mathcal{S}(B_j)$ is as desired. Since $E_1 \cap E_2$ is a generalized peak set we see that $B_j|_{\{x\}} \neq \{0\}$ for $x \in E_j \backslash (E_1 \cap E_2)$ so that Theorem 7.3(iii) gives

$$B_j = I_{E_1 \cap E_2} \quad \text{in } C(E_j).$$

If $\mu \in A^\perp$ then

$$\mu|_{E_j} \in (A|_{E_j})^\perp \subset B_j^\perp = \{\mu \in M(E_j)\text{: supp } \mu \subset E_1 \cap E_2\} \quad (j = 1, 2).$$

Thus $\mu \in A^\perp$ implies supp $\mu \subset E_1 \cap E_2$ so that Theorem 7.5 shows $K \subset E_1 \cap E_2$ and

$$I_{E_1 \cap E_2} \subset A.$$

The case $E_1 = X$ and $E_2 = \varnothing$ amounts to a theorem of Hoffman and Wermer [**130**].

7.9. Corollary. *If* re A *is uniformly closed in* $C_{\mathbb{R}}(X)$ *then* $A = C(X)$.

Proof. We have $A(Z_A)|_{S_A} = \{u|_{S_A};\, u = \text{re } f,\ f \in A\}$ so the hypothesis says $X \cup -i\varnothing$ is an interpolation pair, giving

$$I_\varnothing = C(X) \subset A.$$

Let $E_1 \triangle E_2 = (E_1 \backslash E_2) \cup (E_2 \backslash E_1)$.

7.10. Corollary. *If* E_j $(j = 1, 2)$ *are M-hulls with* $E_1 \cup -iE_2$ *an A-interpolation pair then*

$$\mu \in A^\perp \textit{ implies } |\mu|(E_1 \triangle E_2) = 0.$$

Proof. If $E = E_1 \cup E_2$ then E is an M-hull and so $\mu \in A^\perp$ implies

$$\mu|_E \in (A|_E)^\perp.$$

By applying Theorem 7.8 to E, we see supp $\mu|_E \subset E_1 \cup E_2$ so that $|\mu|(E_1 \triangle E_2) = 0$.

If we combine Theorem 7.6 and Corollary 7.10 we have a characterization of the split faces of Z_A.

7.11. Corollary. *Let A be a function algebra with complex state space Z_A. The following are equivalent*:

(i) $E_j (j=1,2)$ *is an M-hull and* $E_1 \cup -iE_2$ *is an A-interpolation pair,*
(ii) E_j *is an M-hull and* $\mu \in A^\perp$ *implies* $|\mu|(E_1 \triangle E_2)=0$;
(iii) $E_j = F_j \cap X$ *and* $F = \operatorname{co}(F_1 \cup -iF_2)$ *is a split face of* Z_A.

Proof. Corollary 7.10 gives (i) implies (ii). If (ii) holds and $\mu = \mu_1 - i\mu_2$ with $\tilde{\mu} = \mu_1 \oplus \mu_2$ on $X \cup -iX$ as in Theorem 3.1 then $\tilde{\mu} \in A(Z_A)^\perp$ implies $\mu \in A^\perp$ so that

$$\int_F \theta f \, d\tilde{\mu} = \int_{E_1 \cap E_2} \operatorname{re} f \, d\mu_1 + \int_{E_1 \cap E_2} \operatorname{im} f \, d\mu_2 = \operatorname{re} \int_{E_1 \cap E_2} f \, d\mu = 0.$$

Theorem 7.6 gives (iii) implies (i).

If we apply Corollary 7.11 to the case $E_1 = E$ and $E_2 = \emptyset$ then we obtain a refined version of Corollary 7.9.

7.12. COROLLARY. *The following are equivalent*:
(i) *E is an M-hull and* $(\operatorname{re} A)|_E$ *is closed in* $C_{\mathbb{R}}(E)$;
(ii) *E is an M-hull and* $A|_E = C(E)$;
(iii) $E = F \cap X$ *where* $F \subset S_A$ *is a split face of* Z_A.

7.13. COROLLARY. *The following are equivalent*:
(i) $\operatorname{co}(F \cup -iS_A)$ *is split in* Z_A;
(ii) $\mu \in A^\perp$ *implies* $\operatorname{supp} \mu \subset E = F \cap X$;
(iii) *the essential set* $K \subset E$.

Proof. The case $F = F_1$ and $S_A = F_2$ in Corollary 7.11 gives the equivalence of (i) and (ii) since (ii) implies E is an M-hull. The equivalence of (ii) and (iii) is Theorem 7.5(ii).

7.14. COROLLARY. *If A is a function algebra in* $C(X)$ *then A is reflexive if and only if X is finite and* $A = C(X)$.

Proof. Assume A is reflexive. If $x \in \operatorname{ext} S_A$ then $\{x\}$ is an M-hull with $A|_{\{x\}} = C(\{x\})$ so that Corollary 7.12 shows $\{x\}$ is split in Z_A. But then the complementary face is compact. The intersection of all the complementary faces of extreme points of Z is a compact face disjoint from $\operatorname{ext} Z_A$ and hence empty. But if x_0 is an accumulation point of $\operatorname{ext} Z_A$ then x_0 must belong to each complementary face. It follows that $\operatorname{ext} Z_A$ is finite and hence a finite dimensional simplex. Thus A is self-adjoint and equals $C(X)$. Since $f^2 \in A$ whenever $f \in A$ it must be the case that $X = \operatorname{ext} S_A$.

A theorem of Sidney and Stout [**194**] shows that the equivalence of (i) and (ii) in Corollary 7.12 holds for arbitrary closed sets E, not just M-hulls.

7.15. THEOREM. *If A is a function algebra and* $(\operatorname{re} A)|_E$ *is closed then* $A|_E = C(E)$.

Proof. Let $B = (A|_E)^-$. Thus if $g \in B$ then there is a sequence $(f_n)_{n=1}^{\infty}$ in A such that $f_n|_E \to g$ uniformly on E. Hence $\operatorname{re} f_n|_E \to \operatorname{re} g$ uniformly so that there is an $f \in A$ with $\operatorname{re} f|_E = \operatorname{re} g$. This shows $\operatorname{re} B = (\operatorname{re} A)|_E$ is closed in $C(E)$ and hence, by Corollary 7.9 $B = C(E)$. Thus $\operatorname{re} A|_E = C_{\mathbb{R}}(E)$ and $A|_E$ is dense in $C(E)$. Therefore $(A|_E)^{\perp} = A^{\perp} \cap M(E) = \{0\}$ and it suffices to show there is a c such that, for $\pi_1\mu = \mu|_E$ and $\pi_2\mu = \mu|_{X\setminus E}$

$$\|\pi_1 m\| \le c\|\pi_2 m\| \quad \text{for all } m \in A^{\perp}.$$

Given $m \in A^{\perp}$ and $\varepsilon > 0$ there is a Borel function ψ on E with range in $[0, 2\pi)$ and

$$\mathrm{d}|m| = \mathrm{e}^{i\psi}\,\mathrm{d}m.$$

By Lusin's Theorem there is $u \in C_{\mathbb{R}}(E)$ such that $\|u\| \le 2\pi$ and

$$\int_E |\mathrm{e}^{i\psi} - \mathrm{e}^{iu}|\,\mathrm{d}|m| < \varepsilon.$$

Since the map $f \to \operatorname{re} f|_E$ is a real bounded linear operator from A onto $(\operatorname{re} A)|_E$ there is a constant K (not dependent on u) and an $f \in A$ such that $\operatorname{re} f|_E = u$ and $\|f\| \le K\|u\|$. Now

$$0 = m(\mathrm{e}^{if}) = \pi_1 m(\mathrm{e}^{if}) + \pi_2 m(\mathrm{e}^{if}).$$

Thus

$$|\pi_1 m(\mathrm{e}^{if})| \le \mathrm{e}^{2\pi K}\|\pi_2 m\|.$$

On the other hand, with $f|_E = u + iv$,

$$\begin{aligned}|\pi_1 m(\mathrm{e}^{if})| &= \left|\int_E \mathrm{e}^{i(u+iv)}\,\mathrm{d}m\right| = \left|\int_E \mathrm{e}^{-v}\,\mathrm{e}^{i\psi}\,\mathrm{d}m - \int_E \mathrm{e}^{-v}(\mathrm{e}^{i\psi} - \mathrm{e}^{iu})\,\mathrm{d}m\right| \\ &\ge \int_E \mathrm{e}^{-v}\,\mathrm{d}|m| - \int_E \mathrm{e}^{-v}\,|\mathrm{e}^{i\psi} - \mathrm{e}^{iu}|\,\mathrm{d}|m| \\ &\ge \mathrm{e}^{-2\pi K}(\|\pi_1 m\| - \varepsilon).\end{aligned}$$

Thus

$$\|\pi_1 m\| \le \mathrm{e}^{4\pi K}\|\pi_2 m\|.$$

The preceding results indicate the importance of identifying the measures in $A^{\perp}$. If we consider the disc algebra $A(T)$ (Example 6.12) we are impressed by the special role of the (normalized) Lebesgue measure σ on T. It is the *unique* representing measure on T for the multiplicative state L defined by

$$L(f) = \tilde{f}(0); \qquad \tilde{f} \text{ the holomorphic extension of } f.$$

More generally, if S_A is a simplex then each multiplicative state has a unique (boundary) representing measure. This is, for example, the case whenever A is a Dirichlet algebra (re A dense in $C_{\mathbb{R}}(X)$), where in fact S_A is a Bauer simplex. The abstract F. and M. Riesz Theorem (Theorem 7.17 below) says that if σ is a *unique* representing measure on X for a multiplicative state in S_A then each $m \in A^{\perp}$ has Lebesgue decomposition $m_a + m_s$ with respect to σ such that $m_a, m_s \in A^{\perp}$.

7.16. PROPOSITION. *Let A be a function algebra and let ϕ be a multiplicative linear functional in S_A. Then ϕ has a unique representing measure σ on X if and only if for all $u \in C_{\mathbb{R}}(X)$*

$$\int_X u \, d\sigma = \sup\left\{\int_X \text{re} \, f \, d\sigma : f \in A \text{ and } \text{re} \, f \le u\right\}.$$

Proof. Consider u on $X \subset S_A$ and let $\check{u}$ be the lower envelope of u on S_A. Then by Theorem 1.6.1

$$\check{u}(\phi) = \sup\{\theta f(\phi) : \theta f \le u \quad \text{on} \quad X \subset S_A\} = \inf\left\{\int_X u \, d\nu : \nu \sim \phi\right\}.$$

But $\theta f = \text{re} \, f$ on $X \subset S_A$ so that uniqueness of σ implies the formula for $\int_X u \, d\sigma$. Conversely,

$$\int_X u \, d\sigma = \check{u}(\phi)$$

for all $u \in C_{\mathbb{R}}(X)$ shows the uniqueness of σ.

7.17. THEOREM. *Let A be a function algebra with σ a unique representing measure for the multiplicative state $\phi \in S_A$. Then $m \in A^{\perp}$ if and only if $m_a, m_s \in A^{\perp}$ where $m = m_a + m_s$ is the Lebesgue decomposition of m with respect to σ.*

Proof. Let $m \in A^{\perp}$ with m_s singular to σ and $dm_a = \rho \, d\sigma$ where ρ is a σ-integrable Borel function on X. Let $h \in A$ and $\varepsilon > 0$ be given. Choose k so large that if

$$B = \{x \in X : |\rho|(x) \ge k\}$$

then

$$2\|h\| \int_B |\rho| \, d\sigma < \frac{\varepsilon}{4}. \tag{1}$$

Let E be a Borel set such that

$$\sigma(E) = 0 = |m_s|(X \backslash E).$$

Then choose K compact, n positive, $\delta > 0$ and U open satisfying

$$K \subset E \subset U$$

$$\|h\| |m_s|(E \backslash K) < \frac{\varepsilon}{4} \tag{2}$$

$$(|m_s|(K))\|h\| \, \mathrm{e}^{-n} < \frac{\varepsilon}{4} \tag{3}$$

$$k\|h\|\sqrt{2}(1-\mathrm{e}^{-\delta})^{1/2} < \frac{\varepsilon}{4} \tag{4}$$

$$n\sigma(U) < \delta.$$

Let $u \in C_{\mathbb{R}}(X)$ with $u|_K \equiv 1$, $0 \leq u \leq 1$ and supp $u \in U$. Then

$$\int u \, \mathrm{d}\sigma \leq \sigma(U),$$

so that

$$\int -nu \, \mathrm{d}\sigma \geq -n\sigma(U) > -\delta.$$

Choose, by Proposition 7.16, $f \in A$ with (by adding an imaginary constant)

$$\sigma(\mathrm{im}\, f) = 0$$

$$\mathrm{re}\, f \leq -nu$$

$$\sigma(\mathrm{re}\, f) > -\delta.$$

Let $g = \mathrm{e}^f$. Then

$$|g| = \mathrm{e}^{\mathrm{re}\, f} \leq 1$$

and

$$|g| \leq \mathrm{e}^{-n} \quad \text{on } K. \tag{5}$$

Thus

$$\left(\int |1-g| \, \mathrm{d}\sigma\right)^2 \leq \int |1-g|^2 \, \mathrm{d}\sigma = \int (1-2\,\mathrm{re}\, g + |g|^2) \, \mathrm{d}\sigma \leq 2\int (1-\mathrm{re}\, g) \, \mathrm{d}\sigma.$$

But since σ is multiplicative, $\sigma(g) = \mathrm{e}^{\sigma(f)} = \mathrm{e}^{\sigma(\mathrm{re}\, f)} = \mathrm{re}\, \sigma(g) = \sigma(\mathrm{re}\, g)$. Hence

$$\left(\int |1-g| \, \mathrm{d}\sigma\right)^2 \leq 2(1-\mathrm{e}^{\sigma(\mathrm{re}\, f)}) \leq 2(1-\mathrm{e}^{-\delta}). \tag{6}$$

Now

$$0 = \int gh \, \mathrm{d}m = \int_X h \, \mathrm{d}m_a - \int_X (1-g)h\rho \, \mathrm{d}\sigma + \int_E gh \, \mathrm{d}m_s \tag{7}$$

with

$$\left|\int_E gh \, \mathrm{d}m_s\right| \leq \|h\| |m_s|(E \backslash K) + (|m_s|(K))\|h\| \, \mathrm{e}^{-n} < \frac{\varepsilon}{2}$$

by (5), (2) and (3).

Likewise

$$\left|\int_X (1-g)h\rho \, \mathrm{d}\sigma\right| \leq 2\|h\| \int_B |\rho| \, \mathrm{d}\sigma + k\|h\| \int_X |(1-g)| \, \mathrm{d}\sigma < \frac{\varepsilon}{2}$$

by (1), (6) and (4).

Therefore (7) gives

$$\left|\int_X h \, \mathrm{d}m_a\right| < \varepsilon.$$

Since ε is arbitrary, $m_a \in A^\perp$ and thus so is $m_s = m - m_a$.

7.18. COROLLARY. *Let A be the disc algebra on T and σ normalized Lebesgue measure on T. If $m(f) = 0$ for all $f \in A$ such that $\sigma(f) = 0$ then m is absolutely continuous with respect to σ.*

Proof. Let $M = \{f \in A: \sigma(f) = 0\}$. Then, since $\sigma(z^k) = 0$ for all $k = 1, 2, \ldots$ we have A spanned by M and the constants. Thus if $m \in M^\perp$ and $c = m(1)$, $m - c\sigma \in A^\perp$ so that $(m - c\sigma)_s = m_s \in A^\perp$ by the theorem. Thus $\hat{m}_s(k) = 0$ for all $k \leq 0$. If $\hat{m}_s(k) = 0$ for all $k \leq n$ then consider $\mathrm{d}\nu = (\bar{z})^{n+1} \, \mathrm{d}m_s$. We have for $j \geq 1$

$$\int_T z^j \, \mathrm{d}\nu(z) = \int (\bar{z})^{n+1-j} \, \mathrm{d}m_s(z) = \hat{m}_s(n+1-j) = 0$$

so that $\nu \in M^\perp$. But the above then yields $\nu_s = \nu \in A^\perp$. Hence

$$0 = \int_T \mathrm{d}\nu = \int_T (\bar{z})^{n+1} \, \mathrm{d}m_s(z) = \hat{m}_s(n+1).$$

We conclude $\hat{m}_s(n) = 0$ for all n so that $m_s = 0$.

7.19. COROLLARY (RUDIN–CARLESON THEOREM). *Let E be a subset of T with $\sigma(E) = 0$. Then E is an M-hull for the disc algebra A on T. In particular E is a peak set and a full exact ρ-interpolation set for any dominator ρ.*

Proof. Let $m \in A^\perp$. Then $\mathrm{d}m = h \, \mathrm{d}\sigma$ for a σ-integrable h on T, by the preceding corollary. Thus $m|_E = 0$. Then Theorem 7.1 shows E is an M-hull and the rest follows from Corollary 2.3 and Corollary 6.10.

We can apply Corollary 7.18 to obtain another example of non-exact interpolation. Let

$$B = \left\{ f \in A : f(1) = \int_T f \, \mathrm{d}\sigma \right\}$$

so that B is the sub*algebra* of $C(T)$ given by

$$B = A \cap \ker (\varepsilon_i - \sigma\rangle$$

and

$$B^\perp = A^\perp + \langle \varepsilon_1 - \sigma \rangle.$$

Let $E = \{-1, 1\}$.

7.20. EXAMPLE. *With B and E as above, E is an approximate, but not exact, full interpolation set for B in the uniform norm.*

Proof. If $\mu \in B^\perp$ then

$$\mu = m + \alpha(\varepsilon_1 - \sigma); \qquad m \in A^\perp$$

and hence $\mathrm{d}m = h \, \mathrm{d}\sigma$ with $\int_T h \, \mathrm{d}\sigma = 0$. Thus

$$\mu = h_0 \, \mathrm{d}\sigma + \alpha\varepsilon_1; \qquad \int_T h_0 \, \mathrm{d}\sigma = -\alpha.$$

If $N = M(E)$ then clearly $B^\perp \cap N = \{0\}$ and

$$\|\pi_1 \mu\| = |\alpha|.$$

Since ε_1 and $h_0 \, \mathrm{d}\sigma$ are mutually singular

$$\|\pi_2 \mu\| = \|h_0 \, \mathrm{d}\sigma\| \geq \left| \int_T h_0 \, \mathrm{d}\sigma \right| = |\alpha| = \|\pi_1 \mu\|.$$

This shows E is a full approximate interpolation set. If $f \in C(E)$ is $z|_E$ then there is no extension g (even in $C(T)$) satisfying $\|g\| = 1 = \int_T g \, \mathrm{d}\sigma$.

As noted after Theorem 7.1 the extension constant $e(B, E)$ must be 0 or ≥ 1. Here it is one and exact interpolation fails.

Notes

General reference books for function algebras are Browder [**57**], Gamelin [**114**], Leibowitz [**150**] and Stout [**204**].

The fact that, for function algebras, generalized peak sets are interpolation sets was shown by Bear [**36**] and by Hoffman and Singer [**129**]. In Theorem 7.1 the equivalence of (i) and (ii) is given by Glicksberg [**118**] using arguments of Bishop [**38**], as indicated in the notes to Sections 2 and 4; the equivalence of (iv) and (v) is due to Ellis [**102**]. For Dirichlet algebras Glicksberg [**118**] has shown that interpolation sets are generalized peak sets.

Proposition 7.2 is due to De Branges [**76**]. The antisymmetric decomposition for function algebras was introduced by Bishop [**38**], where he shows that maximal antisymmetric sets are generalized peak sets. The notion of essential set was introduced and developed by Bear [**35**]. Theorem 7.6 and Corollaries 7.11–7.14 are due to Briem [**52**] and Ellis [**102, 103**].

Corollary 7.9 is a theorem of Hoffman and Wermer [**130**] (see also Browder [**56**]); the generalization of the Hoffman–Wermer Theorem given in Theorem 7.15 is due to Sidney and Stout [**194**]. Theorem 7.17 and Corollary 7.18 are versions of the F. and M. Riesz Theorem, and for the history of these results we refer to the book by Stout [**204**]. Example 7.20 is due to Glicksberg [**118**] and Gamelin [**113**].

8. FACIAL TOPOLOGIES AND DECOMPOSITIONS

We recall from Chapter 3 that the facial topology on ext K (K compact convex) has as its closed sets all subsets of ext K of the form $(\text{ext } K) \cap F$, F a closed split face of K. This topology is Hausdorff if and only if K is a Bauer simplex, in which case $A(K) \simeq C_{\mathbb{R}}(X_0)$, $X_0 = \text{ext } K$. We shall interpret this in the context of $A(Z_A)$ where A is a (closed, separating with constants) subspace of $C_{\mathbb{C}}(X)$ and $Z_A = \text{co}\,(S_A \cup -iS_A)$ is the complex state space.

We denote $\partial Z_A = \text{ext } Z_A$ $(= \partial_A X \cup -i\partial_A X)$ with $\partial_A X = \text{ext } S_A$. We define the closed sets of the *symmetric facial topology* on ∂Z_A as those of the form $\partial Z_A \cap F$, F a symmetric split face of Z_A.

Since the collection of symmetric split faces is closed under intersections and finite convex hulls this is indeed a topology on ∂Z_A. Now F is a symmetric split face if and only if $F \cap X \subset S_A$ is an M-hull for A. Thus a set $E_0 \subset \partial_A X$ is (relatively) closed in $\partial_A X$ if and only if

$$E_0 = \partial_A E \ (= \partial_A X \cap E)$$

for an M-hull E of A. For this reason we call the restriction of the symmetric facial topology to $\partial_A X$ the *M-topology*. This is of course weaker than the restriction of the facial topology of ∂Z_A to $\partial_A X$.

8.1. THEOREM. *$\partial_A X$ is compact in the M-topology.*

Proof. The M-topology is weaker than the facial topology on $\partial_A X$ in S_A which is compact by Theorem 3.1.1.

We next develop the notion of central functions and the result of Theorem 3.1.4 to the present context. Let $\theta A = A(Z_A)$ and let f be a real function on

∂Z_A. We say f is *symmetric* if $f(x)=f(-ix)$ for all $x \in \partial_A X$. If u is a real function on $\partial_A X$ then we denote the symmetric extension to ∂Z_A by $\tilde{u}$. We see that f is facially continuous if and only if there is an $h \in A$ such that for $x \in \partial_A X$

$$\text{re } h(x)=f(x)$$

$$\text{im } h(x)=f(-ix)$$

and for each $a \in A$ there is a $b \in A$ such that

$$(\text{re } h)(\text{re } a)=\text{re } b$$

$$(\text{im h})(\text{im } a)=\text{im } b$$

pointwise on $\partial_A X$.

It is easy to see that if f is facially continuous and symmetric then f is continuous in the symmetric facial topology. Likewise, if u is continuous on $\partial_A X$ in the M-topology then $\tilde{u}$ is symmetric facially continuous on ∂Z_A.

As in Section 7 we denote

$$A_0=\{u \in A\colon u \text{ real-valued}\}.$$

Finally, observe that if f is real-valued on $\partial_A X$ then f is real-valued on X.

8.2. THEOREM. *Let u_0 be a real-valued function on $\partial_A X$. The following are equivalent*:

(i) *u_0 is continuous in the M-topology,*

(ii) *$\tilde{u}_0=\theta h|_{\partial Z_A}$; $h=u+iu$, $u \in A_0$ and θh central in $A(Z_A)$,*

(iii) *$u_0=u|_{\partial X}$ for $u \in A_0$ such that for each $a \in A$ there is a $b \in A$ with $ua=b$ pointwise on ∂X.*

Proof. If u_0 is M-topology continuous then $\tilde{u}_0$ is symmetric facially continuous, and hence facially continuous, on ∂Z_A. Thus there is an $h \in A$ with $\theta h|_{\partial Z_A}=\tilde{u}_0$ and the symmetry of $\tilde{u}_0$ shows $h=u+iv$ with $u=v$ on $\partial_A X$. But then $h-ih \in A$ and is real-valued on $\partial_A X$, hence on X. This shows $u=v \in A_0$ and thus, (i) implies (ii). If (ii) holds the above formulas show

$$u(\text{re } a)=\text{re } b$$

$$u(\text{im } a)=\text{im } b$$

on ∂X so that $ua=b$ on ∂X. This in turn shows that $\theta(u+iu)$ is central in $A(Z_A)$ so that $\tilde{u}_0$ is symmetric facially continuous and hence u_0 is M-topology continuous.

8.3. COROLLARY. *The following are equivalent*:

(i) *The M-topology on $\partial_A X$ is Hausdorff*;

(ii) *$\partial_A X$ is compact and A is isometric to $C(\partial_A X)$.*

Proof. If (i) holds then the M-topology, the facial topology and the w^*-topology on $\partial_A X \subset S_A$ coincide. Thus S_A is a Bauer simplex with $A(S_A)$ isometric to $C_{\mathbb{R}}(\partial S_A) = C_{\mathbb{R}}(\partial_A X)$. Now $f \in A$ implies re $f|_{\partial X}$ is continuous and hence continuous in the M-topology. Then Theorem 8.2 shows re $f \in A$ so that A is self-adjoint, so that in particular re $A|_{\partial X}$ is closed in $C_{\mathbb{R}}(\partial X)$. But re $A|_{\partial X}$ is always dense in $A(S_A) = C_{\mathbb{R}}(\partial X)$ so that re $A|_{\partial X} = C_{\mathbb{R}}(\partial X)$ and (ii) follows. The converse is clear.

8.4. COROLLARY. *Let A be a function algebra.*

(i) *A real function u_0 on ∂X is M-topology continuous if and only if $u_0 = u|_{\partial X}$ for some $u \in A_0$.*

(ii) *The M-topology is Hausdorff if and only if $A = C(X)$.*

Proof. Theorem 8.2 shows (i). For (ii) we note that Corollary 8.3 gives A isometric to $C(\partial X)$ and in particular, A is self-adjoint. Then the Stone–Weierstrass Theorem gives $A = C(X)$.

8.5. THEOREM. *Let A be a function algebra. The following are equivalent*:

(i) *θf is central in $A(Z_A)$;*

(ii) *$f = u + iv$; $u, v \in A_0$ and $u - v \equiv 0$ on the essential set K.*

Proof. If (i) holds we can apply the multiplicative formulas for central functions with $a = 1$ to obtain $b_1 \in A$ such that on ∂X

$$u = \text{re } b_1$$
$$v = \text{im } b_1.$$

Thus b_1 is real and so $b_1 \in A_0$. If $a = -i$ we obtain $b_2 \in A$ with

$$0 = \text{re } b_2$$
$$-v = \text{im } b_2$$

on ∂X. Then b_2 is imaginary so that $ib_2 \in A_0$ and equals v on ∂X. Therefore $b_1 - b_2 = f$ on X and it follows that $b_1 = u$ on X, $ib_2 = v$ on X, giving $u, v \in A_0$. Let $w = u - v$. We show θw is central in $A(Z_A)$. Given $a \in A$ choose $b \in A$ such that

$$u(\text{re } a) = \text{re } b$$
$$v(\text{im } a) = \text{im } b$$

on ∂X. Then $b - va = w(\text{re } a)$ on ∂X. But $b - va \in A$ and hence $b - va \in A_0$. Thus

$$(\theta w)(\theta a) = \theta(b - va) = \begin{cases} w(\text{re } a) & \text{on } \partial X \\ 0 & \text{on } -i\partial X. \end{cases}$$

Thus θw is central. Now

$$-i\partial X \subset \{z \in \partial Z_A : \theta w(z) = 0\} = \partial Z_A \cap F$$

for a closed split face F of Z. But $-i\partial X \subset F$ implies

$$F = \text{co}\,(F_1 \cup -iS_A): \qquad F_1 = F \cap S_A.$$

Then Corollary 7.13 gives $K \subset F_1 \cap X$. But then

$$0 = \theta w|_K = w|_K.$$

If (ii) holds and $a \in A$ is given let

$$b = u(\text{re}\, a) + iv(\text{im}\, a) \quad \text{on } X.$$

Then $b \in C(X)$ and $b = ua$ on K. But $ua \in A$ and so Theorem 7.5(v) gives $b \in A$. It follows that θf is central.

8.6. COROLLARY. *The function algebra A is antisymmetric if and only if the only central functions in $A(Z_A)$ are constants.*

In general we say the centre of $A(K)$ is *trivial* if the only central functions are constants.

We next consider abstract versions in $A(K)$ of the Šilov decomposition for function algebras. We first define the Bishop decomposition for function algebras and then give $A(K)$ analogues of both decompositions.

Thus, let A be a function algebra and let $\mathcal{B}$ denote the collection of maximal (with respect to inclusion) antisymmetric subsets of X. Since each $\{x\}$ is antisymmetric and the closure of the union of all antisymmetric sets containing x is again antisymmetric, we see that $\mathcal{B}$ is a partition of X into closed subsets. $\mathcal{B}$ is called the *Bishop decomposition* of A.

8.7. THEOREM.

(i) *If $E \in \mathcal{B}$ then $E \subset E_0$ for $E_0 \in \mathcal{S}$;*

(ii) *each $E \in \mathcal{B}$ is a generalized peak set;*

(iii) $a = \{f \in C(X)\colon f|_E \in A|_E$ *for all* $E \in \mathcal{B}\}$.

Proof. (i) is the conclusion of Theorem 7.3(ii). For (ii), let E_1 denote the intersection of all peak sets containing E. Then E_1 is a generalized peak set and hence an M-hull. If E is a proper subset of E_1 then the maximality of E shows there is an $f \in A$, $f|_{E_1}$ real and not constant. But $f|_E$ *is* constant, say $f|_E = c$, so that for $\varepsilon > 0$ small,

$$g = 1 - \varepsilon(f - c)^2$$

is a real peak function on E_1 with

$$E \subset \{x \in E_1 \colon g(x) = 1\} \triangleq E_2.$$

But then E_2 is an M-hull for $A|_{E_1}$ and hence an M-hull for A. It follows that E_2 is a generalized peak set such that

$$E \subset E_2 \subset E_1.$$

Since this contradicts the minimality of E_1, it must be the case that $E = E_1$. The proof of (iii), using Proposition 7.2, is identical to that of Theorem 7.3(iii).

We see that in both $\mathcal{S}$ and $\mathcal{B}$ the elements are M-hulls and therefore correspond to (symmetric) split faces of Z_A. If K is a compact convex set we let $\mathcal{F}$ be the collection of non-empty closed split faces of K. A sub-collection of $\mathcal{F}$ is called a *decomposition of K* if it consists of pairwise disjoint split faces covering ∂K $(=\text{ext } K)$. Let $\mathcal{F}'$ be the sub-collection of sets $F \in \mathcal{F}$ such that $f|_F$ is constant whenever f is a central function in $A(K)$. We denote by $\mathcal{F}_S$ the collection of maximal (with respect to inclusion) elements in $\mathcal{F}'$.

8.8. THEOREM. *The collection $\mathcal{F}_S$ is a decomposition of K with restriction to ∂K consisting of the maximal sets of constancy in ∂K of central functions in $A(K)$.*

Proof. If F_i $(i = 1, 2) \in \mathcal{F}_S$ and $F_1 \cap F_2 \neq \varnothing$ and $f \in A(K)$ is constant on F_1 and F_2 then f is constant on $F_1 \cup F_2$, hence on co $(F_1 \cup F_2)$ which is also a split face. It follows that the elements of $\mathcal{F}_S$ are pairwise disjoint. If $x \in \partial K$ let

$$E = \{y \in \partial K : f(x) = f(y) \quad \text{for all } f \text{ central in } A(K)\}.$$

Since each such f is facially continuous, E is facially closed in ∂K. Thus $E = \partial K \cap F$, $F \in \mathcal{F}$. Since $E = \partial F$ $(=\text{ext } F)$, if f is central then $f|_E$ constant implies $f|_F$ is constant so that $F \in \mathcal{F}'$. If $F \subset F_1 \in \mathcal{F}'$ then the definition of E shows that $\partial F_1 \subset \partial F$. Hence F is maximal and so, $\mathcal{F}_S$ covers ∂K.

The collection $\mathcal{F}_S$ is termed the *abstract Šilov decomposition* of K. If A is a function algebra then the abstract Šilov decomposition of Z_A determines the Šilov decomposition of A in the following sense.

8.9. THEOREM. *Let A be a function algebra with complex state spa᾽e Z_A. Then $E \in \mathcal{S}$, the Šilov decomposition of A if and only if*

$$E = F \cap X \subset S_A$$

for $F \in \mathcal{F}_S$, the abstract Šilov decomposition for Z_A.

Proof. Given $E \in \mathcal{S}$ then E is an M-hull so that, with $\partial E = E \cap \partial X \subset S_A$,

$$E = \overline{\text{co}}\, (\partial E).$$

Let $x \in \partial E \subset \partial Z_A$ and let $x \in F \in \mathcal{F}_S$. If θf is central in $A(Z_A)$ then Theorem 8.5 shows $\operatorname{re} f = u \in A_0$. Hence u is constant on E so that $\theta f|_{\partial E} = u|_{\partial E}$ is constant. It follows from Theorem 8.8 that $\partial E \subset F \cap X$ and hence $E \subset F \cap X \subset S_A$. Conversely, $u \in A_0$ implies (by Theorem 8.5) that $\theta(u+iu)$ is central and hence constant on F. Thus u is constant on $F \cap X$ so that $F \cap X \subset E$. This shows $E = F \cap X$.

We shall see that the abstract Šilov decomposition determines $A(K)$ in a fashion analogous to Theorem 7.3(iii) for function algebras. We develop this property in conjunction with the *abstract Bishop decomposition* which is described as follows. We let, for $\mathcal{F}$ the closed non-empty split faces of K,

$$\mathcal{F}'' = \{F \in \mathcal{F}; \text{ the centre of } A(F) \text{ is trivial}\}.$$

The elements of $\mathcal{F}''$ are called *abstractly antisymmetric*. Let $\mathcal{F}_B$ denote the maximal elements of $\mathcal{F}''$.

8.10. THEOREM. *The collection $\mathcal{F}_B$ is a decomposition of K such that for each $F \in \mathcal{F}_B$, $F \subset F_0 \in \mathcal{F}_S$.*

Proof. First, let F_i $(i = 1, 2) \in \mathcal{F}_B$ with $F_1 \cap F_2 \neq \emptyset$. Let $F = \operatorname{co}(F_1 \cup F_2) \in \mathcal{F}$. Now, central functions in $A(F)$ restrict to central functions in $A(F_i)$ and since F_i is antisymmetric it follows that F is as well. This shows elements of $\mathcal{F}_B$ are pairwise disjoint. If $x \in \partial K$ let F_x be the smallest element of $\mathcal{F}$ containing x. Then F_x must be antisymmetric, for if f is central in $A(F_x)$ there is a split face G of F_x such that

$$\partial F_x \cap G = \{y \in \partial F_x : f(x) = f(y)\}.$$

Hence $x \in G = F_x$ so that f is constant and $F_x \in \mathcal{F}''$. Now, if

$$\mathcal{F}_x = \{G : x \in G \in \mathcal{F}''\}$$

then $\mathcal{F}_x$ is non-empty. Thus, let F be the smallest element in $\mathcal{F}$ containing $\bigcup \mathcal{F}_x$. We show F is antisymmetric. Let f be central in $A(F)$ and let

$$E = \{y \in \partial F : f(y) = f(x)\}.$$

If $G \in \mathcal{F}_x$ then $f|_G$ is central in $A(G)$ so f is constant on G and $\partial G = \partial F \cap G \subset E$. But $E = \partial F \cap G_0$ for G_0 a split face of F which contains ∂G for all $G \in \mathcal{F}_x$. Thus $G_0 = F$ and so f is constant on F. Clearly F is maximal in $\mathcal{F}''$ and $F \subset F_0 \in \mathcal{F}_S$.

As in the Šilov decomposition for function algebras, we can associate the abstract Bishop decomposition with the Bishop decomposition of A.

8.11. THEOREM. *Let A be a function algebra with Bishop decomposition $\mathcal{B}$. Then $E\in\mathcal{B}$ if and only if $E=F\cap X\subset S_A$ for some $F\in\mathcal{F}_B$, the abstract Bishop decomposition for Z_A.*

Proof. Let $E\in\mathcal{B}$ so that E is an M-hull with $F_0=\overline{\text{co}}\,(E)$ and $G=\text{co}\,(F_0\cup -iF_0)\in\mathcal{F}$, the split faces of Z_A. Now G is the complex state space of $A|_E$ so that if h is central in $A(G)$ then $h=\theta f|_G$, $f\in A$, and Theorem 8.5 shows $f=u+iv$ with $u|_E$, $v|_E\in(A|_E)_0$ and $u=v$ on the essential set of $A|_E$. But E antisymmetric for A implies that $u|_E$, $v|_E$ are constant functions and that $A|_E$ is essential. Thus $u=v$ on E and θf is constant on G. Thus $G\in\mathcal{F}''$ so that

$$E=G\cap X\subset F\cap X,\qquad F\in\mathcal{F}_B.$$

Now let $F=\text{co}\,(F_1\cup -iF_2)\in\mathcal{F}_B$ with $F_i\subset S_A$. Let $E_i=F_i\cap X$. Theorem 7.6 shows E_i is an M-hull and hence so is $E=E_1\cup E_2$. Let G be the symmetric split face of Z_A such that $G\cap X=E$. We show E is antisymmetric for A. Let $f\in A$ such that $f|_E\in(A|_E)_0$. Then $(f+if)|_E$ satisfies Theorem 8.5(ii) so that $\theta(f+if)|_G$ is central in $A(G)$. Thus its restriction to F is constant, say, c. Then

$$\theta(f+if)|_{E_1}=\text{re}\,f|_{E_1}=\alpha=\text{re}\,f|_{E_2}=\theta(f+if)|_{-iE_2}$$

so that $f|_E=\text{re}\,f|_E=\alpha$. Hence we have, since E is antisymmetric,

$$F\cap X=F_1\cap X=E_1\subset E\subset E_0\in\mathcal{B}.$$

We now take up the question of to what degree these abstract decompositions determine $A(K)$. Let $X=(\text{ext}\,K)^-$ and consider $A=A(K)|_X$ as a subspace of $C_{\mathbb{R}}(X)$ with state space K. As usual, for subsets C of $M_{\mathbb{R}}(X)$, $\partial_A C$ or ∂C denotes the intersection of C with the boundary measures in $M_{\mathbb{R}}(X)$. We continue to use ∂E for $(\text{ext}\,K)\cap E$ for subsets $E\subset K$.

We first establish the abstract analogue of Theorem 7.6.

8.12. THEOREM. *If $\mu\in\partial(\text{ext}\,A_1^{\perp})$ then* supp $\mu\subset F\cap X$ *for some* $F\in\mathcal{F}_B$.

Proof. Let F_0 be the smallest element in $\mathcal{F}$ containing supp μ. We show F_0 is antisymmetric. Let f be central in $A(F_0)$ and assume $0\le f\le 1$. Then for each $a\in A(F_0)$ there is a $b\in A(F_0)$ with $af=b$ on $F_0\cap X$ and hence on supp μ. This gives

$$\int_X af\,\mathrm{d}\mu=\int_X b\,\mathrm{d}\mu=0$$

so that $f\,\mathrm{d}\mu$ and $(1-f)\,\mathrm{d}\mu\in A_1^{\perp}$. Since μ is extreme in $A_1^{\perp}$ and $\mathrm{d}\mu=f\,\mathrm{d}\mu+(1-f)\,\mathrm{d}\mu$, we have $f\,\mathrm{d}\mu=\lambda\,\mathrm{d}\mu$, λ a constant. Then $f=\lambda$ a.e. $(\mathrm{d}|\mu|)$

so that $f \equiv \lambda$ on supp μ. If G is the split face of F_0 such that

$$\partial F_0 \cap G = \{x \in \partial F_0 : f(x) \leq \lambda\}$$

then $f \leq \lambda$ on G so that supp $\mu \subset G$, implying $G = F_0$. Thus $f \leq \lambda$ on F_0. Similarly $f \geq \lambda$ on F_0 and hence f is constant, showing F_0 is antisymmetric.

We need to show next that if $f \in C(X)$ is annihilated by $\partial(\text{ext } A_1^{\perp})$ then f is annihilated by all $\mu \in \partial A^{\perp}$. To accomplish this it is necessary to introduce a variant of the Choquet ordering on a set. If W is a compact convex set partially ordered by $\succ\cdot$ then we say a subset T is *hereditary-up* if

$$\eta \in T \quad \text{and} \quad \nu \succ\cdot \eta \quad \text{implies } \nu \in T.$$

We say T is *extremal* if

$$x = \lambda y + (1-\lambda) z \in T; \quad y, z \in W, \quad \text{implies } y, z \in T.$$

For example, T may be a union of faces. A function f on W is *isotonic* if

$$\nu \succ\cdot \eta \quad \text{implies } f(\nu) \geq f(\eta).$$

8.13. LEMMA. *Let $(W, \succ\cdot)$ be as above with the continuous convex isotonic functions separating the points of W. If T is a closed extremal hereditary-up subset of W then there is an element of T which is maximal with respect to $\succ\cdot$ and belongs to* ext W.

Proof. Let

$$\mathscr{E} = \{E \subset T : \varnothing \neq E \text{ is closed, extremal and hereditary-up}\}.$$

Since the properties of elements of $\mathscr{E}$ are preserved under intersections every sub-collection with the finite intersection property has non-empty intersection. Thus there is a minimal (with respect to inclusion) element $E_0 \in \mathscr{E}$. If f is continuous, convex and isotonic and if E_{00} is the subset of E_0 where f attains its maximum on E_0, then E_{00} is non-empty, extremal and, since E_0 is hereditary-up, E_{00} is as well. Thus $E_{00} \in \mathscr{E}$. Therefore $E_{00} = E_0$ and must be a singleton by the point separation hypothesis. It follows that this point must be maximal for $\succ\cdot$ and extremal, hence extreme in W.

For $f \in C(X)$ let $\hat{f}$ and $\check{f}$ be the upper and lower envelopes of f on K as in Chapter 1, Section 6.

8.14. PROPOSITION. *Let $X = (\text{ext } K)^{-}$ and $f \in C(X)$. If $\mu(f) = 0$ for all $\mu \in \partial(\text{ext } A_1^{\perp})$ then $\mu(f) = 0$ for all $\mu \in \partial A^{\perp}$.*

Proof. Define $\succ\cdot$ on $A_1^{\perp}$ by

$$\nu \succ\cdot \eta \quad \text{if } \nu^{+}(p_1) + \nu^{-}(p_2) \geq \eta^{+}(p_1) + \eta^{-}(p_2)$$

for all p_1, p_2 continuous convex functions on K. If p is continuous, convex and non-negative then the maps

$$\mu \to \mu^+(p), \qquad \mu \to \mu^-(p)$$

are continuous and convex on $A_1^\perp$. By considering the pairs p, 0 and 0, p they are seen to be isotonic as well. Since $\{p_1 - p_2 : p_j \text{ continuous, convex on } K\}$ is dense in $C_\mathbb{R}(K)$ we see that the elements of $A_1^\perp$ are separated by the continuous convex isotonic functions. Now, given $f \in C(X)$, let $\mu \in \partial A_1^\perp$ with $\mu(f) > 0$. We show $\nu(f) > 0$ for some $\nu \in \partial(\text{ext } A_1^\perp)$. Since μ is a boundary measure, μ^+, μ^- are maximal so that

$$\mu^+(f) = \mu^+(\check{f}) \quad \text{and} \quad \mu^-(-f) = \mu^-(-f)^\vee.$$

Therefore there exist p_1, p_2 continuous and convex on K with

$$p_1 \le f, \qquad p_2 \le -f$$

and

$$\mu^+(p_1) + \mu^-(p_2) \ge \tfrac{1}{2}(\mu^+(f) + \mu^-(-f)) = \tfrac{1}{2}\mu(f) > 0.$$

Let

$$\alpha = \sup\{\nu^+(p_1) + \nu^-(p_2) : \nu \in A_1^\perp\}$$

and

$$T = \{\nu \in A_1^\perp : \nu^+(p_1) + \nu^-(p_2) = \alpha\}.$$

Then T is closed and hereditary-up in $A_1^\perp$. Furthermore, T is extremal since $\alpha > 0$ implies $\|\nu\| = 1$ for all $\nu \in T$. But then if $\nu \in T$ and

$$\nu = \tfrac{1}{2}(\nu_1 + \nu_2); \qquad \nu_i \in A_1^\perp \quad (i = 1, 2)$$

then $\|\nu_i\| = 1$. Since these measures annihilate the function 1,

$$\|\nu^+\| = \tfrac{1}{2} = \|\nu^-\| \quad \text{and} \quad \|\nu_i^+\| = \tfrac{1}{2} = \|\nu_i^-\|.$$

If B_1, B_2 is the Jordan decomposition of X for ν then

$$1 = 2\nu^+(B_1) \le \nu_1^+(B_1) + \nu_2^+(B_1) \le 1$$

so that

$$\nu^+ = \tfrac{1}{2}(\nu_1^+ + \nu_2^+) \quad \text{and} \quad \nu^- = \tfrac{1}{2}(\nu_1^- + \nu_2^-).$$

Thus

$$\alpha = \tfrac{1}{2}(\nu_1^+(p_1) + \nu_1^-(p_2)) + \tfrac{1}{2}(\nu_2^+(p_1) + \nu_2^-(p_2)) \le \alpha$$

and it follows that $\nu_1, \nu_2 \in T$.

Therefore Lemma 8.13 yields $\nu \in (\text{ext } A_1^\perp) \cap T$ and ν maximal for $\succ\cdot$. The latter clearly implies that ν^+ and ν^- are maximal measures so that

$$\nu \in \partial(\text{ext } A_1^\perp).$$

Finally $\nu \in T$ gives

$$\nu(f)=\nu^+(f)+\nu^-(-f)\geq \nu^+(p_1)+\nu^-(p_2)=\alpha>0.$$

If $X=(\text{ext } K)^-$, let $A=A(K)|_X\subset C_{\mathbb{R}}(X)$ and denote

$$L=\{f\in C_{\mathbb{R}}(X): \hat{f}=f=\check{f} \quad \text{on } X\subset K\}$$

where $\hat{f}$ and $\check{f}$ are the upper and lower envelopes of f relative to K.

8.15. PROPOSITION.

$$A=\{f\in L: \mu(f)=0 \textit{ for all } \mu\in\partial(\text{ext } A_1^\perp)\}.$$

Proof. Clearly A is contained in the set on the right. Suppose then, $f\in L$ and f is annihilated by all $\mu\in\partial(\text{ext } A_1^\perp)$. Then Proposition 8.14 shows $\mu(f)=0$ for all $\mu\in\partial A^\perp$. We prove that $\mu(f)=0$ for all $\mu\in A^\perp$, implying that $f\in A^{\perp\perp}=A$. Thus, given $\mu\in A^\perp$ with $\mu\neq 0$ we have $\mu=\mu^+-\mu^-$ and we can assume $\|\mu^+\|=\mu^+(1)=\mu^-(1)=\|\mu^-\|=1$. Now choose ν_1 and ν_2 maximal on K with

$$\nu_1>\mu^+ \quad \text{and} \quad \nu_2>\mu^-.$$

If $a\in A(K)$ then

$$\nu_1(a)-\nu_2(a)=\mu^+(a)-\mu^-(a)=\mu(a)=0$$

so that $\nu=\nu_1-\nu_2\in\partial A^\perp$. Also

$$\nu_1(f)=\nu_1(\hat{f})\leq\mu^+(\hat{f})=\mu^+(f)=\mu^+(\check{f})\leq\nu_1(\check{f})=\nu_1(f)$$

so that $\nu_1(f)=\mu^+(f)$ and similarly $\nu_2(f)=\mu^-(f)$. Consequently

$$\mu(f)=\mu^+(f)-\mu^-(f)=\nu(f)=0.$$

8.16. THEOREM. *Let $X=(\text{ext } K)^-$ and $A=A(K)|_X\subset C_{\mathbb{R}}(X)$. If $f\in L$ and*

$$f|_{X\cap F}\in A(F)|_{X\cap F} \quad \textit{for all } \mathscr{F}\in\mathscr{F}_B,$$

(the abstract Bishop decomposition of K) then $f\in A$.

Proof. Let $\mu\in\partial(\text{ext } A_1^\perp)$ so that by Theorem 8.12

$$\text{supp } \mu\subset X\cap F \quad \text{for some } F\in\mathscr{F}_B.$$

If $g\in A(F)$ and $g=f$ on $X\cap F$ then choose $h\in A(K)$ with $h|_F=g$

$$\mu(f)=\int_{X\cap F} f\,d\mu=\int_{X\cap F} g\,d\mu=\mu(h)=0.$$

Then Proposition 8.15 shows $f\in A$.

8.17. COROLLARY. *If $X \subset \bigcup \mathscr{F}_B$ then $f \in A$ if and only if $f \in C_{\mathbb{R}}(X)$ and*

$$f|_{X \cap F} \in A(F)|_{X \cap F} \quad \text{for all } F \in \mathscr{F}_B.$$

Proof. Theorem 1.6.1 shows

$$\hat{f}(x) = \text{supp}\,\{\mu(f)\colon \mu \in M_1^+(X) \quad \text{and } \mu \text{ represents } x\}.$$

But $x \in F$ implies supp $\mu \subset F$ whenever μ represents x. It follows that

$$\hat{f}(x) = a(x) = f(x) \quad \text{for} \quad a \in A(F) \quad \text{with} \quad a|_{X \cap F} = f|_{X \cap F}.$$

Similarly $\hat{f}(x) = f(x)$ for all $x \in X$ so that $f \in L$ and Theorem 8.16 applies.

8.18. COROLLARY. *If $f \in C(X)$ then $f \in A$ if and only if*

$$f|_{X \cap F} \in A(F)|_{X \cap F} \quad \text{for all } F \in \mathscr{F}_S.$$

Proof. Since the abstract Bishop decomposition is finer (Theorem 8.10) than $\mathscr{F}_S$, if $f \in L$ and satisfies

$$f|_{X \cap F} \in A(F)|_{X \cap F} \quad \text{for all } F \in \mathscr{F}_S$$

then $f \in A$ by Theorem 8.16. To show $f \in L$ it suffices to show

$$X \subset \bigcup \mathscr{F}_S$$

as in Corollary 8.17. Thus, let $x \in X$ and G be the smallest closed split face of K containing x. If h is central but not constant on G then choose α strictly between the minimum and maximum values of h on G. Then there are split faces G_i $(i = 1, 2)$ of K such that

$$\partial K \cap G_1 = \{y \in \partial K\colon h(y) \leq \alpha\}$$

$$\partial K \cap G_2 = \{y \in \partial K\colon h(y) \geq \alpha\}.$$

Now $K = \text{co}\,(G_1 \cup G_2)$ and $X = (\partial G_1 \cup \partial G_2)^- \subset (\partial G_1)^- \cup (\partial G_2)^-$ so that $x \in G_1$ or $x \in G_2$. But $G_i \cap G$ is a proper split face of K which contradicts the minimality of G. Hence h is constant on G so that

$$x \in G \subset F \in \mathscr{F}_S.$$

8.19. COROLLARY. *K is a simplex if and only if each member of the abstract Bishop (Šilov) decomposition for K is a simplex.*

Proof. If K is a simplex then so is each closed face. Assume then that the members of $\mathscr{F}_B$ are simplexes. We show $\partial A^\perp = \{0\}$. If not, then Proposition 8.14 shows there is a $\mu \in \partial(\text{ext}\, A_1^\perp)$ which, by Theorem 8.12, has supp $\mu \subset F \in \mathscr{F}_B$. But then $\mu \in \partial A(F)^\perp$ so that $\mu = 0$. Therefore K is a simplex.

Finally, if each element of $\mathcal{F}_S$ is a simplex then the same holds for $\mathcal{F}_B$ by Theorem 8.10, and again, K is a simplex.

We conclude with an example showing that in general $\mathcal{F}_B \neq \mathcal{F}_S$. Let $F = \{(0, t) \in l^1 \times \mathbb{R}; |t| \leq 1\}$ and $G = \{(x, 0) \in l^1 \times \mathbb{R}: x \geq 0 \text{ and } \|x\| \leq 1\}$. Let $K = \text{co}\,(F \cup G)$ in $l^1 \times \mathbb{R}$.

8.20. EXAMPLE. *The set K above is a metrizable simplex for which*

$$\mathcal{F}_B \neq \mathcal{F}_S.$$

Proof. It is easy to see that

$$\text{ext}\, K = \{(0, \pm 1), (\delta_n, 0)_{n=1}^{\infty}\}$$

and

$$(\text{ext}\, K)^{-} = (\text{ext}\, K) \cup \{(0, 0)\}.$$

Also, each element of K has a unique representation as a convex combination of an element of G and $(0, \pm 1)$. Hence K is a metrizable simplex. We have

$$A(K) \simeq c_0 \times \mathbb{R}^2; \qquad f \to (a, s, t); \qquad f(x, r) = a \cdot x + rs + t$$

and elements of the form $(a, 0, 0)$ are central. Hence each $\{(\delta_n, 0)\} \in \mathcal{F}_S$. If f is central in $A(K)$ then there is an $h \in A(K)$ such that

$$h = f^2 \quad \text{on} \quad \{(0, 0), (0, \pm 1)\}.$$

Thus f is constant on F and

$$\mathcal{F}_S = \{F, \{(\delta_n, 0)\}_{n=1}^{\infty}\}.$$

On the other hand, every element of $A(F)$ is central (since F is a Bauer simplex) so that $(0, \pm 1) \in \mathcal{F}_B$ and $\mathcal{F}_B$ is finer than $\mathcal{F}_S$.

Notes

Theorems 8.1, 8.2, 8.4 and Corollaries 8.3, 8.5, 8.6 are given in Ellis [**102**] and Hirsberg [**127**]. The study of facial decompositions for compact convex sets was introduced by Ellis [**104**], where the results 8.7 to 8.11 are proved. Theorems 8.12, 8.16 and their corollaries are due to Briem [**53**], and Example 8.20 is due to Asimow.

9. LINDENSTRAUSS SPACES

A real or complex Banach space A is called a *Lindenstrauss space* if A^* is isometric to a concrete $L^1(\mu)$ space for some measure space $(X, \mathcal{S}, \mu)$. We shall restrict our attention here to the case where A is a (close separating, with constants) subspace of $C(X)$. If A is a real subspace of $C_{\mathbb{R}}(X)$ then A^* is an L^1 space if and only if the state space S_A is a simplex, as per the discussion in Chapter 2, Section 6. We now have the tools at hand to characterize the complex Lindenstrauss function spaces using a general characterization due to Effros.

The set-up here is the same as in Section 1 where U denotes the unit ball of A^* (w^*-topology). We are interested in the subspace of $C(U)$ consisting of the T-homogeneous functions, that is, functions f satisfying $f(tx) = tf(x)$ for all $t \in T$. We define the projection hom of $C(U)$ onto the T-homogeneous functions by

$$\hom(f)(x) = \int_T \bar{t} f(tx)\, \mathrm{d}t$$

where $\mathrm{d}t$ denotes Haar measure on T. It is easy to check that hom is indeed a norm-decreasing projection. We also have use for the adjoint projection in $M(U)$ given by

$$(f, \hom \mu) = (\hom f, \mu); \qquad \mu \in M(U), \qquad f \in C(U).$$

We define U to be a *T-simplex* if whenever λ, μ are maximal measures on U representing the same $L \in U$ then

$$\hom \lambda = \hom \mu.$$

Then the theorem of Effros referred to above says A is a Lindenstrauss space if and only if $U = A_1^*$ is a T-simplex. For the purposes of our discussion here we shall consider this the defining property of Lindenstrauss spaces.

9.1. THEOREM. *Let $1 \in A$ be a separating subspace of $C(X)$. The following are equivalent*:

(i) *A is a Lindenstrauss space*;
(ii) $\partial_A A^\perp = \{0\}$;
(iii) *Z_A is a simplex*;
(iv) *A is self-adjoint and S_A is a simplex*;
(v) *each $E \subset X$ which is a peak set for* re *A is an M-hull*;
(vi) *A is self-adjoint and each $L \in U$ is represented by a unique maximal probability measure on U.*

Proof. (i) implies (ii): Suppose $\nu \in \partial_A A^\perp$ and $\|\nu\| = 1$. Then let $\tilde{\nu} = |\nu| \circ (\sigma_h)^{-1}$ as in Section 1. Thus $\sigma_h(x) = h(x)x$ where $\mathrm{d}\nu = h\, \mathrm{d}|\nu|$ and $|h| = 1$ on X and

$$\int_U f(tx)\, \mathrm{d}\tilde{\nu}(x, t) = \int_X f(h(x)x)\overline{h(x)}\, \mathrm{d}\nu(x).$$

Hence, if f is T-invariant then $\tilde{\nu}(f)=|\nu|(f)$ so that by Proposition 1.4 $\tilde{\nu}$ is maximal. Since $\nu\in A^\perp$ and $\int_X f\,d\nu=\int_U f\,d\tilde{\nu}$ for T-homogeneous f, we have $\tilde{\nu}$ represents 0. Thus hom $\tilde{\nu}=0$. Given $f\in C(X)$, we then extend f to TX by $f(tx)=tf(x)$ and then to $g\in C(U)$ by the Tietze Theorem. If $k=\text{hom } g$ then

$$k(tx)=tk(x)=tf(x)=f(tx)$$

and

$$0=(hom\ \tilde{\nu})(k)=\tilde{\nu}(g)=\int_{TX} tf(x)\,d\tilde{\nu}(x,t)=\int_X f(x)\,d\nu(x)=\nu(f).$$

Thus $\nu=0$.

Now (ii) and (iii) are equivalent since Theorem 3.1 shows $\partial_A A^\perp=\{0\}$ if and only if $\partial A(Z_A)=\{0\}$ which is equivalent by Theorem 2.7.3 to Z_A being a simplex. The equivalence of (iii) and (iv) is immediate from Theorem 0.2. The equivalence of (ii) and (vi) is essentially Theorem 1.5(ii).

(ii) implies (i). Let $\tilde{\mu}$, $\tilde{\nu}$ be maximal on U representing the same $L\in U$. Let μ, ν by $H\tilde{\mu}$, $H\tilde{\nu}$ so that μ, ν are boundary measures with $\mu-\nu\in\partial A^\perp$. Thus $\mu=\nu$ so that if f is T-homogeneous on U

$$\int_{TX} f(tx)\,d\tilde{\mu}(x,t)=\int_X f\,d\mu=\int f\,d\nu=\int_{TX} f(tx)\,d\tilde{\nu}(x,t)$$

so that hom $\tilde{\mu}=\text{hom } \tilde{\nu}$.

(iii) implies (v). Let $F=\overline{\text{co}}\,E\subset S_A$. If E is a peak set for re A then F is an (exposed) face of S_A, hence a face of Z_A. But then F is split in Z_A so that co $(F\cup -iF)$ is split, showing E is an M-hull.

(v) implies (iv). If F_0 is an exposed face of S_A then $E_0=F_0\cap X$ is a peak set for re A. Thus (v) gives that E_0 is an M-hull and hence, co $(F_0\cup -iF_0)$ is split in Z_A. In particular F_0 is split in S_A so that Theorem 2.7.2 shows S_A is a simplex. To show A is self-adjoint we need to prove (Theorem 0.2) that S_A is split in Z_A. We use the set-up of Section 3, identifying $\mu=\mu_1-i\mu_2\in M(X)$ with $\tilde{\mu}=\mu_1\oplus\mu_2\in M_{\mathbb{R}}(X\cup -iX)$ and f with $\theta f\in A(Z_A)$, where

$$(\theta f,\tilde{\mu})=\int_Z(\theta f)\,d\tilde{\mu}=\int_X\theta f\,d\mu_1+\int_{-iX}\theta f\,d\mu_2.$$

Let $\|\theta f\|$, $\|\tilde{\mu}\|$ denote the norms in $A(Z_A)$, $A(Z_A)^*$.

If $\tilde{\mu}\in\partial A(Z_A)^\perp$ then $\tilde{\mu}|_{S_A}=\mu_1$, a boundary measure on $X\subset S_A$. We show $\mu_1\in A(Z_A)^\perp$, that is, $\mu_1\sim 0$. We consider first any support pair (a,x) in $(A(Z_A),A(Z_A)^*)$ with $\|a\|=1$ and $a(x)=\|x\|$. Let ξ be a boundary measure on $X\cup -iX\subset Z_A$ representing x with

$$\|x\|=\|\xi\|=\|\xi^+\|+\|\xi^-\|$$

where ξ^+, ξ^- is the Jordan decomposition of ξ on Z_A. Then

$$\|\xi\|=\xi^+(a)-\xi^-(a)=\|\xi^+\|+\|\xi^-\|$$

so that (since $\|a\|=1$) $\xi^+(a)=\|\xi^+\|$ and $\xi^-(-a)=\|\xi^-\|$. Then

$$\operatorname{supp}\xi^+\subset\{z\in Z_A\colon a(z)=1\}\doteq F^+$$

$$\operatorname{supp}\xi^-\subset\{z\in Z_A;\ a(z)=-1\}\doteq F^-$$

where F^+, F^- are disjoint exposed faces of Z_A. Then

$$F^+=\operatorname{co}(F_1^+\cup -iF_2^+)$$

$$F^-=\operatorname{co}(F_1^-\cup -iF_2^-)$$

where $F_j^\pm$ $(j=1,2)$ are exposed faces of S_A. The hypothesis then shows that

$$E_j^\pm=\operatorname{co}(F_j^\pm\cup -iF_j^\pm)\quad (j=1,2)$$

is split in Z_A. Thus if f is the function on Z_A which is $\theta(1)$ on E_1^+ and 0 elsewhere then $h_1^+\doteq\hat{f}$ is affine on Z_A, $0\le h_1^+\le 1$ and satisfies (since ξ is boundary)

$$\xi(h_1^+)=\xi(f)=\int_{F_1^+}d\xi=\int_{F_1^+}d\xi^+=\xi^+(F_1^+).$$

Also, using Theorem 1.6.1(ix) and the fact that $h_1^+\equiv 0$ on $-iX$,

$$0=\tilde{\mu}(h_1^+)=\mu_1(h_1^+).$$

Similarly there is an affine l.s.c. h_2^+ with

$$\xi(h_2^+)=\xi^+(-iF_2^+)$$

and, since $h_2^+\equiv 0$ on X,

$$\mu_1(h_2^+)=0.$$

Proceeding as above with F_j^- $(j=1,2)$ we have affine l.s.c. h_j^- with

$$\xi^-(h_1^-)=\xi^-(F_1^-)$$

$$\xi^-(h_2^-)=\xi^-(-iF_2^-)$$

$$\mu_1(h_1^-)=0=\mu_1(h_2^-)$$

Now if $\mu_1\sim x_0$ we can assume (replacing μ_1 with a suitable boundary measure on S_A) that $\|\mu_1\|=\|x_0\|$. Given $\varepsilon>0$ we can, by the Bishop–Phelps Theorem (Chapter 1, Section 7), choose a support pair (a,x) with $\|a\|=1$ and $\|x-x_0\|<\varepsilon$. Then using the above ξ,

$$\begin{aligned}\|x\|=\|\xi\|&=\xi^+(F_1^+)-\xi^+(-iF_2^+)+\xi^-(F_1^-)+\xi^-(-iF_2^-)\\&=\textstyle\sum\xi(h_j^\pm)=\sum(\xi-\mu_1)(h_j^\pm)\le 4\|\xi-\mu_1\|=4\|x-x_0\|<4\varepsilon.\end{aligned}$$

Thus $\|x_0\|\le\|x-x_0\|+\|x\|<5\varepsilon$.

We can combine Theorem 4.4 with the above to characterize the symmetric split simplicial faces of Z_A.

9.2. COROLLARY. *Let A be as above and let* $E = k(E)$, $F = \overline{\text{co}}\, E \subset S_A$ *and* $\hat{Z}_E = \text{co}\,(F \cup -iF) \subset Z_A$. *The following are equivalent:*

(i) *E is an M-hull and* $A|_E$ *is a Lindenstrauss space,*
(ii) $\mu \in \partial A^{\perp}$ *implies* $\mu|_E = 0$,
(iii) $\hat{Z}_E$ *is a simplex and a split face of* Z_A.

9.3. COROLLARY. *If* $A \subset C(X)$ *is a Lindenstrauss space and* $\partial_A X = E$ *is closed in X then* $A|_E = C(E)$.

Proof. If $\mu \in (A|_E)^{\perp}$ then $\mu \in \partial A^{\perp}$ so that $\mu|_E = \mu = 0$. Thus $(A|_E)^{\perp\perp} = C(E)$; but E is a boundary so that $A|_E$ is closed.

9.4. COROLLARY. *If A is a function algebra in* $C(X)$ *and a Lindenstrauss space then* $A = C(X)$.

Proof. A is self-adjoint.

9.5. COROLLARY. *If A is a function algebra in* $C(X)$ *such that A and* re *A have the same peak sets then* $A = C(X)$.

Proof. Combine Theorem 9.1(v) and the preceding corollary.

Notes

The underpinning for this section is the characterization of T-simplexes due to Effros [**96**]. The various equivalences of Theorem 9.1 (except for (v)) are due to Hirsberg and Lazar [**128**] and Fuhr and Phelps [**112**]. The equivalence of (v) to the others and the peak set characterization of $C(X)$ is due to Briem [**55**].

CHAPTER 5

Convexity Theory for C^*-Algebras

1. THE STRUCTURE TOPOLOGY AND PRIMITIVE IDEALS IN $A(K)$

Much of the motivation for the development of infinite-dimensional convexity theory during recent years has come from the study of C^*-algebras and W^*-algebras. In this section we will develop the notion of structure topology for the primitive ideal space of $A(K)$ in direct analogy to the Jacobson structure topology for C^*-algebras. In a later section we will tie the abstract and concrete cases together, after we have studied the facial structure of the state space of a unital C^*-algebra.

Let K be a compact convex set. For each $x \in \partial K$ we write F_x for the smallest closed split face of K which contains x. The face F_x is called a *primitive face* of K and its annihilator $(F_x)_\perp = \{f \in A(K) : f(y) = 0, \forall y \in F_x\}$ is called a *primitive ideal* in $A(K)$.

If F is a closed split face of K then the set $F_\perp$ is called an *M-ideal* (or *near lattice ideal*). Since F is evidently the closed convex hull of all the primitive faces of K which are contained in F, dually $F_\perp$ is the intersection of all primitive ideals containing $F_\perp$.

If J is an M-ideal in $A(K)$ we will write $\mathscr{h}(J)$ for the set of all primitive ideals of $A(K)$ which contain J; $\mathscr{h}(J)$ is called the *hull* of J. If $E = \{I_\alpha\}$ is a set of primitive ideals in $A(K)$ then we write $\mathscr{k}(E)$ for the largest M-ideal contained in $\bigcap_\alpha I_\alpha$. Dually, $\mathscr{k}(E) = F_\perp$ where F is the smallest closed split face of K which contains $\bigcup_\alpha (I_\alpha)^\perp$, where $I_\alpha^\perp = \{x \in K : f(x) = 0, \forall f \in I_\alpha\}$. The M-ideal $\mathscr{k}(E)$ is called the *kernel* of E. It is clear that for all such J and E we have

$$\mathscr{k}(\mathscr{h}(J)) \supseteq J, \qquad \mathscr{h}(\mathscr{k}(E)) \supseteq E.$$

Denote by Prim $A(K)$ the set of all primitive ideals in $A(K)$ and call a subset E of Prim $A(K)$ *structurally closed* if $E = \mathscr{h}(J)$ for some M-ideal J. If E is structurally closed then

$$E \subseteq \mathscr{h}(\mathscr{k}(E)) = \mathscr{h}(\mathscr{k}(\mathscr{h}(J))) \subseteq \mathscr{h}(J) = E,$$

so that $E = \mathscr{h}(\mathscr{k}(E))$. Moreover, if $E = \{I_\alpha\}$ is structurally closed then E contains every primitive ideal which contains $\mathscr{k}(E)$, and consequently $\mathscr{k}(E) = \bigcap_\alpha I_\alpha$.

Let E_1, E_2 be structurally closed with $E_j = \mathscr{h}(J_j)$, $j = 1, 2$. Then, if $J = J_1 \cap J_2$, J is an M-ideal and $\mathscr{h}(J) \supseteq E_1 \cup E_2$. However if $(F_x)_\perp \in \mathscr{h}(J)$ then $F_x \subseteq (J_1 \cap J_2)^\perp = \text{co}\,(J_1^\perp \cup J_2^\perp)$, and since $x \in J_1^\perp \cup J_2^\perp$ we must have either $F_x \subseteq J_1^\perp$ or $F_x \subseteq J_2^\perp$; thus $\mathscr{h}(J) \subseteq E_1 \cup E_2$. Therefore $E_1 \cup E_2$ is structurally closed. If $\{E_\alpha\}$ is a family of structurally closed sets with $E_\alpha = \mathscr{h}(J_\alpha)$ for each α, then if $J = (\bigcap_\alpha J_\alpha^\perp)_\perp$ we see that J is an M-ideal and $\bigcap_\alpha E_\alpha = \mathscr{h}(J)$, using the facts that $F = (F_\perp)^\perp$, $I = (I^\perp)_\perp$ for every closed split face F of K and every M-ideal I in $A(K)$. Hence we have shown that the structurally closed sets induce a topology on Prim $A(K)$ called the *structure topology*. This topology will only be Hausdorff when K is a Bauer simplex (see Corollary 1.2), but it is always T_0. In fact if I, J are primitive ideals with $I \neq J$ then either $I \not\supseteq J$ or $J \not\supseteq I$; if $I \not\supseteq J$ then $\mathscr{h}(J)$ is a structurally closed set containing J but not I.

1.1. THEOREM. *Let K be a compact convex set and let f*: Prim $A(K) \to \mathbb{R}$ *be a bounded structurally continuous function. Then there exists a central function g in $A(K)$ such that*

$$f((F_x)_\perp) = g(x), \qquad x \in \partial K.$$

Proof. If f is constantly equal to 1, we can put $g = 1$. Therefore, without loss of generality, we assume that $0 \leq f \leq 1$. Let n be a natural number and define

$$U_j = f^{-1}\left(\left(\frac{j-1}{n}, \frac{j+1}{n}\right)\right) \quad \text{for } j = 0, \ldots, n.$$

Then, if X denotes Prim $A(K)$ with the structure topology, $X \backslash U_j$ is closed and so $J_j = \bigcap\{I : I \in X \backslash U_j\}$ is an M-ideal. Now we have $J_0 + J_1 + \cdots + J_n = \{\bigcap_{j=0}^n J_j^\perp\}_\perp = A(K)$, because if $F = \bigcap_{j=0}^n J_j^\perp \neq \phi$ there would exist a primitive face $G \subseteq F$ and hence a primitive ideal $G_\perp \in \bigcap_{j=0}^n (X \backslash U_j) = \varnothing$.

If $h \in A(K)$ satisfies $0 \leq h \leq 1$, and if $\varepsilon > 0$, then we will prove by induction on n that there exist $h_j \in J_j$, $j = 0, 1, \ldots, n$, such that $h = \sum_{j=0}^n h_j$ and $-\varepsilon < h_j < 1 + \varepsilon$ for $0 \leq j \leq n$. This is obviously true for $n = 0$; suppose it is true with $(n-1)$ replacing n and for any $\varepsilon > 0$. If $J'_{n-1} = J_{n-1} + J_n$ then J'_{n-1} is an M-ideal and $J_0 + J_1 + \cdots + J_{n-2} + J'_{n-1} = A(K)$, so that there exist $h_j \in J_j$, $0 \leq j \leq n-2$, $h'_{n-1} \in J'_{n-1}$ with $h = h_0 + \cdots + h_{n-2} + h'_{n-1}$ and $-\delta \leq h_j$, $h'_{n-1} \leq 1 + \delta$, $0 \leq j \leq n-2$, for some $\delta \in (0, \varepsilon)$. Let $F_{n-1} = J_{n-1}^\perp$, $F_n = J_n^\perp$ so that $(J'_{n-1})^\perp = F_{n-1} \cap F_n$ and $h'_{n-1} = 0$ on $F_{n-1} \cap F_n$. Clearly we can define $h_{n-1} \in A(\text{co}\,(F_{n-1} \cup F_n))$ so that $h_{n-1} = h'_{n-1}$ on F_n and $h_{n-1} = 0$ on F_{n-1}; in particular we have $0 \leq h_{n-1} \leq h'_{n-1}$ on co $(F_{n-1} \cup F_n)$. Since co $(F_{n-1} \cup F_n)$ is a closed split face of K we can, by Theorem 2.10.5, extend h_{n-1} to K so that $h_{n-1} \in A(K)$ and $0 \leq h_{n-1} \leq h'_{n-1} + \frac{1}{2}(\varepsilon - \delta)$. Now if we put $h_n = h'_{n-1} - h_{n-1}$ we see that $h_{n-1} \in J_{n-1}$, $h_n \in J_n$, $0 \leq h_{n-1} < 1 + \varepsilon$, $-\varepsilon < h_n < 1 + \varepsilon$. Moreover, $h = \sum_{j=0}^{n} h_j$ and so the inductive proof is concluded.

Define $u_n = \sum_{j=0}^{n} jh_j/n \in A(K)$, and let $x \in \partial K$ be fixed. For $j = 0, \ldots, n$ we either have $(F_x)_\perp \in U_j$, in which case $|f((F_x)_\perp) - j/n| \leq 1/n$, or we have $(F_x)_\perp \in X \backslash U_j$, in which case $(F_x)_\perp \supseteq J_j$ so that $h_j(x) = 0$; the first possibility can occur for at most two values of j. Therefore

$$|f((F_x)_\perp)h(x) - u_n(x)| = \left| \sum_{j=0}^{n} f((F_x)_\perp) - j/n)h_j(x) \right| \leq \frac{2(1+\varepsilon)}{n}.$$

Since $u_n \in A(K)$ for each n, it follows that $\{u_n\}$ is a Cauchy sequence converging to some $u \in A(K)$ such that $u(x) = f((F_x)_\perp)h(x)$ for all $x \in \partial K$. In the particular case when $h = 1$ we obtain a $g \in A(K)$ such that $g(x) = f((F_x)_\perp)$ for all $x \in \partial K$. Therefore, in general, we have $u(x) = g(x)h(x)$ for all $x \in \partial K$, and consequently we deduce that g belongs to the centre of $A(K)$.

In the special case when every $x \in \partial K$ is a split face (for example when K is a simplex or the state space of a function algebra) the primitive faces of K coincide with the extreme points of K. It is evident in this case that Prim $A(K)$ with the structure topology is homeomorphic to ∂K with the facial topology, and so Theorem 1.1 coincides with Theorem 3.1.4 in this situation.

Notes

The description of the structure topology for Prim $A(K)$ given above follows the paper of Alfsen and Effros [**8**], and differs somewhat from that given in Alfsen's book [**5**] where Størmer's axiom for K is assumed. Theorem 1.1 is due to Alfsen *et al.* [**7, 8**] and is an abstract version of a special case of a theorem of Dauns and Hofmann [**71**]. The proof given above is due to Elliott and Olesen [**99**]. A much more general version of 1.1, which applies in any real Banach space, appears in [**8**].

For the connections between 1.1 and spectral theory for $A(K)$ we refer to the paper of Edwards [**86**].

2. THE IDEAL STRUCTURE OF A UNITAL C^*-ALGEBRA AND THE FACIAL STRUCTURE OF ITS STATE SPACE

Let A be a unital C^*-algebra, let $K = \{\phi \in A^* : \phi(1) = 1\}$, with the w^*-topology, be the state space of A and let A_h be the Hermitian elements of A. Then A_h is a partially ordered real Banach space, for the ordering of positivity of operators, or equivalently for the positive cone $\{xx^* : x \in A\}$. Moreover, 1 is an order unit defining the norm in A_h, and so 2.2.3 shows that A_h is isometrically and order isomorphic to $A(K)$. For the basic theory of C^*-algebras we will refer to Dixmier [**81**], or Sakai [**188**].

If $F \subseteq K$ and if $M \subseteq A$ we will write $M^\perp = \{x \in K : a(x) = 0, \forall a \in M\}$, $F_0 = \{x \in A : \phi(x) = 0, \forall \phi \in F\}$, $F_\perp = F_0 \cap A_h$, $L_F = \{a \in A : a^*a \in F_\perp\}$; we note that $L_F \subseteq F_0$ by the Cauchy–Schwarz inequality. Note that $(F_\perp)^+$ is a face of $A(K)^+$. If N is a closed face of $A(K)^+$ and if $L = \{a \in A : a^*a \in N\}$ then L is a closed left ideal in A. In fact $a \in L$ implies that $\lambda a \in L$ for $\lambda > 0$, and $a, b \in L$ implies that $0 \leq (a+b)^*(a+b) = 2(a^*a + b^*b) - (a-b)^*(a-b) \leq 2(a^*a + b^*b) \in N$, so that $a + b \in L$. Finally if $b \in A$ and $a \in L$ we have $(ba)^*ba \leq a^*\|b\|^2 a \in N$, and hence $ba \in L$. It follows that L_F is a closed left ideal in A. We now prove that there is a one-to-one correspondence between the closed left ideals of A and the closed faces F of K.

2.1. THEOREM. *F is a closed face of K if and only if $F = L^\perp$ for some closed left ideal L in A; in this case we may take $L = L_F$.*

Proof. Let L be a closed left ideal in A, and let $F = L^\perp$. Let $\phi \in (L^+)^\perp$ and let $a \in L$. Then $a^*a \in L^+$ so that $\phi(a^*a) = 0$, whence $\phi(a) = 0$ by the Cauchy–Schwarz inequality for positive linear functionals. Therefore $L^\perp = (L^+)^\perp$, and hence F is a face of K.

Conversely, let F be a closed face of K. Then L_F is a closed left ideal, by the previous discussion. Let $M = A^{**}F \equiv \{\phi \in A^* : \phi(a^{**}) = \Psi(b^{**}a^{**})$, $\forall a^{**} \in A^*$, some $b^{**} \in A^{**}$, some $\Psi \in F\}$. M is a linear subspace of the normal dual space A^* of the W^*-algebra A^{**}, and $AF \subseteq A^{**}F \subseteq \overline{AF}^{w^*}$, by the w^*-density of A in A^{**}. We will prove (i) $M \cap K = F$, and (ii) M is w^*-closed. From this it will follow that $F = (L_0)^\perp$, and that M is the w^*-closure of AF so that $L_0 = \{a \in A : \phi(ba) = 0, \forall b \in A, \forall \phi \in F\} = L_F$.

(i) For any subset E of K the *carrier* of E is defined to be the smallest (Hermitian) projection e in A^{**} such that $\phi_e = \phi$, for all $\phi \in E$, where $\phi_e(a) = [\phi(eae)/\phi(e)]$ for all $a \in A$, and $\phi(e) \neq 0$. When $\phi \in K$ let F_ϕ denote the smallest norm-closed face of K containing ϕ. If e is the carrier of $\{\phi\}$ then the set $G_\phi = \{\theta_e : \theta \in K\}$ is norm-closed and convex (using the fact that $((\theta_e)_e = \theta_e)$, and is a face of K containing ϕ since if $\lambda\phi_1 + (1-\lambda)\phi_2 \in G_\phi$, with $0 < \lambda < 1$ and $\phi_1, \phi_2 \in K$, we have $\phi_j(e) = 1$ and it follows using the Cauchy–Schwarz inequality that $(\phi_j)_e = \phi_j$ for $j = 1, 2$. Hence we have $F_\phi \subseteq G_\phi$. To obtain the reverse inclusion, since each $\Psi \in G_\phi$ has carrier $\leq e$, it is sufficient to show that every $\Psi \in K$ belongs to F_ϕ whenever carrier $\{\Psi\} \leq e$. Since the only elements of A^{**} that we need to consider belong to $eA^{**}e$ we can assume that $e = 1$, and show that $F_\phi = K$.

Corresponding to ϕ, let ρ denote the usual GNS representation of A^{**} as a weakly-closed algebra of operators on a Hilbert space H, with cyclic vector

z. Then, since carrier $\phi = 1$, $\rho(A^{**})'z$ is strongly dense in H, where $\rho(A^{**})'$ is the commutant of $\rho(A^{**})$.

Let $s \in \rho(A^{**})'$ with $\|s\| \leq 1$, and let $t = (1 - s^*s)^{1/2}$. If θ is defined for all $b \in A^{**}$ by $\theta(b) = (\rho(b)sz, sz)$ we clearly have $\theta \geq 0$, and further $\phi(b) - \theta(b) = (\rho(b)z, z) - (\rho(b)sz, sz) = (\rho(b)t^2z, z) \geq 0$ for all $b \geq 0$, so that $\theta \leq \phi$. Therefore F_ϕ contains a multiple of θ. However, since the mapping $x \rightsquigarrow (\rho(\cdot)x, x)$ is strongly continuous, and since $\rho(A^{**})'z$ is strongly dense in H, and also since every positive normal functional on A^{**} has the form $\Psi(a) = (\rho(a)x, x)$ for some $x \in H$, it follows that $F_\phi = K$. We conclude that $F_\phi = G_\phi$.

Let e' be the carrier of F, and let $\{e_\alpha\}$ be a maximal orthogonal family of projections in A^{**} that are carriers of elements of F. Put $e'' = \sup_\alpha \{e_\alpha\}$ and $f = e' - e'' \geq 0$. If $f \neq 0$ then $\phi(f) > 0$ for some $\phi \in F$, since otherwise we would have $\phi_{e''} = \phi_{e'} = \phi$, for all $\phi \in F$, which contradicts the definition of e'. Since $\phi_{1-e''}(f) = [\phi(f)/\phi(1-e'')] > 0$ we can assume that the carrier g of ϕ is disjoint from e''. This contradicts the maximality of $\{e_\alpha\}$, and therefore $e' = e''$.

Suppose that $\theta_{e'} = \theta$ for some $\theta \in K$. Since θ is a normal functional on A^{**} we have $\theta(e') = \sum_\alpha \theta(e_\alpha)$, and so $\theta(e') = \sum_j \theta(e_j)$ for a countable subfamily $\{e_j\}$ of the $\{e_\alpha\}$. Because $\theta(e' - \sum e_j) = 0$ and the carrier h of θ is $\leq e'$ we see that $h \leq \sum e_j$. If $\phi_j \in F$ are such that e_j is the carrier of ϕ_j then $\phi \equiv \sum 2^{-j}\phi_j$ belongs to F with carrier $\sum 2^{-j}e_j \geq h$. The first part of the proof of (i) now gives $\theta \in F$. Consequently we obtain $F = \{\theta_{e'} : \theta \in K, \theta(e') \neq 0\}$, the reverse inclusion following by the corresponding argument for F_ϕ given above. The preceding result gives $M = \{a^{**}e'\theta e' : \theta \in K,\ a^{**} \in A^{**}\}$. Now if $\Psi = a^{**}e'\theta e' \in K$ we have $\Psi e' = \Psi = \Psi^* = e'\Psi = e'\Psi e' = \Psi_{e'} \in F$, and hence $M \cap K = F$, and the proof of (i) is complete.

(ii) Let Σ^* denote the closed unit ball of A^*. We need to show that $M \cap \Sigma^*$ is w^*-compact. Suppose that $\{\phi_\alpha\}$ is a w^*-convergent net in $M \cap \Sigma^*$ with limit ϕ. The polar decomposition (cf. Sakai [**188**, 1.14.4], Dixmier [**81**, 12.2.4]) enables us to write $\phi_\alpha = w_\alpha \eta_\alpha$, with $w_\alpha \in A^{**}$, $\eta_\alpha \geq 0$, $\|\eta_\alpha\| = \|\phi_\alpha\|$ and $\eta_\alpha = w_\alpha^*\phi_\alpha$. In particular we have $\eta_\alpha \in M \cap (\Sigma^*)^+ = \text{co}\,(F \cup \{0\})$. We can therefore, by choosing a subnet if necessary, assume that $\{\eta_\alpha\}$ is w^*-convergent to some $\eta \in \text{co}\,(F \cup \{0\})$. We have, for each $a \in A$, $|\phi_\alpha(a)|^2 \leq \eta_\alpha(w_\alpha w_\alpha^*)\eta_\alpha(a^*a)$, so that $|\phi(a)|^2 \leq \eta(a^*a)$ and consequently $|\phi(b)|^2 \leq \eta(b^*b)$ for all b in A^{**}. Now let $\phi = w\zeta$, $\zeta = w^*\phi$ be the polar decomposition of ϕ. Then $|\zeta(a)|^2 = |\phi(w^*a)|^2 \leq \eta(a^*ww^*a) \leq \eta(a^*a)$ for all $a \in A^{**}$. Moreover, by choosing a subnet such that $\|\phi_\beta\| \to \|\phi\|$, we see that $\eta(1) = \lim \eta_\beta(1) = \lim \|\phi_\beta\| = \|\phi\| = \|\zeta\| = \zeta(1)$. If $a \in A_h$ we obtain, for all $\lambda \in \mathbb{R}$,

$$\lambda^2 + 2\lambda\zeta(a) + \zeta(a)^2 = |\zeta(a+\lambda)|^2 \leq \eta((a+\lambda)^2) = \lambda^2 + 2\lambda\eta(a) + \eta(a^2)$$

and this implies that $\zeta = \eta$. Consequently $\phi = w\eta \in M \cap \Sigma^*$, and so $M \cap \Sigma^*$ is w^*-compact.

2.2. COROLLARY. *Every closed face of the state space K of the C^*-algebra A is semi-exposed.*

Proof. We prove directly that F is a semi-exposed face of K if and only if $F = L_F^\perp$, and the result then follows from 2.1.

Let F be a semi-exposed face of K, so that $F = (F_\perp)^\perp = ((F_\perp)^+)^\perp$ by 2.8. If $a \in (F_\perp)^+$ then $a^2 \le \|a\|a$, so that $\phi(a^2) = 0$ for all ϕ in F, and hence $a \in L_F$. Therefore we have $(F_\perp)^+ \subseteq L_F \subseteq F_0$, and it follows immediately that $F = (L_F)^\perp$.

Conversely, let $F = L_F^\perp$ and let $\phi \in K \backslash F$. There exists an $a \in L_F$ with $\phi(a) \neq 0$, and hence $\phi(a^*a) > 0$ by the Cauchy–Schwarz inequality. Since L_F is a left ideal a^*a belongs to $(L_F)^+$, and hence F is semi-exposed.

Our next investigation in this section is the connection between closed two-sided ideals in A and closed split faces of K. A subset E of K is *invariant* if for each $\phi \in E$ and each $a \in A$ with $\phi(a^*a) > 0$ the functional ϕ_a belongs to E, where $\phi_a(b) = (\phi(a^*ba)/\phi(a^*a))$ for each $b \in A$.

2.3. THEOREM. *Let F be a closed face of the state space K of a unital C^*-algebra A. Then the following statements are equivalent.*

(i) *F is invariant*;
(ii) *F_0 is a two-sided ideal in A*;
(iii) *$F_\perp$ is positively generated*;
(iv) *F is a split face of K.*

Proof. (iv) implies (iii). This follows from (2.10.5).

(ii) implies (iv). We have $F = (F_\perp)^\perp$ by Corollary 2.2. Let $\phi \in K$ and let π denote the representation of A induced by ϕ, with cyclic vector ξ, in the sense of the Gelfand–Neumark–Segal construction. If H denotes the Hilbert space associated with π let L be the closed linear subspace of H spanned by $\{\pi(a)x : a \in F_0, x \in H\}$, and let M be the orthogonal complement of L. Since F_0 is a two-sided ideal L is invariant under $\pi(A)$, and hence also is M; write π_1, π_2 for the restrictions of π to L and M respectively, and write ξ_1, ξ_2 for the projections of ξ onto L and M respectively. For $j = 1, 2$ define $\phi_j(a) = (\pi_j(a)\xi_j, \xi_j)$, where $a \in A$. Then ϕ_1 and ϕ_2 are clearly positive linear functionals on A and, using orthogonality, we have for $a \in A$

$$\phi_1(a) + \phi_2(a) = (\pi_1(a)\xi_1, \xi_1) + (\pi_2(a)\xi_2, \xi_2) = (\pi_1(a)\xi_1 + \pi_2(a)\xi_2, \xi_1 + \xi_2)$$
$$= (\pi(a)\xi, \xi) = \phi(a).$$

Similarly we obtain $\phi_2(a) = 0$ whenever $a \in F_0$, and also that

$$\|\phi_1\| = \|\xi_1\|^2 = \|\phi_1|F_0\|.$$

Suppose now that $\phi = \phi_1' + \phi_2'$ where $\phi_2' = 0$ on F_0 and $\|\phi_1'\| = \|\phi_1'|F_0\|$. Then if $\{u_\alpha\}$ is an approximate identity for the ideal F_0, and if π' denotes the representation associated with ϕ_1' and with cyclic vector ξ', $\pi'(u_\alpha)\xi'$ converges to ξ'. Consequently $\phi_1'(xu_\alpha)=(\pi'(u_\alpha)\xi', \pi'(x^*)\xi')$ converges to $(\xi', \pi'(x^*)\xi')=\phi_1'(x)$ for each $x \in A$, and similarly $\phi_1(xu_\alpha)$ converges to $\phi_1(x)$. Since ϕ_2 and ϕ_2' vanish on F_0, and $xu_\alpha \in F_0$ for each α, it follows that $\phi_1=\phi_1'$ and $\phi_2=\phi_2'$.

Finally the set $\{\Psi \in K : \|\Psi\|=\|\Psi|F_0\|\}$ is convex because $\|\Psi|F_0\|=\lim \Psi(u_\alpha)$ (cf. Dixmier [**80**, 2.1.5]), and it is therefore evident that F is a split face of K.

(iii) implies (ii). As in the proof of 2.2 we have $(F_\perp)^+ \subseteq L_F \subseteq F_0$. Since F_0 is self-adjoint it follows that $F_0=\mathrm{lin}_{\mathbb{C}}\,(F_\perp)^+ \subseteq L_F$, so that $F_0=L_F$. Hence F_0 is a self-adjoint left ideal and therefore a two-sided ideal.

(i) implies (ii). We know that $F=L_F^\perp$. Let $u \in A$ be unitary, so that $\phi_u \in F$ whenever $\phi \in F$. Hence if $a \in L_F$ it follows that $u^*au \in L_F$, and therefore $au=u(u^*au) \in L_F$ since L_F is a left ideal. Since A is the complex linear span of its unitary elements we deduce that L_F is a right ideal also. In particular $A_h \cap L_F$ is a strongly Archimedean ideal because $A_h/(A_h \cap L_F)$ is isometrically order-isomorphic to $(A/L_F)_h$. Therefore $A_h \cap L_F=F_\perp$ (cf. Chapter 2 Section 8), and also $L_F=F_0$.

(ii) implies (i). Let $F=I^\perp$, where I is a closed two-sided ideal in A, and let $\phi \in F$ and $a \in A$ with $\phi(a^*a)>0$. Then if $b \in I$ we have $a^*ba \in I$, so that $\phi_a(b)=0$. Hence $\phi_a \in F$ and F is invariant.

2.4. COROLLARY. *If A is a unital C^*-algebra with state space K then $\phi \in \partial K$ is a split face of K if and only if ϕ is an algebra homomorphism. Moreover, the following statements are equivalent*:

(i) *A_h is a vector lattice*;

(ii) *K is a simplex*;

(iii) *A is commutative.*

Proof. If ϕ is an algebra homomorphism then $\phi^{-1}(0)$ is a closed two-sided ideal, and hence $\phi=(\phi^{-1}(0))^\perp$ is a split face of K by Theorem 2.3. Conversely, if ϕ is a split face of K then $\phi^{-1}(0)$ is a two-sided ideal, by Theorem 2.3. It is elementary now to verify that ϕ is an algebra homomorphism.

(i) implies (ii) is trivial. If A is commutative then the Gelfand–Neumark Theorem shows that $A=C(X)$, and hence that $A_h=C_{\mathbb{R}}(X)$, for some compact Hausdorff space X; consequently we obtain (iii) implies (i). Finally, if K is a simplex then each $\phi \in \partial K$ is a split face of K, and hence ϕ is an algebra homomorphism. Therefore $\phi(ab)=\phi(a)\phi(b)=\phi(ba)$ for all $a, b \in A$ and $\phi \in \partial K$, so that $ab=ba$ by the Krein–Milman Theorem; consequently (ii) implies (iii).

State spaces of unital C^*-algebras provide some interesting examples concerning split faces.

2.5. EXAMPLE. *Let A denote the C^*-algebra $B(\mathscr{C}^2)$ (the bounded linear operators on $\mathscr{C}^2$) and let K be the state space of A and $F = L^{\perp}$, where L is the two-sided ideal of compact operators in A. Then F is a closed split face of K and there exist $f, g \in A(K)^+$, $h \in A(F)^+$ with $h \le f, g$ on F, and such that no extension $h' \in A(K)$ of h exists satisfying $0 \le h' \le f, g$ on K.*

Proof. By 2.3, F is a closed split face of K. Let $\{\mathbf{e}_n\}$ be the coordinate basis for $\mathscr{C}^2$ and for $n \in \mathbb{N}$, define $f(\mathbf{e}_{2n-1}) = 0$, $f(\mathbf{e}_{2n}) = \mathbf{e}_{2n}$, $k(\mathbf{e}_{2n-1}) = n^{-1}\mathbf{e}_{2n-1} + (n^{-1} - n^{-2})^{1/2}\mathbf{e}_{2n}$, $k(\mathbf{e}_{2n}) = (n^{-1} - n^{-2})\mathbf{e}_{2n-1} - n^{-1}\mathbf{e}_{2n}$.

It is easily checked that f and $g = f + k$ are projections with $k \in L$ while $f, g \notin L$. We have $0 \le f, g$ on K and $0 \le g \le f, g$ on F. Put $h = g|F$ and suppose that there exists an $h' \in A(K)^+$ extending h with $h' \le f, g$. If q denotes the range projection for h' we have $q \le f, g$, so that the intersection M of the ranges of f and g contains the range of q. However, inspection shows that $M = \{0\}$, giving a contradiction.

The above example should be compared with 2.10.5. In Chapter 2 we saw that every closed split face is semi-exposed; the following example answers a natural question in this connection.

2.6. EXAMPLE. *Let H be a non-separable Hilbert space, let $A = B(H)$ and let K be the state space of A. Then if L denotes the compact operators on H, $F = L^{\perp}$ is a closed split face of K which is not an intersection of exposed split faces of K.*

Proof. Since L is a minimal closed two-sided ideal in A, F is a maximal proper closed split face of K. It will therefore be sufficient to show that F is not exposed. Suppose that there exists an $f \in A(K)$ with $f \in F_{\perp}$, $f(x) > 0$ for all $x \in K \backslash F$. Then f identifies with a compact Hermitian operator T on H, and since H is non-separable there exists a $\xi \in H$ with $\|\xi\| = 1$ and $T\xi = 0$. But then if ϕ is defined on A by $\phi(a) = (a\xi, \xi)$ we have $\phi \in K$ and $f(\phi) = (T\xi, \xi) = 0$. However, if $a \in L$ satisfies $a\eta = (\eta, \xi)\xi$ for $\eta \in H$, then $\phi(a) = 1$ so that $\phi \in K \backslash F$, giving a contradiction.

We now give a generalization of Corollary 4.9.5.

2.7. THEOREM. *If a complex unital Banach algebra A is a complex Lindenstrauss space then $A = C(X)$ for some compact Hausdorff space X.*

Proof. Let e denote the algebraic identity of A and let S denote the state space of A. The second Banach dual A^{**} of A is also a complex Banach algebra with respect to the Arens product and has unit $\hat{e}$, the canonical embedding of e in A^{**}. Let S^{**} denote the state space of A^{**} relative to $\hat{e}$.

Since A is a complex Lindenstrauss space A^{**} is linearly isometric to the space $C(Y)$ for some compact Hausdorff space Y. Let $T: A^{**} \to C(Y)$ implement this isometry, and let $u = T\hat{e}$. The Bohnenblust–Karlin Theorem [**43**] shows that $\hat{e}$ is an extreme point of $(A^{**})_1$, and so u is an extreme point of $(C(Y))_1$. Hence $|u| = 1$ on Y.

Let S' denote the state space of $C(Y)$ and define $\rho: (C(Y))^* \to A^{***}$ by $\rho = T^* \circ \tau$, where $\tau: (C(Y))^* \to (C(Y))^*$ is defined by $\tau(\phi)(f) = \phi(f\bar{u})$ for $\phi \in (C(Y))^*$ and $f \in C(Y)$. Now τ is a linear isometry onto $(C(Y))^*$ and hence ρ is a linear isometry onto A^{***}. If $\phi \in S'$ then $\|\phi\| = 1 = \phi(1)$ so that $\|\rho(\phi)\| = 1$, and also $\rho(\phi)(\hat{e}) = T^*(\tau(\phi))\hat{e} = \tau(\phi)(T\hat{e}) = \phi(u\bar{u}) = 1$, giving $\rho(\phi) \in S^{**}$. Conversely, if $\psi \in S^{**}$ we have $\|\psi\| = 1 = \psi(\hat{e})$ and $(\rho^{-1}\psi)(1) = (\tau^{-1}(T^{*-1}\psi))(1) = (T^{*-1}\psi)(u) = \psi(T^{-1}u) = \psi(\hat{e}) = 1$, so that $\rho^{-1}\psi \in S'$.

We have shown that $\rho|S'$ is an affine isomorphism between the simplex S' and S^{**}, and therefore S^{**} is a simplex. Similarly $\rho|\text{co}\,(S' \cup -iS')$ is an affine isomorphism between the simplex $\text{co}\,(S' \cup -iS')$ and $Z^{**} = \text{co}\,(S^{**} \cup -iS^{**})$ so that S^{**} is a split face of Z^{**}. Therefore A^{**} is a B^*-algebra (see 4.0.2), and Corollary 2.4 shows that A^{**} is commutative. The subalgebra A of A^{**} is consequently a function algebra, and so the result follows from Corollary 4.9.5.

Notes

Theorem 2.1 is due, independently, to Effros [**92**, 5.1] and Prosser [**176**, 5.11]; our treatment follows the proof of Prosser. Corollary 2.2 was noted by Chu [**66**], and the particular case for extreme points follows from a result of Glimm [**120**, Theorem 9]. The connection between closed two-sided ideals in a C^*-algebra and the split faces of the state space as presented in 2.3 was developed in the papers of Effros [**92**], Størmer [**206**] and Alfsen and Andersen [**6**]. The implication (ii)$\Rightarrow$(iv) in Theorem 2.3 can also be proved by a method of Smith and Ward [**197**, 5.3] by which they show that the M-ideals and the closed two-sided ideals coincide for any C^*-algebra. The equivalence of (i) and (iii) in Corollary 2.4 was first proved by Sherman [**192**].

Example 2.5 is due to Stefánson [**200**], Example 2.6 is due to Andersen and Theorem 2.7 to Ellis [**105**].

3. COMMUTATIVITY AND ORDER IN C^*-ALGEBRAS

The relationship between commutativity and order properties in C^*-algebras has been considered by many authors; in this section we discuss some of their results.

Let V be a real-linear subspace of the bounded self-adjoint operators on a Hilbert space H. Then if $a \in V$ we denote by $|a|$ the positive square root of a^2. Even if $|a|$ belongs to V it is not in general the lattice modulus of a in V unless the operators in V commute.

3.1. THEOREM. *Let V be a linear space of bounded linear self-adjoint operators on a Hilbert space H. Then if $|a| \in V$ whenever $a \in V$, and if V has the Riesz decomposition property, V is commutative.*

Proof. Let L be the closure of V in $B(H)$ for the strong (or weak) operator topology. Then $|a| \in L$ whenever $a \in L$ (cf. Kadison [**137**]). If $P = \{a \in L : 0 \leq a, \|a\| \leq 1\}$ then its strong closure Q is compact for the weak operator topology. If ∂Q is commutative then so is Q by the Krein–Milman Theorem, and hence so is V because V is positively generated since $|a| \geq a, -a$ for each $a \in V$.

Before showing that ∂Q is commutative we will prove that V is a vector lattice with $|a|$ equal to the lattice modulus of a, for each $a \in V$. First suppose that $a, b \in V^+$ and that $ab = 0$. If $x \in V$ and $x \leq a, b$ then, by the Riesz interpolation property, there exists $y \in V$ with $0, x \leq y \leq a, b$. If $z^{1/2}$ denotes the positive square root of any $z \in V^+$ we have $0 \leq a^{1/2}ya^{1/2} \leq a^{1/2}ba^{1/2} = 0$ since $a^{1/2}$ is the limit of certain polynomials in a, and $ab = 0$. Therefore $0 = a^{1/2}ya^{1/2} = (a^{1/2}y^{1/2})(a^{1/2}y^{1/2})^*$, so that $a^{1/2}y^{1/2} = 0$ and hence $ay = 0$. Moreover, we have $0 \leq y^2 \leq y^{1/2}ay^{1/2} = 0$ so that $y^2 = 0 = y$. Hence every lower bound x for a and b belongs to $-V^+$, and so $\inf(a, b) = 0$.

Now for any $a \in V$ we have $|a| + a, |a| - a \geq 0$, and $(|a| + a)(|a| - a) = 0$, and therefore the preceding argument gives $\inf(|a| + a, |a| - a) = 0$. Therefore $\sup(a, -a) = |a| - \inf(|a| + a, |a| - a) = |a|$, and V is a vector lattice with the required property. The strong continuity of the modulus operation now shows that L is an M-space with $|a|$ the lattice modulus of a for each $a \in L$. Since Q is compact for the weak operator topology, and is also a directed set, Q contains a greatest element so that L is a $C_{\mathbb{R}}(X)$ space, for some compact Hausdorff space X.

The extreme points of Q correspond to characteristic functions of open and closed sets in X. Hence if $e_1, e_2 \in \partial Q$ then $e_1 \wedge e_2$, $e_1 - e_1 \wedge e_2$ and $e_2 - e_1 \wedge e_2$ all belong to ∂Q and are disjoint. Suppose that $a, b \in Q$ and $a \wedge b = 0$. Then we have

$$(a + b)^2 = (a \vee b + a \wedge b)^2 = (a \vee b - a \wedge b)^2 = |a - b|^2 = (a - b)^2,$$

so that $ab + ba = 0$. But then $a^2b^2 = -abab = b^2a^2$ and so the positive square root a of a^2 also commutes with b^2. Applying this argument again gives $ab = ba$, so that $ab = 0 = ba$. It follows that $e_1 \wedge e_2$, $e_1 - e_1 \wedge e_2$ and $e_2 - e_1 \wedge e_2$ commute, and hence e_1 and e_2 commute.

The above result enables us to extend Corollary 2.4 to the non-unital case immediately.

3.2. COROLLARY. *Let A be a C^*-algebra. Then A is commutative if and only if A_h is a simplex space.*

3.3. COROLLARY. *Let A be a C^*-algebra. Then A is commutative if and only if it has the property that $0 \leq a \leq b$ in A implies that $a^2 \leq b^2$.*

Proof. Of course $|a|$ belongs to A_h whenever $a \in A_h$. The result will follow if we can show that A_h is a vector lattice; for this it is sufficient, as in the proof of 3.1, to show that $a, b \in A_h^+$ with $ab = 0$ implies that $\inf(a, b) = 0$.

Suppose that $x \in A_h$ with $x \leq a, b$. Then we have $0 \leq b - x \leq 2a + b - x$ and $0 \leq a - x \leq a + 2b - x$. Hence we obtain $(b - x)^2 \leq (2a + b - x)^2$, so that $2a^2 - ax - xa \geq 0$, and consequently $x^2 \leq (2a - x)^2$. It now follows that $|x| \leq 2a - x$. [In fact if S, T are positive operators on a Hilbert space with $S^2 \leq T^2$ then, for $\lambda > 0$, $(\lambda I + S + T)(\lambda I + T - S) = (\lambda^2 I + 2\lambda T + T^2 - S^2) + i(iTS - iST) = X + iY$, where X and Y are Hermitian. Moreover X is positive and invertible and since $X^{-1/2} Y X^{-1/2}$ is self-adjoint the spectrum of $I - iX^{-1/2} Y X^{-1/2}$ lies in $\{z \in \mathbb{C} : \mathrm{re}\, z = 1\}$ and consequently $(X + iY)^{-1} = X^{-1/2}(I + iX^{-1/2} Y X^{-1/2})^{-1} X^{-1/2}$ exists. Therefore it follows that the spectrum of $(T - S)$ lies in the positive half-line, giving $S \leq T$.] Similarly we obtain $|x| \leq 2b - x$, and the two inequalities together give $0 \leq \frac{1}{2}(|x| + x) \leq a, b$.

As in the proof of Theorem 3.1 the fact that $ab = 0$ now implies that $|x| + x = 0$, and hence $x \leq 0$. Consequently we obtain $0 = \inf(a, b)$ as required.

Notes

An extensive investigation of the relationship between order properties and commutativity in C^*-algebras was carried out by Topping [**208**], where, in particular, he proved Theorem 3.1. Corollary 3.3 was originally proved by Ogasawara [**166**].

4. THE CENTRE OF A C^*-ALGEBRA, WEAKLY CENTRAL ALGEBRAS, PRIME ALGEBRAS

In Section 3.1 we introduced the abstract notion of centre for $A(K)$ spaces. We now show that this notion coincides for A_h with the algebraic centre, that is those elements in A_h which commute with all elements of the algebra. In carrying out this process we investigate the concrete C^*-algebra version of the Dauns–Hofmann Theorem that we studied in Section 5.1.

If A is a unital C^*-algebra with state space K, and if $\phi \in \partial K$, then the usual GNS construction gives an irreducible *-representation π_ϕ of A, and the kernel of π_ϕ is called a *primitive ideal* in A. The space of all primitive ideals in A will be denoted by Prim A, and will be endowed with the Jacobson (or hull-kernel) topology (see Dixmier [**81**]).

4.1. THEOREM. *If K is the state space of a unital C^*-algebra then the following statements hold.*

(i) *I is a primitive ideal in A if and only if $I^\perp = \{x \in K; a(x) = 0, \forall a \in I\}$ is a primitive face of K.*

(ii) *K satisfies Størmer's axiom.*

(iii) *If $f: \mathrm{Prim}\, A \to \mathbb{R}$ is continuous for the Jacobson topology then there exists an $a \in A_h$, with a in the algebraic centre of A, such that $f(\ker \pi_\phi) = \phi(a)$ for all $\phi \in \partial K$.*

(iv) *The algebraic centre of A_h coincides with the centre of $A(K)$.*

Proof. Let $\phi \in \partial K$, $I = \ker \pi_\phi$ so that I is a primitive ideal in A, and let $F = I^\perp$. Then, if $\Psi \in \partial F$, Ψ is the w^*-limit of states of the form $x \rightsquigarrow \phi(y^*xy)$ (see Dixmier [**81**, Corollary 3.4.3]), and since the primitive face F_ϕ containing ϕ is invariant we see that $\Psi \in F_\phi$. Therefore we have $F = F_\phi$, a primitive face of K. Conversely, if $\phi \in \partial K$ then $I = \ker \pi_\phi$ is a primitive ideal, and again we obtain $F_\phi = I^\perp$.

(ii) Let $\{F_\alpha\}$ be a family of closed split faces of K and let $\{I_\alpha\}$ be the family of closed two-sided ideals in A with $F_\alpha = I_\alpha^\perp$ (see 2.3). Then if $I = \bigcap_\alpha I_\alpha$ and if $E = \bigcup_\alpha \partial F_\alpha$ we clearly have

$$\overline{\mathrm{co}} \bigcup_\alpha F_\alpha = \overline{\mathrm{co}} \bigcup_\alpha \Big(\bigcup_{\phi \in E} F_\phi \Big), \quad \text{so that } I = \Big(\overline{\mathrm{co}} \bigcup_\alpha F_\alpha \Big)_0 = \bigcap_{\phi \in E} (F_\phi)_0.$$

Therefore $I^\perp$ contains the set $\overline{\mathrm{co}} \bigcup_\alpha F_\alpha$. Now, if $\Psi \in \partial(\bigcap_{\phi \in E} (\ker \pi_\phi)^\perp)$, Ψ is the w^*-limit of states of the form $(\pi_\phi(\cdot)\xi, \xi)$ where $\phi \in E$ (see Dixmier [**81**, 3.4.2]). Therefore we obtain $\Psi \in \bigcup_{\phi \in E} (\ker \pi_\phi)^\perp$, and consequently $\Psi \in \overline{\mathrm{co}} \bigcup_\alpha F_\alpha$. It follows that $\overline{\mathrm{co}} \bigcup_\alpha F_\alpha = I^\perp$, and hence $\overline{\mathrm{co}} \bigcup_\alpha F_\alpha$ is a closed split face of K.

(iii) By the result of (i) the Jacobson topologies on $\mathrm{Prim}\, A$ and on $\mathrm{Prim}\, A(K)$ are equivalent. Therefore, by Theorem 1.1, there exists an $a \in A_h$, belonging to the centre of $A(K)$, such that $f(\ker \pi_\phi) = \phi(a)$ for all $\phi \in \partial K$. We need to show that a belongs to the algebraic centre of A.

Let I be a primitive ideal in A and let $p \in \partial I^\perp$. Then $\ker \pi_p$ is a primitive ideal containing I, so that $\ker \pi_p$ belongs to the closure of I in $\mathrm{Prim}\, A(K)$ and thus $f(I) = f(\ker \pi_p)$. In particular we see that $p(a - f(I)) = p(a - f(\ker \pi_p)) = 0$ for all $p \in \partial I^\perp$, so that $a - f(I)$ belongs to I by the Krein–Milman Theorem. Now for any $b \in A$ we have $ab - f(I)b \in I$ and $ba - f(I)b \in I$, so that $ab - ba \in I$ for any primitive ideal I. Because A is semi-simple it now follows that $ab = ba$, that is a belongs to the algebraic centre of A.

(iv) Let a belong to the centre of $A(K)$. Then, by Theorem 1.1, the function $f: \mathrm{Prim}\, A \to \mathbb{R}$ defined by $f(\ker \pi_\phi) = \phi(a)$ is continuous. Hence the proof of (iii) shows that a belongs to the algebraic centre of A.

Conversely, let a belong to the algebraic centre of A, $a \in A_h$. We may assume that $\|a\| < \frac{1}{2}$. If $\phi \in \partial K$ then $\delta = \frac{1}{2} - \phi(a) \in (0, 1)$, and we can define $\Psi(b) = \delta^{-1}\phi(b(\frac{1}{2} - a))$ for all $b \in A$. If $b \geq 0$ then we have $0 \leq b(\frac{1}{2} - a) \leq b$ so that $0 \leq \Psi \leq \delta^{-1}\phi$. Therefore we see that $\Psi \in K$ and also $(1 - \delta)^{-1}(\phi - \delta\Psi) \in K$. Since ϕ is extreme in K the equation $\phi = \delta\Psi + (\phi - \delta\Psi)$ implies that $\phi = \Psi$, and hence $\delta\phi(b) = \phi(b(\frac{1}{2} - a)) = (\frac{1}{2} - \phi(a))\phi(b)$. Consequently we obtain $\phi(ba) = \phi(ab) = \phi(a)\phi(b)$, and therefore a belongs to the centre of $A(K)$.

In Chapter 4 we defined the Šilov and Bishop decompositions for a compact convex set K, and we investigated their relationship with function algebras. If K is the state space of a unital C^*-algebra A then these decompositions give rise to natural decompositions for A. Using some of the results developed in this chapter it is not difficult to see that if $\{F_\alpha\}$ denotes the Šilov decomposition for K then $\{(F_\alpha)_0\}$ is the family of closed two-sided ideals in A which are generated by the maximal ideals of the centre of A. Similarly, if $\{F_\beta\}$ denotes the Bishop decomposition for K then $\{(F_\beta)_0\}$ forms the family of closed two-sided ideals I which are minimal subject to the property that A/I has trivial centre.

If K is the state space of a C^*-algebra then the Bishop decomposition need not cover $\overline{\partial K}$ in the C^*-algebra situation, and it is not known in general if the Bishop decomposition determines $A(K)$ amongst Banach subspaces of $C_{\mathbb{R}}(\overline{\partial K})$. However, if some subspace L of $C_{\mathbb{R}}(\overline{\partial K})$ has the form $L = B_h$ for some unital C^*-algebra B containing A as a *-subalgebra, and if L contains

$$\{f \in C_{\mathbb{R}}(\overline{\partial K}) : f|(F_\beta \cap \overline{\partial K}) \in A(K)|(F_\beta \cap \overline{\partial K}), \forall_\beta\}$$

then $L = A_h$ by Glimm's Stone–Weierstrass Theorem [**119**]. In general the best result we have is the following, which shows that the Bishop decomposition always determines the centre of $A(K)$.

4.2. THEOREM. *Let K be a compact convex set satisfying Størmer's axiom and let $\{F_\beta\}$ be the Bishop decomposition for K. Then we have*

$$(\textit{centre } A(K))|\overline{\partial K} = \{f \in C_{\mathbb{R}}(\overline{\partial K}) : f|(F_\beta \cap \overline{\partial K}) \textit{ is constant}, \forall_\beta\}.$$

Proof. If f belongs to the centre of $A(K)$ then certainly f is constant on each F_β because of antisymmetry. Conversely, let $f \in C_{\mathbb{R}}(\overline{\partial K})$ be constant on each $F_\beta \cap \overline{\partial K}$. Since each primitive face of K is contained in some F_β we can define a function g: Prim $A(K) \to \mathbb{R}$ by $g((F_x)_\perp) = f(x)$ for all $x \in \partial K$. If we can show that g is structurally continuous it will follow that $f \in$ (centre $A(K))|\overline{\partial K}$ by Theorem 1.1.

It will be sufficient to show that g is upper semi-continuous for the structure topology. Let $\alpha \in \mathbb{R}$ such that $E = \{(F_x)_\perp : f(x) \geq \alpha\}$ is non-empty.

Since K satisfies Størmer's axiom the set $F=\overline{\mathrm{co}}\,\{\bigcup F_x : f(x)\geq\alpha\}$ is a closed split face of K; put $J=F_\perp$. Certainly we have $(F_x)_\perp\supseteq J$ for each $(F_x)_\perp\in E$. On the other hand, if $(F_x)_\perp\supseteq J$ then F_x is contained in F. Now the set $G=\{y\in\overline{\partial K}: f(y)\geq\alpha\}$ is compact and contains all ∂F_u with $f(u)\geq\alpha$, and consequently $G\supseteq\partial F$ by Milman's Theorem. Hence we have $f(x)\geq\alpha$ so that $(F_x)_\perp$ belongs to E. We have therefore shown that $E=h(J)$, and thus E is structurally closed and g is upper semi-continuous as required.

A C^*-algebra A is said to be *weakly central* if whenever I, J are maximal two-sided ideals in A, and Z is the centre of A, then $I\cap Z=J\cap Z$ implies that $I=J$. A compact convex set K is *weakly central* if whenever F and G are minimal closed split faces of K, and $\mathfrak{z}$ is the centre of $A(K)$, then $F_\perp\cap\mathfrak{z}=G_\perp\cap\mathfrak{z}$ implies that $F=G$. We show next that the Bishop decomposition always determines $A(K)$ when K is weakly central.

4.3. THEOREM. *If K is weakly central then the Bishop and Šilov decompositions for K coincide. Consequently if A is a W^*-algebra, or any weakly central unital C^*-algebra, the Bishop decomposition for the state space of A determines A.*

Proof. Let F be a face in the Šilov decomposition for K, and let $\mathfrak{z}$ be the centre of $A(K)$. Then if G and H are distinct minimal closed split faces of F we have $G_\perp\cap\mathfrak{z}=H_\perp\cap\mathfrak{z}=F_\perp\cap\mathfrak{z}$ since each $f\in\mathfrak{z}$ is constant on F. Therefore $H=G$ because K is weakly central and this contradiction shows F has no non-trivial, mutually disjoint, closed split faces. Hence $A(F)$ must have trivial centre, and consequently the Bishop and Šilov decompositions for K coincide.

The preceding general theory now shows that the Bishop decomposition determines $A(K)$ and hence A, whenever K is the state space of a weakly central unital C^*-algebra. Since every W^*-algebra is weakly central (cf. Misonou [**160**]) the result for W^*-algebras now follows.

4.4. THEOREM. *A compact convex set K is weakly central if and only if for every closed split face F of K the centre of $A(F)$ equals the restriction to F of the centre of $A(K)$. Consequently a unital C^*-algebra A is weakly central if and only if for every closed two-sided ideal I in A the centre of A/I is the quotient of the centre of A.*

Proof. Suppose first that centre $A(F)=(\text{centre } A(K))|F$ for every closed split face F of K. Let G and H be minimal closed split faces of K, so that, in particular G and H are disjoint. Let $F=\mathrm{co}\,(G\cup H)$ and let $f\in A(F)$ such that $f(x)=0$ for all $x\in G$ and $f(y)=1$ for all $y\in H$. Clearly f is central in

$A(F)$ and so there exists a g central in $A(K)$ with $g|F=f$. But then g belongs to $G_\perp$ but not to $H_\perp$, which shows that K is weakly central.

Conversely, suppose that K is weakly central. Let F be a closed split face of K, and let $q:A(K)\to A(F)$ be the restriction (quotient) map. If $\mathfrak{z}=$ centre $A(K)$ and $\mathfrak{z}'=$ centre $A(F)$ let $\tilde{q}:\text{Max }\mathfrak{z}'\to\text{Max }\mathfrak{z}$ denote the map $\tilde{q}(N)=q^{-1}(N)\cap\mathfrak{z}$, where N belongs to the maximal ideal space of $\mathfrak{z}'$; $q(N)$ belongs to Max $\mathfrak{z}$ because $N=\{f\in\mathfrak{z}': f(x)=0\}$ for some $x\in\partial F$ so that $\tilde{q}(N)=\{g\in\mathfrak{z}: g(x)=0\}$ is a maximal ideal in $\mathfrak{z}$. Let $N_1=\{f\in\mathfrak{z}': f(x_1)=0\}$ be distinct from N, with $x_1\in\partial F$. Then there exist disjoint closed split faces G and G_1 of F containing x and x_1 respectively. But then x and x_1 cannot belong to the same member of the Šilov decomposition for K, arguing as in Theorem 4.3. Hence there exists a $g\in\mathfrak{z}$ with $g(x)\neq g(x_1)$, and consequently $\tilde{q}(N)\neq\tilde{q}(N_1)$, which shows that $\tilde{q}$ is an injection.

It is easily seen that $\tilde{q}$ is continuous. In fact let $N_\alpha=\{f\in\mathfrak{z}': f(x_\alpha)=0\}$ be a net in Max $\mathfrak{z}'$ converging to $N=\{f\in\mathfrak{z}': f(x)=0\}$, that is $f(x_\alpha)\to f(x)$ for all $f\in\mathfrak{z}'$. But if $g\in\mathfrak{z}$ we have $g|F\in\mathfrak{z}'$ so that $g(x_\alpha)\to g(x)$, which means that $\tilde{q}(N_\alpha)\to\tilde{q}(N)$. This argument shows that, under $\tilde{q}$, Max $\mathfrak{z}'$ is homeomorphic to a closed subset of Max $\mathfrak{z}$.

Let $f\in\mathfrak{z}'$ and let g be a continuous extension of $f\circ\tilde{q}^{-1}$ to Max $\mathfrak{z}$, so that $g\in\text{Max }\mathfrak{z}$. If $x\in\partial F$ we have $g(x)=g(\{h\in\mathfrak{z}: h(x)=0\})=g(\tilde{q}(\{h\in\mathfrak{z}': h(x)=0\}))=f(x)$, so that g is an extension of f.

We now consider the C^*-algebraic counterpart of Theorem 5.1. A C^*-algebra A is called a *prime* algebra if whenever I, J are closed two-sided ideals in A then $IJ=\{xy: x\in I,\ y\in J\}=\{0\}$ implies that either $I=\{0\}$ or $J=\{0\}$.

4.5. THEOREM. *Let A be a unital C^*-algebra. Then A is a prime algebra if and only if A_h is an antilattice.*

Proof. By Theorem 3.5.1, A_h will be an antilattice if and only if the state space K of A is a prime compact convex set.

Let A be a prime algebra and let F, G be semi-exposed faces of K with $\text{co}\,(F\cup G)=K$. Then, by Theorem 2.1, there exists a closed left ideal L and a closed right ideal R in A with $F=L^\perp$ and $G=R^\perp$. Now we have $RL\subseteq R\cap L\subseteq(\text{co}\,(R^\perp\cup L^\perp))_0=\{0\}$, and therefore AR and LA are closed two-sided ideals with $ARLA=\{0\}$. Hence we obtain either $AR=\{0\}$ or $LA=\{0\}$, that is either $R=\{0\}$ or $L=\{0\}$. Therefore either $F=K$ or $G=K$ so that K is prime, and A_h is an antilattice.

Conversely, suppose that K is prime and that I, J are closed two-sided ideals in A with $IJ=\{0\}$. We must also have $I\cap J=\{0\}$. In fact $I\cap J$ is a closed two-sided ideal containing IJ and, since $(I\cap J)\cap A_h$ is positively generated, we can deduce that $I\cap J=IJ$ if we prove that every $b\in$

$A_h^+ \cap I \cap J$ belongs to IJ. Since $b^{1/2}$ is a uniform limit of polynomials in b, with no constant terms, we have $b^{1/2} \in I \cap J$, and hence $b = b^{1/2}b^{1/2} \in IJ$ as required. But now it follows that $\text{co}\,(I^\perp \cup J^\perp) = (I \cap J)^\perp = K$, and hence either $I^\perp = K$ or $J^\perp = K$. Therefore either $I = \{0\}$ or $J = \{0\}$, so that A is a prime algebra.

4.6. COROLLARY. *If A is a W^*-algebra with state space K then the following statements are equivalent.*

(i) *A is a factor;*
(ii) *K is a prime compact convex set;*
(iii) *A_h is an antilattice.*

Proof. The equivalence of (ii) and (iii) has already been established.

If K is prime then $A(K)$ is an antilattice, and so clearly the centre of $A(K)$ must be the constants. Consequently A is a factor, by Theorem 4.1(iv).

Conversely, suppose that A is a factor. If A is of type II or type III then $\overline{\partial K} = K$ (see Glimm [**119**, p. 239]) and hence K is prime by Corollary 3.5.2. If A is of type I then there exists a $\phi \in \partial K$ associated with a faithful irreducible representation of A. If F is the smallest closed split face of K containing ϕ then every $\Psi \in \partial K$ is the w^*-limit of states of A of the form $x \rightsquigarrow \phi(y^*xy)$ (see Dixmier [**81**, Corollary 3.4.3]), and the invariance of F now gives $F = K$. If I, J are closed two-sided ideals in A with $IJ = \{0\}$ then $K = \text{co}\,(I^\perp \cup J^\perp)$, as in the proof of 4.5, and since either $\phi \in I^\perp$ or $\phi \in J^\perp$ we conclude that either $I^\perp = K$ or $J^\perp = K$. Therefore either $I = \{0\}$ or $J = \{0\}$ so that A is a prime algebra. Consequently A_h is an antilattice by Theorem 4.5.

Notes

The results contained in Theorem 4.1 are due to Størmer [**206**] and to Alfsen and Andersen [**6**, **7**, **16**]. Theorems 4.2 and 4.3 were proved by Ellis [**106**].

The theory of weakly central C^*-algebras was studied by Vesterstrøm in [**211**], and Theorem 4.4 is an extension by Chu [**64**, **67**] of some of Vesterstrøm's results to compact convex sets. Theorem 4.5 is due to Chu [**66**]; it was also proved by Archbold [**20**] using quite different techniques. Corollary 4.6 was first proved by Kadison [**138**]. Some further results of this kind are proved by Dixmier [**82**].

5. UNIT TRACES FOR C^*-ALGEBRAS

We saw in Section 2 that the state space K of a unital C^*-algebra A is never a simplex unless A is commutative. Nevertheless K does possess an algebraically important subset which is a simplex. Let $T(A) = \{\phi \in K : \phi(ab) = \phi(ba),\ \forall a, b \in A\}$ denote the *unit traces* on A. We will require the fact, proved in the next section, that every Hermitian $\Psi \in A^*$ has a unique decomposition $\Psi = \Psi^+ - \Psi^-$ with Ψ^+, Ψ^- positive and $\|\Psi\| = \|\Psi^+\| + \|\Psi^-\|$; this result is due to Grothendieck [**123**]. We will prove that $\text{lin}_{\mathbb{R}}\, T(A)$ is an L-space with $\Psi^+ = \sup(\Psi, 0)$, $\Psi^- = \sup(-\Psi, 0)$ for each $\Psi \in \text{lin}_{\mathbb{R}}\, T(A)$.

5.1. THEOREM. *The unit traces $T(A)$ on a unital C^*-algebra A form a compact simplex.*

Proof. Let $V = \text{lin}_{\mathbb{R}}\, T(A)$ and let $\phi \in V$. If $u \in A$ is unitary define Ψ^+ and Ψ^- belonging to A^* by $\Psi^+(a) = \phi^+(u^*au)$, $\Psi^-(a) = \phi^-(u^*au)$ for each $a \in A$, where ϕ^+ and ϕ^- are as mentioned immediately prior to the theorem. We have, for each $a \in A$, $\phi(a) = \phi(auu^*) = \phi(u^*au) = \Psi^+(a) - \Psi^-(a)$, and also $\|\Psi^+\| + \|\Psi^-\| \leq \|\phi^+\| + \|\phi^-\| = \|\phi\| \leq \|\Psi^+\| + \|\Psi^-\|$. The uniqueness of the decomposition $\phi = \phi^+ - \phi^-$ now implies that $\Psi^+ = \phi^+$ and $\Psi^- = \phi^-$, that is $\phi^+(a) = \phi^+(u^*au)$ and $\phi^-(a) = \phi^-(u^*au)$ for all $a \in A$ and all unitaries $u \in A$. Therefore $\phi^+(ua) = \phi^+(u^*uau) = \phi^+(au)$, and since every $a \in A$ is a linear combination of unitaries it follows that $\phi^+ \in V$, and similarly $\phi^- \in V$. Consequently V is positively generated and the theorem will follow if we show that $\inf(p, q)$ exists in V for each pair p, q in V^+, or equivalently that $[0, p]$ is lattice-ordered for all p in V^+.

Firstly, let $p \in (A^*)^+$ and let H_p be the Hilbert space associated with p in the usual Gelfand–Neumark–Segal construction, and let L_p denote the representation of A on H_p induced by the left regular representation of A. [H_p is the completion of A/N_p for the norm induced by the inner product $\langle \tilde{a}, \tilde{b} \rangle = p(b^*a)$, where $N_p = \{x \in A : p(x^*x) = 0\}$ and $\tilde{x}$ denotes the image of $x \in A$ in A/N_p under the quotient map.] If $q \in A^*$ with $0 \leq q \leq p$ then the Cauchy–Schwarz inequality for q gives $q(b_1{}^*a_1) = q(ba)$ whenever $\tilde{b}_1 = \tilde{b}$ and $\tilde{a}_1 = \tilde{a}$, and hence there exists a $D_q \in B(H_p)$ such that $q(b^*a) = \langle \tilde{a}, D_q\tilde{b} \rangle$ for all $a, b \in A$. Since $0 \leq q \leq p$ it follows easily that $0 \leq D_q \leq I$, where I is the identity operator on H_p. Moreover $\langle \tilde{c}, D_qL_p(a)\tilde{b} \rangle = \langle \tilde{c}, D_q\widetilde{ab} \rangle = q(b^*a^*c) = \langle \widetilde{a^*c}, D_q\tilde{b} \rangle = \langle L_p(a)^*\tilde{c}, D_q\tilde{b} \rangle = \langle \tilde{c}, L_p(a)D_q\tilde{b} \rangle$, and hence D_q belongs to the commutant $L_p(A)'$ of $L_p(A) = \{L_p(a) \in B(H_p) : a \in A\}$. Conversely, if $D \in L_p(A)'$ and $0 \leq D \leq I$ we can define $q(a) = \langle \tilde{a}, D\tilde{1} \rangle$, and it is easy to check that $0 \leq q \leq p$ and that $D = D_q$. From this reasoning it follows that $\{q \in A^* : 0 \leq q \leq p\}$ is order isomorphic to $\{D \in L_p(A)' : 0 \leq D \leq I\}$.

Now take $p \in V^+$ and $q \in V$ with $0 \leq q \leq p$. In this case the right regular representation of A induces a $*$-representation R_p of A on H_p and the involution in A induces an isometry J on H_p such that $J\tilde{a} = \tilde{a}^*$ for $a \in A$ and $\langle x, y \rangle = \langle Jy, Jx \rangle$ for x, y in H_p. For all, a, b, c in A we have $\langle \tilde{a}, R_p(b)D_q\tilde{c} \rangle = \langle \widetilde{ab^*}, D_q\tilde{c} \rangle = q(c^*ab^*) = q(b^*c^*a) = \langle \tilde{a}, D_q\widetilde{cb} \rangle = \langle \tilde{a}, D_qR_p(b)\tilde{c} \rangle$, so that $D_q \in R_p(A)'$. Conversely if $D'_q \in L_p(A)' \cap R_p(A)'$ where $q' \in A^*$ and $0 \leq q' \leq p$, then $q'(ba) = \langle \tilde{a}, D'_q\widetilde{b^*} \rangle = \langle \tilde{a}, D'_qR_p(b^*)\tilde{1} \rangle = \langle \tilde{a}, R_p(b^*)D'_q\tilde{1} \rangle = \langle \widetilde{ab}, D'_q\tilde{1} \rangle =$

$q'(ab)$, that is $q' \in V$. Therefore we have shown that $\{q \in V: 0 \le q \le p\}$ is order isomorphic to $\{D \in L_p(A)' \cap R_p(A)': 0 \le D \le I\}$.

If we can show that $L_p(A)' \cap R_p(A)'$ is a commutative W^*-algebra then the result will follow. Let $S \in L_p(A)'$ and $T \in R_p(A)'$ be Hermitian. Then $JS\tilde{1} = S\tilde{1}, JT\tilde{1} = T\tilde{1}$ and so there exist x_n, y_n in A_h with $\tilde{x}_n \to S\tilde{1}$ and $\tilde{y}_n \to T\tilde{1}$. For each $a \in A$ we have $S\tilde{a} = SL_p(a)\tilde{1} = L_p(a)S\tilde{1} = \lim R_p(x_n)\tilde{a}$, and $T\tilde{a} = \lim L_p(y_n)\tilde{a}$ similarly. Since each pair $R_p(a)$ and $L_p(b)$ commute it follows that $\langle ST\tilde{a}, \tilde{b}\rangle = \langle T\tilde{a}, S\tilde{b}\rangle = \lim \langle L_p(y_n)\tilde{a}, R_p(x_n)\tilde{b}\rangle = \lim \langle R_p(x_n)\tilde{a}, L_p(y_n)\tilde{b}\rangle = \langle TS\tilde{a}, \tilde{b}\rangle$ so that $ST = TS$.

The first part of the above proof shows that, for $p \in K$, L_p is an irreducible *-representation of A if and only if $p \in \partial K$, whereas the second part of the proof shows that, for $p \in T(A)$, L_p is a factor representation (i.e. $L_p(A)' \cap L_p(A)''$ is trivial) if and only if $p \in \partial T(A)$. If $p \in \partial K \cap T(A)$ then N_p is a two-sided ideal and also, by Theorem 2.1, $\{p\} = N_p^\perp$ so that p is a split face of K; consequently the set $\partial K \cap T(A)$ coincides with the characters of A. In general $\partial K \cap T(A)$ is not equal to $\partial T(A)$; in fact if A is the C^*-algebra of complex 2×2 matrices then (cf. Størmer [**201**]) the state space of A is affinely isomorphic to a ball in $\mathbb{R}^3$, and the unique trace maps into the centre of the ball. In the case when A is separable a result of Dixmier [**80**] shows that N_p is a primitive ideal whenever $p \in \partial T(A)$; in this case $N_p^\perp$ is a primitive face of K containing p.

Another subset of K which is a simplex, and which is of algebraic interest, is the *commutator face* $\mathrm{Cm}(A)$ of K. Here we define $\mathrm{Cm}(A) = (\mathrm{comm}\, A)^\perp$, where the *commutator ideal* comm A of A is the closed two-sided ideal generated by the commutators $ab - ba$, $a, b \in A$. Since $T(A) = \{ab - ba: a, b \in A\}^\perp$ it is clear that $T(A) \supseteq \mathrm{Cm}(A)$. In the case when $A = B(\mathbb{C}^n)$, $T(A)$ is a singleton whereas $\mathrm{Cm}(A)$ is empty, because A is a simple algebra.

5.2. THEOREM. *The commutator face* $\mathrm{Cm}(A)$ *of a unital C^*-algebra A is the largest closed split face of the state space K which is a simplex. Moreover,* $\mathrm{Cm}(A)$ *is the closed convex hull of the characters of A.*

Proof. By Theorem 2.3, $\mathrm{Cm}(A)$ is a closed split face of K; also $\mathrm{Cm}(A)$ can be identified with the state space of the commutative C^*-algebra $A/\mathrm{comm}\, A$, and hence $\mathrm{Cm}(A)$ is a Bauer simplex. It is evident that $\mathrm{Cm}(A)$ contains all the characters of A. If $x \in A \backslash \mathrm{comm}\, A$ then $\tilde{0} \ne \tilde{x} \in A/\mathrm{comm}\, A$, and since $A/\mathrm{comm}\, A$ is commutative there exists a character Ψ of $A/\mathrm{comm}\, A$ such that $\Psi(\tilde{x}) \ne 0$; hence $\phi(a) = \Psi(\tilde{a})$ defines a character ϕ of A with $\phi(x) \ne 0$. It follows from this reasoning that comm A is precisely the intersection of the kernels of all the characters of A. Since each character of A is a split face of K, and since K satisfies Størmer's axiom, the closed convex hull of the characters is also a closed split face F of K and hence $F = (F_\perp)^\perp = (\text{comm } A)^\perp = \text{Cm}(A)$.

Suppose that G is a closed split face of K which is a simplex. Then, if $\phi \in \partial G$, ϕ is a split face of G and hence ϕ is a split face of K. But then, by Corollary 2.4, ϕ is a character of A so that $\phi \in \text{Cm}(A)$. Consequently $\text{Cm}(A)$ contains G.

Notes

Theorem 5.1 is a result of Thoma [**207**], but the proof given here is due to Effros and Hahn [**97**]. A consequence of 5.1 is that each $\phi \in T(A)$ is the barycentre of a unique maximal measure μ on $T(A)$. This measure μ is the central measure corresponding to ϕ (see Sakai [**188**, Theorem 3.1.18]). For a complete discussion of central measures for compact convex sets and for C^*-algebras we refer to Alfsen [**5**, § 2.8], Sakai [**188**, Chapter 3] and Wils [**214**, **215**, **216**]. The use of Choquet boundary theory in the decomposition of lower semi-continuous semi-infinite traces on separable C^*-algebras is discussed by Davies [**74**].

Another approach to the trace simplex for C^*-algebras is given by Alfsen and Shultz [**10**, § 12], and some further results concerning simplexes of states have been proved by Batty [**33**].

The theory of commutator ideals in C^*-algebras is discussed by Arveson [**21**, § 3.3].

6. THE STATE SPACES OF JORDAN OPERATOR ALGEBRAS AND OF C^*-ALGEBRAS

We firstly consider some recent work of E. M. Alfsen, F. W. Shultz and E. Størmer concerning characterizations and representations of certain Jordan algebras.

Let A be a real Banach space which is a Jordan algebra over $\mathbb{R}$ with identity 1 and such that (i) $\|a \cdot b\| \leq \|a\|\|b\|$, (ii) $\|a^2\| = \|a\|^2$ and (iii) $\|a^2\| \leq \|a^2 + b^2\|$ for all a, b in A; A is called a *JB-algebra*. The connection between $A(K)$ spaces and JB-algebras is demonstrated in the following result.

6.1. THEOREM. *If A is a* JB*-algebra then A can be identified with an $A(K)$ space such that $A(K)^+ = A^2 = \{a^2 : a \in A\}$, $1(x) = a$ for all $x \in K$, and $0 \leq a^2 \leq 1$ whenever $-1 \leq a \leq 1$. Conversely, if $A(K)$ is a Jordan algebra over $\mathbb{R}$ with identity* 1, *and if $0 \leq a^2 \leq 1$ whenever $\|a\| \leq 1$, then $A(K)$ is a* JB-*algebra.*

Proof. Let A be a JB-algebra. If $a \in A$ then (cf. Jacobson [**133**, p. 36]) the polynomials in a form an associative subalgebra, and hence their closure

$C(a)$ is a commutative Banach algebra. Let $a, b \in A^2\backslash\{0\}$ and put $\alpha = \|a\|$, $\beta = \|b\|$. Since $\|a/\alpha\| = 1$ we can find $d \in C(a)$ such that $1 - a/\alpha = d^2$, by the Binomial Theorem. Hence if $a = c^2$ and $f = \sqrt{\alpha}d$ we have $\alpha - a = \alpha d^2 = f^2$, so that $\|\alpha - a\| = \|f^2\| \leq \|c^2 + f^2\| = \alpha$; similarly we obtain $\|\beta - b\| \leq \beta$ so that $\|\alpha + \beta - (\alpha + b)\| \leq \alpha + \beta$, or $\|1 - (a + b)/(\alpha + \beta)\| \leq 1$. As above we can find a $g \in A$ with $g^2 = (a + b)/(\alpha + \beta)$, and hence $a + b \in A^2$. This shows that A^2 is a convex cone.

If $a^2 = -b^2$ then $\|a\|^2 = \|a^2\| \leq \|a^2 + b^2\| = 0$, and thus A^2 is a proper cone. Moreover, the preceding discussion implies that $A^2 = \{a \in A : \|a\| \geq \|\|a\| - a\|\}$ from which it follows that A^2 is norm-closed. Suppose that $-1 \leq a \leq 1$ and write $1 - a^2 = (1 - a) \cdot (1 + a) = c^2 \cdot d^2$, for some elements c, d in $C(a)$. Since $C(a)$ is associative we have $a^2 + (c \cdot d)^2 = 1$, so that $0 \leq a^2 \leq 1$. Further we obtain

$$\|a\|^2 = \|a^2\| \leq \|a^2 + (c \cdot d)^2\| = 1.$$

On the other hand $1 + a \in A^2$ and $1 - a \in A^2$ whenever $\|a\| \leq 1$. Therefore we have $\{a \in A : \|a\| \leq 1\} = \{a \in A : -1 \leq a \leq 1\}$, and the first half of the theorem is proved.

Conversely, let $A(K)$ be a Jordan algebra over $\mathbb{R}$ with identity 1 such that $0 \leq a^2 \leq 1$ whenever $\|a\| \leq 1$. Now if $\|a\| \leq 1$ and $\|b\| \leq 1$ we have $\|\frac{1}{2}(a + b)\| \leq 1$ and $\|\frac{1}{2}(a - b)\| \leq 1$ so that $0 \leq [\frac{1}{2}(a + b)]^2 \leq 1$ and $0 \leq [\frac{1}{2}(a - b)]^2 \leq 1$. Therefore $-1 \leq [\frac{1}{2}(a + b)]^2 - [\frac{1}{2}(a - b)]^2 \leq 1$ and consequently $\|a \cdot b\| = \|[\frac{1}{2}(a + b)]^2 - [\frac{1}{2}(a - b)]^2\| \leq 1$. It follows that $\|a \cdot b\| \leq \|a\| \; \|b\|$ for all a, b in $A(K)$.

In particular we see that $\|a^2\| \leq \|a\|^2$ for all $a \in A$. However, if $\|a^2\| \leq 1$ then $0 \leq a^2 \leq 1$ so that $a = \frac{1}{2}[a^2 + 1 - (a - 1)^2] \leq \frac{1}{2}(a^2 + 1) \leq 1$; similarly $-a \leq 1$ and so $\|a\| \leq 1$. Therefore we obtain $\|a^2\| = \|a\|^2$ for all a in A. Since all squares belong to $A(K)^+$ we clearly have $\|a^2\| \leq \|a^2 + b^2\|$ for all a, b in A. Hence $A(K)$ is a JB-algebra.

An immediate consequence of the above result is that every JB-algebra has the property that $\sum_{j=1}^{n} a_j^2 = 0$ implies that $a_j = 0$ for $j = 1, \ldots, n$. The basic ideas in the theory developed by Alfsen and Shultz [**10**] are those of P-projections and projective faces.

Let $V = \text{lin}\, K$ be a base-norm space and let $A = A^b(K) = V^*$. Let $P : V \to V$ be a positive projection and let $P^* : A \to A$ be the dual (positive) projection. P is said to be *smooth* if $((\ker P)^0)^+ = (((\ker P)^+)^0)^+$, and similarly for P^*. Since $(\ker P^*)_0 = PV$, which is positively-generated, we have $(\ker P^*)_0 = ((\ker P^*)^+)_0$, and hence P^* is smooth if and only if $(\ker P^*)_0 \cap K$ is a norm semi-exposed face of K. A positive projection $Q : V \to V$ is said to be *quasi-complementary* to P if $(\ker P)^+ = (QV)^+$ and $(\ker Q)^+ = (PV)^+$, and similar definitions apply to P^* and Q^*. It is not difficult to verify (cf. **10**, Theorem 1.8]) that P and Q are smooth and quasi-complementary if and only if P^* and Q^* are smooth and quasi-complementary.

If P is a smooth positive projection with $\|P\| \leq 1$, and if there exists a smooth quasi-complementary projection Q with $\|Q\| \leq 1$, then P and P^* are called *P-projections*. When this situation attains we have $(PV) \cap K = (\ker P^*)_0 \cap K$, which is a (norm) semi-exposed face of K; such faces are called *projective faces* of K. It is easy to see that every split face F of K is projective with associated P-projection satisfying $P(\lambda x + (1-\lambda)y) = \lambda x$ for $0 \leq \lambda \leq 1$, $x \in F$, $y \in F'$.

Let P be a P-projection with quasi-complement P' (necessarily unique) and let $F_P = PV \cap K$ and $F_{P'} = P'V \cap K$; then F_P and $F_{P'}$ are called *quasi-complementary* projective faces. The connection between projective faces and split faces is demonstrated more clearly in the following result.

6.2. THEOREM. *Let F, G be semi-exposed faces of K and let $M = \text{lin}\,(F \cup G)$. Then F and G are quasi-complementary projective faces of K if and only if the following two conditions hold*:

(i) *F and G are split faces of* $\text{co}\,(F \cup G)$ *and the associated projections $\pi: M \to \text{lin}\, F$ and $\pi': M \to \text{lin}\, G$ are continuous*;

(ii) *there is a continuous positive projection ϕ from V onto M with $\phi(K) \subseteq K$ and such that* $\ker \phi \subseteq \tilde{F} \cap \tilde{G}$, *where $\tilde{F} = ((F^0)^+)_0$, and $\tilde{G} = ((G^0)^+)_0$.*

Proof. First let F and G be quasi-complementary projective faces with $F = F_P$ and $G = F_Q$. The restriction of P to M maps F onto itself and G into 0, and is continuous, so that statement (i) follows. Now let $\phi = P + Q: V \to M$. Then, since $PQ = QP = 0$, ϕ is a continuous positive projection onto M. We have $1 - P^*1 \geq 0$ and $P^*(1 - P^*1) = 0$ so that $Q^*(1 - P^*1) = 1 - P^*1 = Q^*1$, since $Q^*P^* = 0$. Therefore, if $x \in K$ we have $\phi(x)(1) = (P^* + Q^*)(1)(x) = 1$, so that $\phi(K)$ is contained in K. Finally, if $\phi(v) = 0$ then $Qv = Q^2 v = Q(-Pv) = 0$; if $f \in (F^0)^+$ then $f \in (\ker Q)^0$ since Q is smooth, and hence $f(v) = 0$ so that $v \in \tilde{F}$. Similarly we have $\ker \phi \subseteq \tilde{G}$.

Conversely suppose that the semi-exposed faces F and G satisfy conditions (i) and (ii). Let $P = \pi \circ \phi$ and $Q = \pi' \circ \phi$, so that P and Q are continuous and positive. We clearly have $\|\phi\| \leq 1$ and, since $\pi, \pi' \geq 0$ and $(\pi + \pi')(\Psi) = \Psi$ for $\Psi \in M^+$, we have $\|\pi(\Psi)\| \leq \|\Psi\|$ and $\|\pi'(\Psi)\| \leq \|\Psi\|$ so that $\|\pi\| \leq 1$ and $\|\pi'\| \leq 1$. Therefore we obtain $\|P\| \leq 1$ and $\|Q\| \leq 1$. We evidently have $PV \cap K = F$ and $QV \cap K = G$. Let $x \in (\ker Q) \cap K$ and write $x = \phi(x) - (x - \phi(x))$. We have $\phi(x) = (\pi + \pi') \circ \phi(x) = Px + Qx = Px$, since $\pi + \pi'$ is the identity operator on $M = \phi(v)$, and therefore $\phi(x) \in (PV)^+$. Since $x - \phi(x) \in \ker \phi \subseteq \tilde{F}$ we have $x \in V^+ \cap \tilde{F} = (PV)^+$, because F is semi-exposed.

Suppose that $Pu = 0$. Then $u - Qu \in (\ker P) \cap (\ker Q) \subseteq \ker \phi$ and hence $u = Qu + (u - Qu) \in \text{lin}\, G + \ker \phi \subseteq \tilde{G}$. The preceding paragraph gave $\tilde{F}(\widetilde{\ker Q})^+$, and $\tilde{G} = (\widetilde{\ker P})^+$ similarly, so we obtain $\ker P \subseteq (\ker P^+)$. It follows that P is smooth, and similarly that Q is smooth. The relationship between P and Q which was found in the preceding paragraph now implies that P and Q are quasi-complementary P-projections, and that F and G are quasi-complementary projective faces of K.

We now consider the relevance of the above concepts to C^*-algebras and W^*-algebras.

6.3. THEOREM. *Let $A = B_h$, where B is a W^*-algebra, and let K be the normal state space of B. Then $P^*: A \to A$ is a P-projection if and only if $P^*a = pap$, $a \in A$, for some projection $p \in B_h$. Every norm-closed face F of K is projective with corresponding projection $p_F \in B_h$ such that $F = \{x \in K: p_F(x) = 1\}$. Moreover, the map $F \rightsquigarrow p_F$ is a bijection of the set of norm-closed faces of K onto the projections in B_h, which map the split faces of K onto the central projections in B_h.*

Proof. We first note that A can be identified with $A^b(K)$, the dual space of the base normed space $V = \text{lin}\, K$. (V is the Hermitian part of the normal dual space of B, see Sakai [**188**]).

Let $p \in B_h$ be a projection and define, for $a \in A$, $P^*a = pap$ and $Q^*a = (1-p)a(1-p)$. Then P^* and Q^* are w^*-continuous, and hence are the duals of projections P, Q: $V \to V$. Clearly we have $\|P\|, \|Q\| \le 1$. Suppose that $P^*a = 0$ for some $a \in A^+$. Then, for $\rho \in K$, the Cauchy–Schwarz inequality gives $|\rho(pa)|^2 \le \rho(a)\rho(pap) = 0$. Hence $pa = 0$ and also $ap = (pa)^* = 0$, so that $Q^*a = a - ap - pa + pap = a$, that is $(\ker P^*)^+ \subseteq (Q^*A)^+$. Similarly we obtain $(\ker Q^*)^+ \subseteq (P^*A)^+$. Now if $a \in (P^*A)^+$ we have $pap = a$, so that $Q^*a = (1-p)a(1-p) = (1-p)pap(1-p) = 0$. Hence $(P^*A)^+ = (\ker Q^*)^+$, and similarly $(Q^*A)^+ = (\ker P^*)^+$.

In order to show that P^* is smooth we need to show that when $\rho \ge 0$ and $\rho \in (\ker^+ P^*)_0$, then $\rho \in (\ker P^*)_0$. Since PV is closed we need to show that $\rho \in PV$. Now if $b = 1 - P^*1$ we see that $b \in \ker^+ P^*$, so that $b(\rho) = 0$, and hence $\|\rho\| = \rho(1) = \rho(P^*1) = \|P\rho\|$, and we can assume $\|\rho\| = 1$ without loss of generality. Note that $\rho(1-p) = \rho(1 - P^*1) = 0$, and so for each $a \in A^+$ we have, using the Cauchy–Schwarz inequality, $\rho(a) = \rho(pap) + \rho(pa(1-p)) + \rho((1-p)ap) + \rho((1-p)a(1-p)) = \rho(pap) = \rho(P^*a))$. Therefore we obtain $\rho = P\rho$ and thus P^*, and similarly Q^*, is smooth. Consequently P^* is a p-projection.

Conversely, let $P^*: A \to A$ be a P-projection with quasi-complement Q^* and let $P^*1 = p$. Let $0 < \lambda < 1$ and $p = \lambda a + (1-\lambda)b$ with $a, b \in [0, 1]$. Then $0 \le \lambda a$, $(1-\lambda)b \le p$, so that $a, b \in (P^*A)^+ = \ker^+ Q^*$, and also $a = P^*a \le p$, $b = P^*b \le p$. But then $\lambda a = p - (1-\lambda)b \ge \lambda p$, and we obtain $a = p = b$. This shows that p is an extreme point of $[0, 1]$ and is therefore a projection in B (see Prosser [**176**, 2.2]). Now if $R^*: A \to A$ is defined by $R^*a = pap$ we know that R^* is also a P-projection with the same range as P^*, namely the order ideal generated by p. The quasi-complement S^* for R^* will be given by $S^*a=(1-p)a(1-p)$ and since $Q^*1=1-p$, we can deduce that $S^*A=Q^*A$. It follows that $\ker^+ R^*=\ker^+ P^*$ and hence that $\ker R^*=\ker P^*$. Consequently we obtain $P^*=R^*$, and the first equivalence of the theorem is established.

Let $P^*:A\to A$ be any P-projection and let $x\in K\cap PV$, so that $(P^*1)(x)=1(Px)=1$. Conversely, let $x\in K$ such that $(P^*1)(x)=1$. Then, if Q^* is the quasi-complementary P-projection for P^*, we have $(Q^*1)(x)=(1-P^*1)(x)=0$ so that $x\in(\ker Q)^+=(PV)^+$. Hence we have shown that the projective face F_p, associated with P, has the form $F_p=\{x\in K:(P^*1)(x)=1\}$. (In particular, F_p is norm-exposed.)

Now if F is any norm-closed face of K there exists a projection $p_F\in B_h$ (the carrier of F) such that $F=\{x\in K: p_F(x)=1\}$ (see proof of Theorem 2.1). The first part of the proof, together with the preceeding remarks, now shows that F is the projective face associated with the P-projection $P_Fa=p_Fap_F$, $a\in A$.

Suppose that $F_p=F_s$ for some P-projection $S: V\to V$. Then P and S have the same range and so arguing as we did above for the equality of P^* and R^*, we have $P=S$. Therefore the map $F\rightsquigarrow p_F$ is a bijection as required.

If F is a split face of K then it is easily seen that the associated P-projection P^* is order-bounded in $A^b(K)=A$, so that $p_F=P^*1$ is central in A, and in B_h by Theorem 4.1(iv) (see also Section 3.8). Conversely, suppose that p_F is central in A. Then if Q^* is the quasi-complement to $P^*:A\to A$, $P^*a=p_Fap_F=p_Fa$, we have $Q^*a=(1-p_F)a$ and so $P^*+Q^*=I$. It follows that $K=\text{co}\,(F_p\cup F_Q)$ and hence, by Theorem 3.17, that F_p and F_Q are complementary split faces of K.

Since the state space K of a unital C^*-algebra A is precisely the normal state space of the W^*-algebra A^{**} the preceding result shows that every norm-closed face of K is projective. The following result proves, and generalizes, the result concerning unique decompositions which was used in the proof of Theorem 5.1.

6.4. THEOREM. *Let K be a compact convex set such that every norm-exposed face is projective. Then each $\rho\in A(K)^*$ has a unique decomposition $\rho=\sigma-\tau$, where $\sigma,\tau\geq 0$ and $\|\rho\|=\|\sigma\|+\|\tau\|$.*

Proof. Let $\rho=\sigma-\tau$ where $\sigma,\tau\geq0$ and $\|\rho\|=\|\sigma\|+\|\tau\|=1$, $\sigma,\tau\neq0$. Choose $f\in A^b(K)$ such that $1=\|f\|=f(\rho)$, and let $g=\frac{1}{2}(1+f)$, $h=1-g$ so that $0\leq g$, $h\leq 1$. Since $1=(g-h)(\rho)=g(\sigma)-g(\tau)-h(\sigma)+h(\tau)=\|\sigma\|+\|\tau\|$, it follows that $g(\sigma)=\|\sigma\|$, $h(\tau)=\|\tau\|$ and $g(\tau)=h(\sigma)=0$. Put $F=\{x\in K: g(x)=0\}$ so that F is a norm-exposed face of K containing $\tau/\|\tau\|$, and let $P:V\to V$ $(V=A(K)^*)$ be the P-projection such that $F=F_p$. Let $k\in A^b(K)$ be defined by $k=(P')^*1$ where P' is the quasi-complement of P. Now

$g=0$ on $F_p=(PV)^+=(\ker P')^+$, and hence because P' is smooth we have $g\in(\ker P')^0=(P')^*A^b(K)$. Therefore, since $0\le g\le 1$, we have $0\le g\le (P')^*1=k\le 1$.

It is evident that $P\tau=\tau$ and hence $P'\tau=0$. Moreover, since $g(\sigma)=\|\sigma\|$, we have $k(\sigma)=\|\sigma\|$ so that $(P^*1)(\sigma)=(1-(P')^*1)(\sigma)=0$. Hence σ belongs to $(\ker P)^+=(\operatorname{im} P')^+$, and therefore $P'\rho=\sigma$. Since P' did not depend on the choice of decomposition $\rho=\sigma-\tau$ it follows that this decomposition is unique.

Continuing with our notation in which $V=\operatorname{lin} K$ is a base normed space with dual space $A=A^b(K)$, we define (V,A) to be in *spectral duality* if for each $a\in A$ and $\lambda\in\mathbb{R}$ there exists a P-projection P, with quasi-complement Q, such that $a(x)\le\lambda$ for $x\in F_p$ and $a(x)>\lambda$ for $x\in F_Q$, and such that P satisfies the following conditions: (i) $P^*a+Q^*a=a$; (ii) whenever S and T are quasi-complementary P-projections on V with $S^*A+T^*a=a$, then P commutes with S and T.

In the particular case where A and V satisfy the hypotheses of Theorem 6.3, (V,A) is in spectral duality. In fact let $a\in A$, $\lambda\in\mathbb{R}$ and let e_λ be the spectral projection for a corresponding to the value λ. Then $e_\lambda\in A$ and commutes with a, so that $e_\lambda a e_\lambda+(1-e_\lambda)a(1-e_\lambda)=a$, and thus condition (i) holds for the P-projection P corresponding to e_λ. Moreover if $p\in A$ is a projection such that $pap+(1-p)a(1-p)=a$ then, multiplying on the left by p and also on the right by p, we see that p commutes with a, and therefore p commutes with e_λ. Evidently now (ii) is also satisfied. Finally, if $x\in F_P$ then $e_\lambda(x)=1$ so that $a(x)=\int\mu\, de_\mu(x)\le\lambda$, while if $x\in F_Q$ we have $e_\lambda(x)=0$ so that $a(x)=\int\mu\, de_\mu(x)>\lambda$.

In the general case where (V,A) are in spectral duality Alfsen and Shultz ([**10**, Theorem 6.8]) have proved that for each $a\in A$ there exists a unique family $\{P^a_\lambda\}$, $\lambda\in\mathbb{R}$, of P-projections in V the spectral resolution for a, such that the family $\{e^a_\lambda\}=\{(P^a_\lambda)^*1\}$ satisfies

$$\text{(i) } e^a_\lambda\le e^a_\mu \quad\text{for } \lambda\le\mu,\qquad \text{(ii) } e^a_\lambda=\inf_{\mu>\lambda}\{e^a_\mu\},\qquad \text{(iii) } \inf_{\lambda\in\mathbb{R}}\{e^a_\lambda\}=0,$$

$\sup_{\lambda\in\mathbb{R}}\{e^a_\lambda\}=1$, and such that $a=\int\lambda\, de^a_\lambda$ in the sense of a norm-convergent Riemann–Stieltjes Integral.

When K is a compact convex set such that $(V,A^b(K))$ are in spectral duality, K is said to be *spectral*. If, in addition, for each $a\in A(K)$ the family $\{e^a_\lambda\}$ consists of upper semi-continuous functions in $A^b(K)$ then K is said to be *strongly spectral*.

6.5. THEOREM.

(i) *Every compact simplex is spectral.*

(ii) *A compact simplex is strongly spectral if and only if it is a Bauer simplex.*

(iii) *The state space of a unital C^*-algebra is strongly spectral.*

Proof. (i) Since every norm-closed face of a compact simplex is split and hence projective, the P-projections in V clearly commute. Hence, given $a \in A^b(K)$ and $\lambda \in \mathbb{R}$ we need to find complementary faces F and F' such that $a(x) \leq \lambda$ on F and $a(x) > \lambda$ on F'. Without loss of generality we can take $\lambda = 0$. Let $a = a^+ - a^-$ be the lattice decomposition of a and let $F = \{x \in K: a^+(x) = 0\}$ and $G = \{x \in K: a^-(x) = 0\}$. Then F and G are split faces of K with $a(x) \leq 0$ for $x \in F$. If we can show that $K = \text{co}\,(F \cup G)$ it will follow that $F' \subseteq G$, and hence that $a(x) = a^+(x) > 0$ for $x \in F'$.

Fix an x in K, and note that $a^+(x) = \sup\{a(y): 0 \leq y \leq x\}$. Thus we can choose $\{y_n\}$ with $0 \leq y_n \leq x$, such that $a^+(x) - a(y_n) < 2^{-n}$. We have, for each n, $a(y_n) - a(y_n \wedge y_{n+1}) = a(y_n \vee y_{n+1}) - a(y_{n+1}) \leq a^+(x) - a(y_{n+1}) < 2^{-n-1}$, and hence $a^+(x) - a(y_n \wedge y_{n+1}) < 2^{-n} + 2^{-n-1}$. By induction we obtain $a^+(x) - a(y_n \wedge y_{n+1} \wedge \cdots \wedge y_{n+k}) < 2^{-n+1}$ for each k. If $w_n = \inf_k \{y_n \wedge \cdots \wedge y_{n+k}\}$ then $\{w_n\}$ converges in the norm of V to $y = \sup_n \{w_n\}$, and hence $a(y) = a^+(x)$ and $0 \leq y \leq x$. Moreover, we have $a^+(y) \geq a(y) = a^+(x) \geq a^+(y)$, so that $a^+(y) = a^+(x)$. Writing $z = x - y$ we see that $a^+(z) = 0$, $x = y + z$, and $a^-(y) = a^+(y) - a(y) = 0$. It follows that $K = \text{co}\,(F \cup G)$.

(ii) Let K be a strongly spectral simplex. For $x \in K$ let μ_x denote the maximal measure on K with resultant x; and for each $a \in A(K)$ define a regular Borel measure μ_x^a on $\mathbb{R}$ by $\int_{\mathbb{R}} \phi \, d(\mu_x^a) = \int_K (\phi \circ a)\, d\mu_x$, for each bounded Borel function ϕ on $\mathbb{R}$. The formula $e_\lambda^a(x) = \mu_x^a((-\infty, \lambda])$, $x \in K$, now gives the spectral resolution for a. (For the precise details see Alfsen–Schultz [**10**].).

For each $x \in K$ define $a^{(2)}(x) = \int \lambda^2 \, d\mu_x^a$, so that $a^{(2)} \in A^b(K)$. For $x \in \partial K$ we have $a^{(2)}(x) = a(x)^2$. On the other hand if $x \notin \partial K$ then for some $a \in A(K)$ the support of μ_x^a will consist of more than one point, so that $a^{(2)}(x) > (\int \lambda \, d\mu_x^a)^2 = a(x)^2$. Evidently then we have $\partial K = \bigcap_{a \in A(K)} \{x \in K: a^{(2)}(x) = a(x)^2\}$, and it will follow that ∂K is closed if we can show that $a^{(2)} \in A(K)$.

For any open interval (α, β) we have

$$\int (\chi_{(\alpha,\beta)} \circ a)\, d\mu_x = \int (\chi_{(\alpha,\beta)} \circ a)\, d\mu_x - \int (\chi_{(-\infty,-\beta]} \circ (-a))\, d\mu_x$$
$$= 1 - e_\alpha^a(x) - e_{-\beta}^{-a}(x),$$

which is l.s.c. in x. Since every open set U in $\mathbb{R}$ is a disjoint union of open intervals we see that $\int (\chi_U \circ a)\, d\mu_x$ is also l.s.c. in x. Finally, the function $\phi(t) = t^2$, $t \in \mathbb{R}$, is the limit of an increasing sequence of non-negative simple functions of the form $\sum_{k=1}^{l} \alpha_k \chi_{U_k}$, with U_k open in $\mathbb{R}$, and therefore $a^{(2)}$ is l.s.c. on K. Carrying out the same procedure with the function $\Psi(t) = \|a\|^2 - t^2$ gives $-a^{(2)}$ l.s.c. and hence $a^{(2)}$ is continuous.

The converse implication of (ii) is just the commutative version of (iii).

(iii) Let K be the state space of a C^*-algebra B, so that K is the normal state space of $A = B^{**}$, in the usual way. We have already noted that, if

$V = A^*$, then (V, A) are in spectral duality, and hence K is spectral. Since $B_h = A(K)$ is closed under composition with continuous real-valued functions ϕ, we can choose a decreasing sequence $\{\phi_n\}$ converging to $\chi_{(-\infty,\lambda]}$ so that $\{\phi_n \circ a\}$ converges to e^a_λ. Hence e^a_λ is u.s.c. in x, and consequently K is strongly spectral.

There exist further geometric characterizations of spectral duality and strongly spectral compact convex sets. We will state these characterizations, without proof, for compact sets K. Two elements $f, g \in A^b(K)^+$ are said to be *orthogonal*, written $f \perp g$ if there exists a projective face F_p of K such that $F = 0$ on F_p while $g = 0$ on the quasi-complementary projective face $F_{p'}$.

6.6. THEOREM. *Let K be a compact convex set. Then*:

(i) *K is spectral if and only if every norm-exposed face of K is projective and every $f \in A^b(K)$ admits a unique decomposition $f = g - h$, where $g, h \in A^b(K)^+$ and $g \perp h$.*

(ii) *K is strongly spectral if K is spectral and, in addition, every $f \in A(K)$ admits a (necessarily unique) decomposition $f = g - h$ with $g, h \in A(K)^+$ and $g \perp h$.*

Alfsen *et al.* [**14**] have developed a much more detailed theory concerning P-projections, JB-algebras and C^*-algebras. There is insufficient space in this book to give a detailed description of their work, so we will just summarize some of their main results.

If K is a compact convex set and if $E \subseteq K$, then we will denote by σ co E the set of all $x \in K$ such that $x = \sum_{n=1}^{\infty} \lambda_n x_n$, where $0 \leq \lambda_n \leq 1$, $x_n \in E$ for each n and the sum is norm-convergent. We will say that K has the *Hilbert ball property* if, for every pair, $x, y \in \partial K$, the smallest face, face (x, y) of K containing x and y is a norm-exposed face of K which is affinely isomorphic to the closed unit ball of some finite or infinite-dimensional Hilbert space H. (If $x = y$ we take dim $H = 0$.)

6.7. THEOREM.

(a) *Let A be a* JB*-algebra. Then there is a unique Jordan ideal J in A such that A/J has a faithful isometric Jordan representation as a Jordan algebra of self-adjoint operators on a Hilbert space. Moreover, every factor representation of A which does not annihilate J maps A onto the exceptional Jordan algebra M_3^8 of all Hermitian* (3×3) *matrices over the Cayley numbers.*

(b) *A compact convex set K is the space of a* JB*-algebra if and only if K has the following properties*: (i) *K has the Hilbert ball property*; (ii) σ co ∂K *is a split face of K*; (iii) *every norm-exposed face of K is projective*; (iv) *every $f \in A(K)$ admits a decomposition $f = g - h$, where $g, h \in A(K)^+$ and $g \perp h$.*

(c) *A compact convex set K is the state space of a Jordan algebra of self-adjoint operators on some Hilbert space (with Jordan product $S \circ T = \frac{1}{2}(ST+TS)$) if and only if K satisfies the conditions of* (b) *and such that* face (x, y) *is a split face of K whenever $x, y \in \partial K$ and* dim (face $(x, y)) = 9$.

The methods that we have touched on in this section have culminated in the solution of Alfsen and Shultz [**12**] of one of the main problems in the subject, namely the characterization of the state spaces of C^*-algebras. The extra ingredient that is required to distinguish the state spaces of C^*-algebras amongst the state spaces of JB-algebras is the concept of orientability.

K is said to have the *3-ball property* if it has the Hilbert ball property with the dimension of each face (x, y) being either three or one. Suppose that $B =$ face (x, y) has dimension three. Using the affine isomorphism between B and the unit ball in $\mathbb{R}^3$, we can define two opposite vector products in B; an *orientation* of B will mean a choice of one of these vector products for B.

Let A be a JB-algebra with state space K and let $a \in A$, $\rho \in K$. Then if $\rho(a^2) \neq 0$ we can define $\rho_a \in K$ by $\rho_a(b) = \rho(aba)/\rho(a^2)$, $b \in A$. Let $\mathcal{S}$ denote the collection {face (ρ, ρ_a)}, where face (ρ, ρ_a) has dimension three. Then K is said to be *orientable* if it is possible to choose orientations for the balls in $\mathcal{S}$ such that, for each $a \in A$, the map $\rho \rightsquigarrow \rho \times \rho_a$ is w^*-continuous wherever it is defined.

6.8. THEOREM. *A* JB-*algebra is isomorphic to the self-adjoint part of a C^*-algebra if and only if its state space K has the* 3-*ball property and is orientable.*

If A is a C^*-algebra then the Jordan product in A_h is determined by the geometry of K. There may however exist different C^*-products in A, and the choice of such products is connected with the orientability of K. For a detailed discussion of these topics, and for the proof of 6.8 we refer to the fundamental paper of Alfsen and Shultz [**12**].

Notes

This section is intended as an introduction to, and a brief survey of, a large body of fundamental work due to Alfsen *et al.* (see [**10**, **11**, **12**, **13**, **14**]).

Theorem 6.1 is due to Alfsen, Shultz and Størmer [**14**, 2.1]. Theorems 6.2 and 6.3 are due to Alfsen and Shultz [**10**, 3.5 and 11.5]. Theorems 6.4 and 6.6 appear in Alfsen and Shultz [**13**, 1.3, 2.2 and 2.4]. Theorem 6.5 was proved in Alfsen and Shultz [**10**, 10.4, 10.9 and 11.6]. Theorem 6.7 gives the main results of Alfsen and Shultz [**11**] and Theorem 6.8 summarizes the main results of Alfsen and Shultz [**12**].

The geometric characterization of the state space of a C^*-algebra given by Alfsen and Shultz in 6.8 represents a formidable achievement. For many years this problem had been one of the most important in the subject, and had occupied the efforts of many mathematicians. In particular the earlier work of Topping [**209**] on Jordan algebras of self-adjoint operators deserves mention. No such geometric characterization of the state spaces of function algebras is known.

An up-dated version of [**12**] has recently been produced: E. M. Alfsen, H. Hanche-Olsen and F. W. Shultz, "State spaces of C^*-algebras".

References

1. E. M. ALFSEN, On the geometry of Choquet simplexes, *Math. Scand.* **15** (1964), 97–110.
2. E. M. ALFSEN, On the decomposition of a Choquet simplex into a direct convex sum of complementary faces, *Math. Scand.* **17** (1965), 169–176.
3. E. M. ALFSEN, Facial structure of compact convex sets, *Proc. London Math. Soc.* (3) **18** (1968), 385–404.
4. E. M. ALFSEN, Un théorème de Weierstrass–Stone pour les sous-espaces vectoriels de $C_{\mathbb{R}}(X)$, *C.R. Acad. Sci. Paris* **271** (1970), 725–726.
5. E. M. ALFSEN, "Compact Convex Sets and Boundary Integrals" (Springer-Verlag, Berlin, 1971).
6. E. M. ALFSEN and T. B. ANDERSEN, Split faces of compact convex sets, *Proc. London Math. Soc.* (3) **21** (1970), 415–442.
7. E. M. ALFSEN and T. B. ANDERSEN, On the concept of centre in $A(K)$. *J. London Math. Soc.* (2) **4** (1972), 411–417.
8. E. M. ALFSEN and E. G. EFFROS, Structure in real Banach spaces, *Annals Math.* **96** (1972), 98–173.
9. E. M. ALFSEN and B. HIRSBERG, On dominated extensions in linear subspaces of $C_c(X)$, *Pacific J. Math.* **36** (1971), 567–584.
10. E. M. ALFSEN and F. W. SHULTZ, Non-commutative spectral theory for affine function spaces on convex sets, *Memoirs Amer. Math. Soc.* **6**, No. 172 (1976).
11. E. M. ALFSEN and F. W. SHULTZ, State spaces of Jordan algebras, *Acta Math.* **140** (1978), 155–190.
12. E. M. ALFSEN and F. W. SHULTZ, State spaces of C^*-algebras, Univ. of Oslo, Preprint Series No. 8 (June 1978).
13. E. M. ALFSEN and F. W. SHULTZ, On non-commutative spectral theory and Jordan algebras, *Proc. London Math. Soc.* (3) **38** (1979), 497–516.
14. E. M. ALFSEN, F. W. SHULTZ and E. STØRMER, A Gelfand–Neumark theorem for Jordan algebras, *Advances in Math.* **28** (1978), 11–56.
15. P. R. ANDENAES, Extreme boundaries and continuous affine functions, *Math. Scand.* **40** (1977), 197–208.
16. T. B. ANDERSEN, On multipliers and order-bounded operators in C^*-algebras, *Proc. Amer. Math. Soc.* **25** (1970), 896–899.

17. T. B. ANDERSEN, On dominated extensions of continuous affine functions on split faces, *Math. Scand.* **29** (1971), 298–306.
18. T. ANDÔ, On fundamental properties of a Banach space with a cone, *Pacific J. Math.* **12** (1962), 1163–1169.
19. T. ANDÔ, Closed range theorems for convex sets and linear liftings, *Pacific J. Math.* **44** (1973), 393–409.
20. R. J. ARCHBOLD, Prime C^*-algebras and antilattices, *Proc. London Math. Soc.* (3) **24** (1972), 669–680.
21. W. B. ARVESON, Subalgebras of C^*-algebras, *Acta Math.* **123** (1969), 141–224.
22. L. ASIMOW, Universally well-capped cones, *Pacific J. Math.* **26** (3) (1968), 421–431.
23. L. ASIMOW, Directed Banach spaces of affine functions, *Trans. Amer. Math. Soc.* **143** (1969), 117–132.
24. L. ASIMOW, Decomposable compact convex sets and peak sets for function spaces, *Proc. Amer. Math. Soc.* **25** (1970), 75–79.
25. L. ASIMOW, Extensions of continuous affine functions, *Pacific J. Math.* **35** (1) (1970), 11–21.
26. L. ASIMOW, Exposed faces of dual cones and peak-set criteria for function spaces, *J. Funct. Anal.* **12** (1973), 456–474.
27. L. ASIMOW, Complementary cones in dual Banach spaces, *Illinois J. Math.* **18** (1974), 657–668.
28. L. ASIMOW, Interpolation in Banach spaces, *Rocky Mtn. J. Math,* **9** (1979), 543–568.
29. L. ASIMOW, Superharmonic dominated interpolation in subspaces of C_0, *Pacific J. Math,* (to appear).
30. L. ASIMOW, Best approximation by gauges on a Banach space, (to appear).
31. L. ASIMOW and A. J. ELLIS, Facial decomposition of linearly compact simplexes and separation of functions on cones, *Pacific J. Math.* **34** (1970), 301–309.
32. L. ASIMOW and A. J. ELLIS, On hermitian functionals on unital Banach algebras, *Bull. London Math. Soc.* **4** (1972), 333–336.
33. C. J. K. BATTY, Simplexes of states of C^*-algebras, (to appear).
34. H. BAUER, Schilowscher Rand und Dirichletsches Problem, *Ann. Inst. Fourier (Grenoble)* **11** (1961), 89–136.
35. H. S. BEAR, Complex function algebras, *Trans. Amer. Math. Soc.* **90** (1959), 383–393.
36. H. S. BEAR, Some boundary properties of function algebras, *Proc. Amer. Math. Soc.* **11** (1960) 1–4.
37. E. BISHOP, A minimal boundary for function algebras, *Pacific J. Math.* **9** (1959), 629–642.
38. E. BISHOP, A generalization of the Stone–Weierstrass theorem, *Pacific J. Math.* **11** (1961), 777–783.
39. E. BISHOP, A general Rudin–Carleson theorem, *Proc. Amer. Math. Soc.* **13** (1962), 140–143.
40. E. BISHOP and K. DE LEEUW, The representation of linear functionals by measures on sets of extreme points, *Ann. Inst. Fourier (Grenoble)* **9** (1959), 305–331.
41. E. BISHOP and R. R. PHELPS, A proof that every Banach space is subreflexive, *Bull. Amer. Math. Soc.* **67** (1961), 97–98.
42. E. BISHOP and R. R. PHELPS, The support functionals of a convex set, Proc. Symp. in Pure Math. **7**, *Convexity* (Amer. Math. Soc., 1963).

43. F. BOHNENBLUST and S. KARLIN, Geometric properties of the unit sphere of Banach algebras, *Ann. of Math.* **62** (1955), 217–229.
44. F. F. BONSALL, Extreme maximal ideals in partially ordered vector spaces, *Proc. Amer. Math. Soc.* **7** (1956), 831–837.
45. F. F. BONSALL, On the representation of points of a convex set, *J. London Math.* **24** (1963), 265–272.
46. F. F. BONSALL and J. DUNCAN, "Numerical Ranges of Operators on Normed Spaces and of Elements of Normed Algebras" (London Math. Soc. Lecture Note Series No. 2, Cambridge University Press, 1971).
47. F. F. BONSALL and J. DUNCAN, "Complete Normed Algebras" (Springer-Verlag, Berlin, 1973).
48. N. BOURBAKI, "Topologie Générale", Ch. IX (3e éd.) (Hermann, Paris, 1961).
49. R. D. BOURGIN and G. A. EDGAR, Non-compact simplexes in spaces with the Radon-Nikodym property, *J. Funct. Anal.* **23** (1976), 162–176.
50. E. BRIEM, Interpolation in subspaces of $C(X)$, *J. Funct. Anal.* **12** (1973), 1–12.
51. E. BRIEM, Facial topologies for subspaces of $C(X)$, *Math. Ann.* **208** (1974), 9–13.
52. E. BRIEM, Split faces associated with function algebras, *J. London Math. Soc.* (2) **9** (1975), 446–450.
53. E. BRIEM, Extreme orthogonal boundary measures for $A(K)$ and decompositions for compact convex sets, Springer Lecture Notes in Mathematics No. 512, (1976), 8–16.
54. E. BRIEM, A characterization of simplexes by parallel faces, (to appear).
55. E. BRIEM, Peak sets for the real part of a function algebra, (to appear).
56. A. BROWDER, On a theorem of Hoffman and Wermer, "Proc. INT. Symp. Function algebras," Scott-Foresman, 1966, 88–89.
57. A. BROWDER, "Introduction to Function Algebras" (Benjamin, New York, 1969).
58. J. BUNCE, The intersection of closed ideals in a simplex space need not be an ideal, *J. London Math. Soc.* (2) **1** (1969), 67–68.
59. L. CARLESON, Representations of continuous functions, *Math. Z.* **66** (1957), 447–451.
60. G. CHOQUET, Existence des représentation intégrales au moyen des points extrémaux dans les cônes convexes, *Seminaire Bourbaki* (Dec. 1956) 139.
61. G. CHOQUET, Le théorème de représentation intégrales dans les ensembles convexes compacts, *Ann. Inst. Fourier* (*Grenoble*) **10** (1960), 333–344.
62. G. CHOQUET, "Lectures in Analysis", Vol. II. (Benjamin, New York, 1969).
63. G. CHOQUET and P. A. MEYER, Existence et unicité des représentations intégrales dans les convexes compact quelconque, *Ann. Inst. Fourier* (*Grenoble*) **13** (1963), 139–154.
64. C. H. CHU, Ideaux premiers dans $A(X)$, *C.R. Acad. Sci. Paris* **275** (1972), 1179–1182.
65. C. H. CHU, Antilattices and prime sets, *Math. Scand.* **31** (1972), 151–165.
66. C. H. CHU, Prime faces in C^*-algebras, *J. London Math. Soc.* (2) **7** (1973), 175–180.
67. C. H. CHU, On convexity theory and C^*-algebras, *Proc. London Math. Soc.* (3) **31** (1975), 257–288.
68. C. H. CHU and J. D. M. WRIGHT, A theory of types for convex sets and ordered Banach spaces, *Proc. London Math. Soc.* (3) **36** (1978), 494–517.

69. F. CUNNINGHAM, JR., E. G. EFFROS and N. M. ROY, *M*-structure in dual Banach spaces, *Israel. J. Math.* **14** (1973), 304–308.
70. P. C. CURTIS and A. FIGA-TALAMANCA, Factorization theorems for Banach algebras, "Proc. Int. Symp. Function Algebras", Scott-Foresman, 1966, 169–185.
71. F. DAUNS and K. H. HOFMANN, Representations of rings by continuous sections, *Memoirs Amer. Math. Soc.* No. 83, (1968).
72. E. B. DAVIES, On the Banach duals of certain spaces with the Riesz decomposition property, *Quart. J. Math. Oxford* **18** (1967), 109–111.
73. E. B. DAVIES, The structure and ideal theory of the predual of a Banach lattice, *Trans. Amer. Math. Soc.* **131** (1968), 544–555.
74. E. B. DAVIES, Decomposition of traces on separable C^*-algebras, *Quart. J. Math. Oxford* (2) **20** (1969), 97–111.
75. M. M. DAY, "Normed Linear Spaces" (Springer-Verlag, Berlin, 1973).
76. L. DE BRANGES, The Stone–Weierstrass theorem, *Proc. Amer. Soc.* **10** (1959), 822–824.
77. J. DENY, Systèmes totaux de fonctions harmoniques, *Ann. Inst. Fourier* (*Grenoble*) **1** (1949), 103–113.
78. J. DIESTEL, "Geometry of Banach Spaces—Selected Topics", Lecture Notes in Mathematics No. 485, (Springer-Verlag, Berlin, 1975).
79. J. DIESTEL and J. J. UHL, JR., "Vector Measures", *Math. Surveys* No. 15 (Amer. Math. Soc., Providence R.I. 1977).
80. J. DIXMIER, Sur les C^*-algèbres, *Bull. Soc. Math. France* **88** (1960), 95–112.
81. J. DIXMIER, "Les C^*-algèbres et Leurs Représentations", (Gauthier-Villars, Paris, 1964).
82. J. DIXMIER, "Ideal center of a C^*-algebra", *Duke Math. J.* **35** (1968), 375–382.
83. G. A. EDGAR, A noncompact Choquet theorem, *Proc. Amer. Math. Soc.* **49** (1975), 354–358.
84. G. A. EDGAR, Extremal integral representations, *J. Funct. Anal.* **23** (1976), 145–161.
85. C. M. EDWARDS, The theory of pure operations, *Commun. Math. Phys.* **24** (1972), 260–288.
86. C. M. EDWARDS, Spectral theory for $A(X)$, *Math. Ann.* **207** (1974), 67–85.
87. D. A. EDWARDS, On the homeomorphic affine embedding of a locally compact cone into a Banach dual space endowed with the vague topology, *Proc. London Math. Soc.* **14** (1964), 399–414.
88. D. A. EDWARDS, Séparation de fonctions réelles définies sur un simplexe de Choquet, *C.R. Acad. Sci. Paris* **261** (1965), 2798–2800.
89. D. A. EDWARDS, On uniform approximation of affine functions on a compact convex set, *Quart. J. Math. Oxford* (2) **20** (1969), 139–142.
90. D. A. EDWARDS, On the ideal centres of certain partially ordered Banach spaces, *J. London Math. Soc.* (2) **6** (1973), 656–658.
91. D. A. EDWARDS and G. F. VINCENT-SMITH, A Weierstrass–Stone theorem for Choquet simplexes, *Ann. Inst. Fourier* (*Grenoble*) **18** (1968), 261–282.
92. E. G. EFFROS, Order ideals in a C^*-algebra and its dual, *Duke Math. J.* **30** (1963), 391–412.
93. E. G. EFFROS, Structure in simplexes, *Acta Math.* **117** (1967), 103–121.

94. E. G. EFFROS, Structure in simplexes II, *J. Funct. Anal.* **1** (1967), 361–391.
95. E. G, EFFROS, On a class of real Banach spaces, *Israel J. Math.* **9** (1971), 430–458.
96. E. G. EFFROS, On a class of complex Banach spaces, *Illinois J. Math.* **18** (1974), 48–59.
97. E. G. EFFROS and F. HAHN, Locally compact transformation groups and C^*-algebras, *Memoirs Amer. Math. Soc.* No. 75 (1967).
98. E. G. EFFROS and J. L. KAZDAN, Applications of Choquet simplexes to elliptic and parabolic boundary-value problems, *J. Diff. Equations* **8** (1970), 95–134.
99. G. A. ELLIOTT and D. OLESEN, A simple proof of the Dauns–Hofmann theorem, *Math. Scand.* **34** (1974), 231–234.
100. A. J. ELLIS, The duality of partially ordered normed linear spaces, *J. London Math. Soc.* **39** (1964), 730–744.
101. A. J. ELLIS, On faces of compact convex sets and their annihilators, *Math. Ann.* **184** (1969), 19–24.
102. A. J. ELLIS, On split faces and function algebras, *Math. Ann.* **195** (1972), 159–166.
103. A. J. ELLIS, On facially continuous functions in function algebras, *J. London Math. Soc.* (2) **5** (1972), 561–564.
104. A. J. ELLIS, Central decompositions and the essential set for the space $A(K)$, *Proc. London Math. Soc.* (3) **26** (1973), 564–576.
105. A. J. ELLIS, Some applications of convexity theory to Banach algebras, *Math. Scand.* **33** (1973), 23–30.
106. A. J. ELLIS, Central decompositions for compact convex sets, *Comp. Math.* **30** (1975), 211–219.
107. A. J. ELLIS, Weakly prime compact convex sets and uniform algebras, *Math. Proc. Camb. Phil. Soc.* **81** (1977), 225–232.
108. A. J. ELLIS, A facial characterization of Choquet simplexes, *Bull. London Math. Soc.* **9** (1977), 326–327.
109. A. J. ELLIS, A characterization of the state spaces of uniform algebras, *Quart. J. Math. Oxford* (2) **29** (1978), 375–383.
110. A. J. ELLIS and A. K. ROY, Dilated sets and characterizations of simplexes, *Inventiones Math.* **56** (1980), 101–108.
111. H. FAKHOURY, Solution d'un problème posé par Effros, *C.R. Acad. Sci. Paris* **269** (1969), 77–79 and 371.
112. R. FUHR and R. R. PHELPS, Uniqueness of complex representing measures on the Choquet boundary, *J. Funct. Anal.* **14** (1973), 1–27.
113. T. W. GAMELIN, Restrictions of subspaces of $C(X)$, *Trans. Amer. Math. Soc.* **112** (1964), 278–286.
114. T. W. GAMELIN, "Uniform Algebras" (Prentice-Hall, Englewood Cliffs, N.J. 1969).
115. A. GLEIT, On the structure topology of simplex spaces, *Pacific J. Math.* **34** (1970), 389–405.
116. A. GLEIT, On the existence of simplex spaces, *Israel J. Math.* **9** (1971), 199–209.
117. A. GLEIT, A characterization of *M*-spaces in the class of separable simplex spaces, *Trans. Amer. Math. Soc.* **169** (1972), 25–33.
118. I. GLICKSBERG, Measures orthogonal to algebras and sets of antisymmetry, *Trans. Amer. Math. Soc.* **105** (1962), 415–435.
119. J. GLIMM, A Stone–Weierstrass theorem for C^*-algebras, *Ann. Math.* **72** (1960), 216–244.

120. J. GLIMM, Type I C^*-algebras, *Ann. Math.* **73** (1961), 572–612.
121. A. GOULLET DE RUGY, "Geométrie des Simplexes", Centre de Documentation Universitaire, Paris (1968).
122. J. GROSBERG and M. KREIN, Sur la décomposition des fonctionelles en composantes positives, *C.R. (Doklady) de l'Acad. Sci. de l'URSS* **25** (1939), 723–726.
123. A. GROTHENDIECK, Un résultat sur le dual d'une C^*-algèbre, *J. Math. Pures Appl.* (9) **36** (1957), 97–108.
124. R. HAYDON, A new proof that every Polish space is the extreme boundary of a simplex, *Bull. London Math. Soc.* **7** (1975), 97–100.
125. B. HIRSBERG, "A Measure Theoretic Characterization of Parallel and Split Faces and their Connections with Function Spaces and Algebras", Aarhus Univ. Various Publications Series No. 16 (1970).
126. B. HIRSBERG, *M*-ideals in complex function spaces and algebras, *Israel J. Math.* **12** (1972), 133–146.
127. B. HIRSBERG, Note sur les représentations intégrales des formes linéaires complexes, *C.R. Acad. Sci. Paris* **274** (1972), 1222–1224.
128. B. HIRSBERG and A. J. LAZAR, Complex Lindenstrauss spaces with extreme points, *Trans. Amer. Math. Soc.* **186** (1973), 141–150.
129. K. HOFFMAN and I. M. SINGER, Maximal algebras of continuous functions, *Acta Math.* **103** (1960), 217–241.
130. K. HOFFMAN and J. WERMER, A characterization of $C(X)$, *Pacific J. Math.* **12** (1962), 941–944.
131. R. B. HOLMES, "Geometric Functional Analysis and its Applications" (Springer-Verlag, Berlin, 1975).
132. O. HUSTAD, A norm-preserving complex Choquet theorem, *Math. Scand.* **29** (1971), 272–278.
133. N. JACOBSON, "Structure and Representations of Jordan Algebras", Amer. Math. Soc. Colloqu. Publ. No. 39 (Providence, R.I., 1968).
134. G. J. O. JAMESON, "Ordered Linear Spaces", Lecture Notes No. 141, (Springer-Verlag, Berlin, 1970).
135. J. JAYNE and C. A. ROGERS, The extremal structure of convex sets, *J. Funct. Anal.* **26** (1977), 251–288.
136. R. V. KADISON, A representation theory for commutative topological algebra, *Memoirs Amer. Math. Soc.* No. 7, (1951).
137. R. V. KADISON, Order properties of bounded self-adjoint operators, *Proc. Amer. Math. Soc.* **2** (1951), 505–510.
138. R. V. KADISON, Irreducible operators algebras, *Proc. Nat. Acad. Science* **43** (1957), 273–276.
139. S. KAKUTANI, Concrete representations of abstract *L*-spaces and the mean ergodic theorem, *Ann. of Math.* **42** (1941), 523–537.
140. S. KAKUTANI, Concrete representations of abstract *M*-spaces, *Ann. of Math.* **42** (1941), 994–1024.
141. D. G. KENDALL, Simplexes and vector lattices, *J. London Math. Soc.* **37** (1962), 365–371.
142. V. KLEE, Separation properties of convex cones, *Proc. Amer. Math. Soc.* **6** (1955), 313–318.
143. M. KREIN and S. KREIN, On an inner characteristic of the set of all continuous functions defined on a bicompact Hausdorff space, *C.R. (Doklady) Acad. Sci. USSR* **27** (1940), 427–430.
144. H. E. LACEY, "The Isometric Theory of Classical Banach Spaces" (Springer-Verlag, Berlin, 1974).

145. A. J. LAZAR, Spaces of affine continuous funtions on simplexes, *Trans. Amer. Math. Soc.* **134** (1968), 503–525.
146. A. J. LAZAR, Sections and subsets of simplexes, *Pacific J. Math.* **33** (1970), 337–344.
147. A. J. LAZAR, The unit ball in conjugate L_1-spaces, *Duke Math. J.* **39** (1972), 1–8.
148. A. J. LAZAR, Extreme boundaries of convex bodies in l_2, *Israel J. Math.* **20** (1975), 369–374.
149. A. J. LAZAR and J. LINDENSTRAUSS, Banach spaces whose duals are L_1-spaces and their representing matrices, *Acta. Math.* **126** (1971), 165–194.
150. G. M. LEIBOWITZ, "Lectures on Complex Function Algebras" (Scott-Foresman, Glenview, 1969).
151. A. LIMA, Complex Banach spaces whose duals are L_1-spaces, *Israel J. Math.* **24** (1976), 59–72.
152. A. LIMA, Intersection properties of ball and subspaces in Banach spaces, *Trans. Amer. Math. Soc.* **227** (1977), 1–62.
153. J. LINDENSTRAUSS, Extension of compact operators, *Memoirs Amer. Math. Soc.* No. 48 (1964).
154. J. LINDENSTRAUSS, G. OLSEN and Y. STERNFELD, The Poulsen simplex, *Ann. Inst. Fourier* (*Grenoble*) **28** (1978), 91–114.
155. W. LUSKY, The Gurarij spaces are unique, *Arch. Math.* **27** (1976), 627–635.
156. W. LUSKY, A note on the paper "The Poulsen simplex" of Lindenstrauss, Olsen and Sternfeld, *Ann. Inst. Fourier* (*Grenoble*) **28** (1978), 233–243.
157. W. A. J. LUXEMBURG and A. C. ZAANEN, "Riesz spaces I" (North Holland, Amsterdam, 1971).
158. P. MANKIEWICZ, A remark on Edgar's extremal integral representation theorem, *Studia Math.* **63** (1978), 259–265.
159. J. N. MCDONALD, Compact convex sets with the equal support property, *Pacific J. Math.* **37** (1971), 429–443.
160. Y. MISONOU, On a weakly central algebra, *Tôhoku Math. J.* **4** (1952), 194–202.
161. G. MOKOBODZKI, Balayage défini par un cône convexe de fonctions numériques sur un espace compact, *C.R. Acad. Sci. Paris* **245** (1962), 803–805.
162. R. T. MOORE, Hermitian functionals on B-algebras and duality characterizations of C^*-algebras, *Trans. Amer. Math. Soc.* **162** (1971), 253–266.
163. R. NAGEL, "Ideal Theorie in Geordnetan Lokalkonvexen Vektorraümen" (Dissertation Eberhard-Karls-Universität, Tübingen 1969).
164. I. NAMIOKA, Separate continuity and joint continuity, *Pacific J. Math.* **51** (1974), 515–531.
165. K. F. NG, The duality of partially ordered Banach spaces, *Proc. London Math. Soc.* **19** (1969), 269–288.
166. T. OGASAWARA, A theorem on operator algebras, *J. Hiroshima Univ.* **18** (1955), 307–309.
167. F. PERDRIZET, Espaces de Banach ordonnés et idéaux, *J. Math. Pures et Appl.* **49** (1970), 61–98.
168. A. L. PERESSINI, "Ordered Topological Vector Spaces" (Harper and Row, New York, 1967).
169. R. R. PHELPS, "Lectures on Choquet's Theorem", Van Nostrand, Princeton (1966).
170. R. R. PHELPS, Support cones in Banach spaces and their applications, *Adv. Math.* **13** (1974), 1–19.

171. R. R. PHELPS, Dentability and extreme points in Banach spaces, *J. Funct. Anal.* **16** (1974), 78–90.
172. R. R. PHELPS, The Choquet representation in the complex case, *Bull. Amer. Math. Soc.* **83** (1977), 299–312.
173. E. T. POULSEN, A simplex with dense extreme points, *Ann. Inst. Fourier (Grenoble)* **11** (1961), 83–87.
174. A. DE LA PRADELLE, A propos du mémoire de G. F. Vincent-Smith sur l'approximation des fonctions harmoniques, *Ann. Inst. Fourier (Grenoble)* **19** (2) (1969), 355–370.
175. A. DE LA PRADELLE, Approximation des fonctions harmoniques a l'aide d'un théorème de G. F. Vincent-Smith, Sem. Théor. Potent. M. Brelot, G. Choquet, J. Deny, 1969/70, 13, Paris (1971), 3101–3116.
176. R. T. PROSSER, On the ideal structure of operator algebras, *Memoirs Amer. Math. Soc.* No. 45 (1963).
177. M. A. RIEFFEL, Dentable subsets of Banach spaces, with applications to a Radon Nikodym theorem, "Proc. Conf. Functional Analysis" (Thompson Book Co., Washington, D.C. 1967), 71–77.
178. F. RIESZ, Sur la decomposition des operations lineaires, *Atti. del Congresso Bologna* **3** (1928), 143–148.
179. F. RIESZ, Sur quelques notions fondamentales dans la théorie générale des operations linéaires, *Ann. of Math.* **41** (1940), 174–206.
180. M. ROGALSKI, Etude du quotient d'un simplexe par une face fermée et application à un théorème de Alfsen; quotient par une relation d'équivalence, Sem. Théor. Potent. M. Brelot, G. Choquet et J. Deny, 1967/68, 12, Sec. Math. Paris (1969), 2/01–2/25.
181. M. ROGALSKI, Caractérisation des simplexes par des propriétés portant sur les faces fermées et sur les ensembles compacts de points extrémaux, *Math. Scand.* **28** (1971), 159–181.
182. W. ROTH, A general Rudin-Carleson theorem in Banach spaces, *Pacific J. Math.* **73** (1977), 197–214.
183. W. ROTH, A stability theorem for the Choquet ordering in $C(X)$, *Math. Scand.* **43** (1978), 92–98.
184. A. K. ROY, Closures of faces of compact convex sets, *Ann. Inst. Fourier (Grenoble)* **25** (1975), 221–234.
185. L. ROYDEN, "Real Analysis" (Macmillan, New York, 1968).
186. W. RUDIN, Boundary values of continuous analytic functions, *Proc. Amer. Math. Soc.* **7** (1956), 808–811.
187. W. RUDIN, "Real and Complex Analysis" (McGraw-Hill, New York, 1974).
188. S. SAKAI, "C^*-algebras and W^*-algebras" (Springer-Verlag, Berlin, 1971).
189. H. H. SCHAEFER, "Banach Lattices and Positive Operators" (Springer-Verlag, Berlin, 1974).
190. Z. SEMADENI, Free compact convex sets, *Bull. Acad. Sci. Pol.* **13** (1965), 141–146.
191. Z. SEMADENI, "Banach Spaces of Continuous Functions" (Monografie Matematyczne, **55**, Warsaw, 1971).
192. S. SHERMAN, Order in operator algebras, *Amer. J. Math.* **73** (1951), 227–232.
193. F. W. SHULTZ, On normed Jordan algebras which are Banach dual spaces, *J. Funct. Anal.* **31** (1979), 360–376.
194. S. J. SIDNEY and E. L. STOUT, A note on interpolation, *Proc. Amer. Math. Soc.* **19** (1968), 380–382.
195. A. M. SINCLAIR, The states of a Banach algebra generate the dual, *Proc. Edinburgh Math. Soc.* (2) **17** (1971), 341–344.

196. R. R. SMITH, An addendum to "*M*-ideal structure in Banach algebras", *J. Funct. Anal.* **32** (1979), 269–271.
197. R. R. SMITH and J. D. WARD, *M*-ideal structure in Banach algebras, *J. Funct. Anal.* **27** (1978), 337–349.
198. P. J. STACEY, Type I points in a compact convex set, *J. London Math. Soc.* (2) **10** (1975), 306–308.
199. P. J. STACEY, Standard Choquet simplices with prescribed extreme and Šilov boundaries, (to appear).
200. J. STEFÁNSON, On a problem of J. Dixmier concerning ideals in a von Neumann algebra, *Math. Scand.* **24** (1969), 111–112.
201. Y. STERNFELD, Characterization of Bauer simplices and some other classes of Choquet simplices by their representing matrices, (to appear).
202. M. H. STONE, Applications of the theory of Boolean rings to general topology, *Trans. Amer. Math. Soc.* **41** (1937), 375–481.
203. M. H. STONE, A general theory of spectra II, *Proc. Nat. Acad. Sci. U.S.A.* **27** (1941), 83–87.
204. E. L. STOUT, "The Theory of Uniform Algebras" (Bogden and Quigley, Tarrytown-on-Hudson, 1971).
205. E. STØRMER, Positive linear maps of operator algebras, *Acta Math.* **110** (1963), 233–278.
206. E. STØRMER, On partially ordered vector spaces and their duals with applications to simplexes and C^*-algebras, *Proc. London Math. Soc.* (3) **18** (1968), 245–265.
207. E. THOMA, Über unitäre Darstellungen abzählberer, diskreter Gruppen, *Math. Annalen* **153** (1964), 111–138.
208. D. M. TOPPING, Vector lattices of self-adjoint operators, *Trans. Amer. Math. Soc.* **115** (1965), 14–30.
209. D. M. TOPPING, Jordan algebras of self-adjoint operators, *Memoirs Amer. Math. Soc.* No. 53 (Providence R.I., 1965).
210. N. TH. VAROPOULOS, Ensembles pics et ensembles d'interpolation pour les algèbres uniformes, *C.R. Acad. Sci. Paris* **272** (1971), 866–867.
211. J. VESTERSTRØM, On the homomorphic image of the center of a C^*-algebra, *Math. Scand.* **29** (1971), 134–136.
212. G. F. VINCENT-SMITH, Uniform approximation of harmonic functions, *Ann. Inst. Fourier* (*Grenoble*) **19** (2) (1969–70), 339–353.
213. B. C. VULIKH, "Introduction to the Theory of Partially Ordered Vector Spaces" (Wolters-Noordhoff, Groningen, 1967).
214. W. WILS, "Désintégration centrales des formes positives sur les C^*-algèbres", *C.R. Acad. Sci. Paris* **267** (1968), 810–812.
215. W. WILS, 'Désintégration centrale dans une partie convexe compacte d'un espace localement convexe', *C.R. Acad. Sci. Paris* **269** (1969), 702–704.
216. W. WILS, The ideal center of partially ordered vector spaces, *Acta Math.* **127** (1971), 41–77.
217. Y. C. WONG and K. F. NG, "Partially Ordered Topological Vector Spaces" (Clarendon Press, Oxford, 1973).
218. K. YOSIDA, On vector lattice with a unit, *Proc. Imp. Acad. Tokyo* **17** (1941), 121–124.

Index of Definitions

Subject Index

D

E

F

G

H

I

J

K

L

M

N

O

P

Q

R

S

T

U

V

W